Inflammation in Psychiatry

Modern Trends in Pharmacopsychiatry

Vol. 28

Series Editor

B.E. Leonard Galway

Inflammation in Psychiatry

Volume Editors

A. Halaris Maywood, Ill.
B.E. Leonard Galway

16 figures, 4 in color, 2013

KARGER Basel · Freiburg · Paris · London · New York · New Delhi · Bangkok · Beijing · Tokyo · Kuala Lumpur · Singapore · Sydney

Modern Trends in Pharmacopsychiatry
(Formerly published as 'Modern Problems in Pharmacopsychiatry')

Prof. Angelos Halaris
Department of Psychiatry and
Behavioral Neuroscience
Loyola University School of Medicine
and Medical Center
Maywood, Ill. (USA)

Prof. Brian E. Leonard
Department of Pharmacology
National University of Ireland
Galway (Ireland)

Library of Congress Cataloging-in-Publication Data

Inflammation in psychiatry / volume editors, A. Halaris, B.E. Leonard.
p. ; cm. -- (Modern trends in pharmacopsychiatry, ISSN 1662-2685 ; v. 28)
Includes bibliographical references and indexes.
ISBN 978-3-318-02310-7 (hard cover : alk. paper) -- ISBN 978-3-318-02311-4 (e-ISBN)
I. Halaris, Angelos. II. Leonard, B. E. III. Series: Modern trends in pharmacopsychiatry ; v. 28. 1662-2685
[DNLM: 1. Mental Disorders--complications. 2. Inflammation--complications. 3. Mental Disorders--immunology. 4. Neuroimmunomodulation. 5. Stress, Psychological--immunology. W1 MO168P v.28 2013 / WM 140]

616.89--dc23

2012051718

Bibliographic Indices. This publication is listed in bibliographic services, including Current Contents® and Index Medicus.

www.karger.com
Printed in Germany on acid-free and non-aging paper (ISO 9706) by Kraft Druck, Ettlingen
ISSN 1662–2685
e-ISSN 1662–4505
ISBN 978–3–318–02310–7
e-ISBN 978–3–318–02311–4

Contents

Preface

Although the establishment of the field of Immunology dates back many decades, psychoimmunology is generally viewed as a relatively recent discovery in medicine. This discovery is based on the evidence that the immune, endocrine and neurotransmitter systems are interdependent in health and disease states. They are invariably dysfunctional in all major psychiatric disorders. The basic concepts of psychoimmunology have depended largely on knowledge of the changes in these interactive and interdependent systems because of recent technical advances in immunology and molecular neurobiology. For example, the immunotransmitters, cytokines, that play a critical role in the pathophysiology of mental illnesses were only discovered about 30 years ago following the development of sensitive assay techniques, such as ELISA, that permitted their detection.

The fundamental concept that endogenous factors are responsible for cognitive, affective and behavioral dysfunction can be traced back to antiquity. Hippocrates, about 400 BC, formulated the concept of black bile which he held responsible for melancholia, and was associated with the autumn season – shades of seasonal affective disorder perhaps? Some 500 years later, the Roman physician Galen wrote a thesis on black bile ('De atrabile') which described its pathological properties that are now identified as melancholic depression. Undoubtedly the most well-known thesis on melancholia was published in the late Middle Ages by Robert Burton of Oxford (1577–1640) and titled 'The Anatomy of Melancholy.' In his thesis, Burton described 'melancholia flatulosa', a disease caused by black bile that 'affects the brain and causes fear, sorrow, dullness and heaviness which troubles the imagination'. Like many ancient commentators, melancholia was considered to be a disease of scholars due to their 'sedentary life and excessive exertion of the brain'.

Clearly, the concept of disease has changed decidedly from the prescientific to the modern, scientific era. In the former era, medicine was essentially viewed in a holistic way. With the development of scientific medicine in the 19th and 20th centuries, specialized disciplines came to dominate Medicine and the holistic, integrated aspects were often ignored. Nowhere is this more apparent than in psychoimmunology. To the immunologist, the brain was considered to be a privileged organ uninfluenced by the immune system; to the psychiatrist, the immune system was restricted to the peripheral organs and therefore of no importance to psychiatry.

How relevant are such views to the subject of inflammation? These reductionist views have been largely rejected by the discoveries in the past 25 years. Inflammation is involved in most aspects of mental health and mental diseases. It is now apparent that chronic, low-grade inflammation not only plays a major role in the pathophysiology of most major psychiatric disorders, but it also has an important evolutionary significance [1–3].

This volume seeks to highlight the intricate link between psychiatry and psychoimmunology and to underscore the central role of inflammation in furthering our understanding of the pathophysiology underlying mental health and mental illness. The choice of specific topics and authors who have contributed to this volume was dictated by the realization that this area of scientific endeavor is rapidly maturing and deserves broader dissemination across disciplines and specialty areas in medicine. Inflammation research is advancing our knowledge of the pathology of depression, schizophrenia, chronic fatigue syndrome, posttraumatic stress disorder and a host of co-morbid conditions, notably diabetes, cardiovascular and cerebrovascular disease. Stress is a major contributor to mental illness. Stress initiates pathological changes by activating the immune and endocrine systems. The possible mechanisms underlying the bidirectionality of co-morbid medical and psychiatric disorders can be viewed as a consequence of inflammatory changes. These emerging novel concepts illustrate how the knowledge of inflammation can enable meaningful integration of psychopathology with physical co-morbidity. Thus, the black bile of Hippocrates can now be identified as a number of factors that are derived from the inflammatory changes that form part of the psychopathology of major psychiatric disorders.

In conclusion, it must be remembered that a theory goes through 3 major stages. First stage it is ignored. Second stage it is vehemently opposed. Third stage it is accepted as being self-evident! We believe psychoimmunology has reached stage 3. Unlike false hopes raised by earlier theories of mental illness, we are confident that this relatively young subspecialty will have staying power. We base this belief on the fact that psychoimmunology provides a previously unsuspected powerful integrative link between autonomic nervous system function and dysfunction, neurochemical transmission, endocrine function and phenomenology. We trust the readers of this volume will agree with us.

Angelos Halaris, Chicago, Ill.
Brian E. Leonard, Galway

References

1 Crespi B, Summers K, Dorus S: Adaptive evolution of genes underlying schizophrenia. Proc Biol Sci 2007;274:2801–2810.

2 McDade TW: Life history theory and immune system: steps towards a human ecological immunology. Am J Phys Anthropol 2003;(suppl 37):100–125.

3 Raison CL, Miller AH: The evolutionary significance of depression in pathogen host defense (PATHOS-D). Mol Psychiatry 2012, Doi:10.1038/mp.2012.2.

Halaris A, Leonard BE (eds): Inflammation in Psychiatry.
Mod Trends Pharmacopsychiatry. Basel, Karger, 2013, vol 28, pp 1–19 (DOI: 10.1159/000343964)

Basic Aspects of the Immunology of Neuroinflammation

Eric S. Wohleb · Jonathan P. Godbout

Department of Neuroscience, Institute for Behavioral Medicine Research, Center for Brain and Spinal Cord Repair, The Ohio State University, Columbus, Ohio, USA

Abstract

Neuroinflammation is used to describe an immune-related process that occurs within the central nervous system (CNS). The objective of this chapter is to introduce basic aspects of neuroinflammation in the context of psychiatric disorders. Inflammatory processes are evident in the CNS with a myriad of stimuli including neurological disease, CNS injury or infection, peripheral infection, and psychological stress. Because the CNS maintains a degree of immune privilege, endothelial cells of the blood-brain barrier (BBB) and resident CNS innate immune cells are integral to the interpretation and propagation of inflammatory signals. For instance, activated CNS immune cells and the BBB coordinate production of cytokines and secondary messengers that act directly to influence neurophysiology. Another key concept is that there is bidirectional communication between the immune system and CNS. Afferent neuro-immune pathways relay the peripheral inflammatory profile to the CNS by secretion of cytokines and direct neuronal activation. Communication through afferent neuro-immune pathways promotes physiological and behavioral responses that are aimed to help clear pathogens from the host. For example, peripheral infection increases circulating IL-1β that induces production of prostaglandins and IL-1β by CNS immune cells that initiate fever and the behavioral symptoms of sickness. Although transient cytokine responses in the brain may be beneficial to the host, prolonged neuroinflammation associated with chronic illness, neurodegenerative disease, stress, and aging impair neuroimmune regulation and negatively affect normal cognitive and behavioral processes. Thus, understanding neuroimmune regulation and mechanisms that mediate neuroinflammation is important because these pathways likely contribute to the pathophysiology of several mental health disorders.

'Neuroinflammation'

The primary objective of this chapter is to introduce the basic aspects of the immunology of neuroinflammation in the context of psychiatric illness. Therefore, it is first important to define 'neuroinflammation'. The term neuroinflammation is used to broadly describe immune-related processes that occur within the CNS. It is important to highlight that neuroinflammation is context dependent, and the inflammatory processes that occur in neurological disease, CNS injury or infection, physiological stress or peripheral infection are unique. The overlap in these conditions is that there is activation of resident CNS innate immune cells that produce cytokines, chemokines and secondary messengers in response to inflammatory stimuli. Thus, there is a degree of interpretation and propagation of inflammatory signals by resident immune cells of the CNS. A significant difference in the examples mentioned above is whether or not the inflammatory exposure is transient or chronic. In general transient inflammatory responses in the CNS are beneficial to the host. Chronic neuroinflammation, however, is maladaptive and profoundly affects neuronal plasticity and brain homeostasis. For instance, chronic exposure to inflammatory cytokines is implicated in exacerbations of pathology associated with neurodegenerative diseases that lead to accelerated cognitive decline [1, 2].

Much of our understanding of neuroinflammatory processes is based on experimental models of neurological disease [3], traumatic CNS injury [4], and CNS infection [5]. For example, seminal findings in the field of neuroimmunology are derived from work using experimental autoimmune encephalomyelitis (EAE) models of multiple sclerosis (MS). In models of MS and other neurological diseases, there is active recruitment of peripheral immune cells (T cells and macrophages) into the brain and spinal cord. In addition, EAE is associated with breakdown of the blood-brain barrier (BBB), edema, axonal damage, and neuronal loss. Similar immune responses are detected in models of CNS infection and injury. Although these models provide insight into the immunological mechanisms that mediate neuroinflammation, these inflammatory processes cause significant and debilitating pathology.

In this context, it is important to highlight that neuroinflammation associated with disease and injury is vastly different from neuroinflammation that is associated with non-pathological conditions. For instance, transient cytokine production induces adaptive and beneficial physiological and behavioral responses that help fight pathogens [6, 7]. Neurological disease, traumatic CNS injury, CNS infection, and autoimmunity induce immune responses that lead to neuronal dysfunction, significant tissue damage, and infiltration of peripheral immune cells which can be categorized as pathological neuroinflammation. Nonpathological neuroinflammation describes conditions that do not exhibit significant breakdown of the BBB, infiltration of peripheral immune cells, or obvious loss of CNS integrity. In consideration of these differences, this chapter will focus on neuroinflammation associated with nonpathological conditions. Some examples of nonpathological neuroinflammation include cytokine-mediated sickness following infection, psychological stress, and aging. Although this chapter will focus on cellular and molecular mechanisms that induce and regulate nonpathological neuroinflammation, it is important to point out that these mechanisms are also relevant during pathological neuroinflammation.

Another objective of this chapter is to discuss CNS components and immune cells that mediate neuroinflammation. The CNS is traditionally described as an immune privileged organ because it lacks the adaptive arm of the immune system and diminished cellular immune function. Even though the CNS has no resident adaptive immune cells (e.g. T cells or B cells) lymphocytes still provide immune surveillance in the CNS [8, 9]. In fact, T cells patrol the CNS in the meninges, choroid plexus, and cerebrospinal fluid [10]. The importance of immune surveillance provided by T cells is exemplified in MS patients that received treatment to block T cells from entering the CNS. In these cases patients had increased incidence of CNS viruses and encephalopathies [11–13]. Thus, the topic of immune privilege in the CNS is complex and is best defined as dynamic and actively regulated [8, 14]. The presence of the BBB supports the immune privileged nature of the CNS by

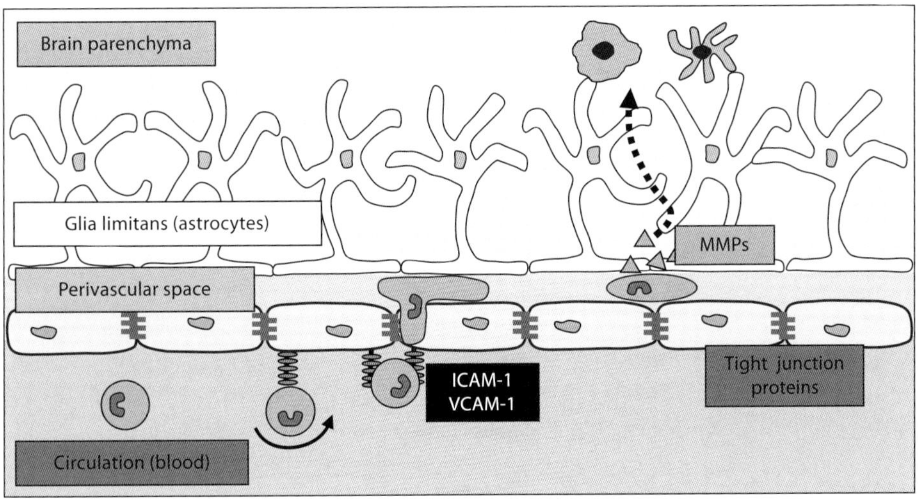

Fig. 1. BBB and immune cell infiltration of the CNS parenchyma.

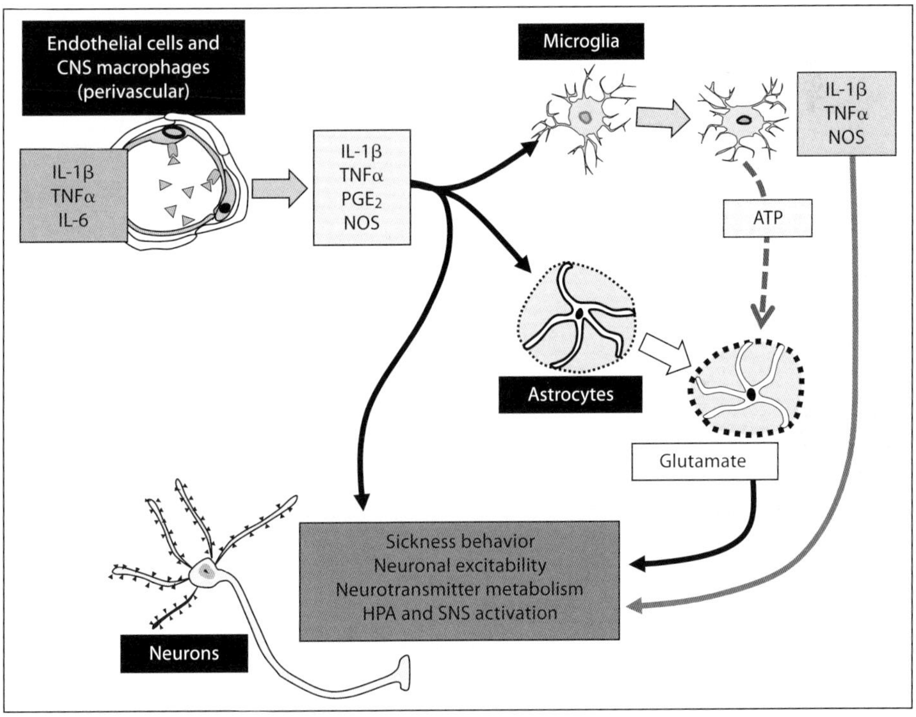

Fig. 2. Cytokines and secondary messenger signaling.

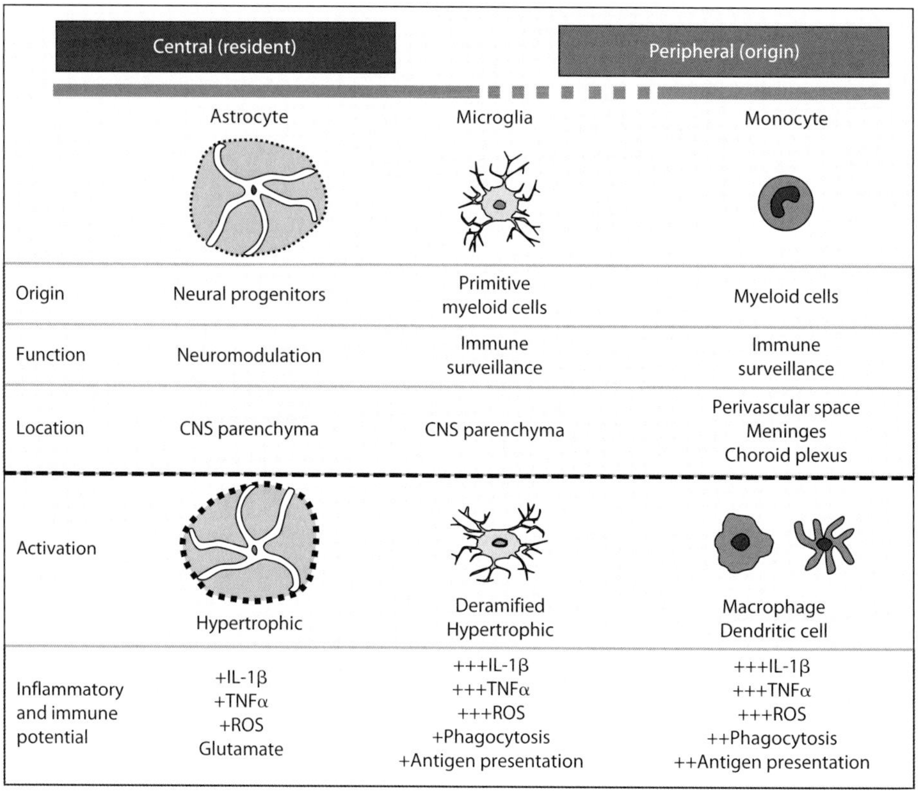

Fig. 3. CNS cells with neuroinflammatory potential.

providing a barrier between the parenchyma and vasculature (fig. 1). Cells that comprise the BBB also facilitate neuroimmune communication (fig. 2). In addition, several CNS cell types have the capacity to initiate and transmit neuroinflammatory signals. Microglia, CNS macrophages, and astrocytes are the primary immune cells that reside in the CNS (fig. 3). Communication between the CNS and immune system is facilitated through humoral and neuronal pathways that transmit inflammatory signals (fig. 4). Collectively, the BBB, resident CNS immune cells, and neuroimmune pathways integrate neuroinflammatory signals and orchestrate responses to promote the organism's health and maintain CNS function.

Role of the Blood-Brain Barrier

The BBB is an important neuroimmune component that consists of endothelial cells, pericytes, and glia. The initial interface between the blood and CNS are endothelial cells. Endothelial cells in the CNS have tight junctions comprised of junctional proteins (e.g. occludins, claudins). These proteins limit infiltration of macromolecules (e.g. cytokines) and cells into the parenchyma (fig. 1). The selectivity of endothelial cells is important because it maintains low cellularity and an optimal balance of ions, proteins, and pH in the interstitial fluid required for neurotransmission [15, 16]. Similar to other CNS functions, the selectivity of the BBB varies depending on the CNS

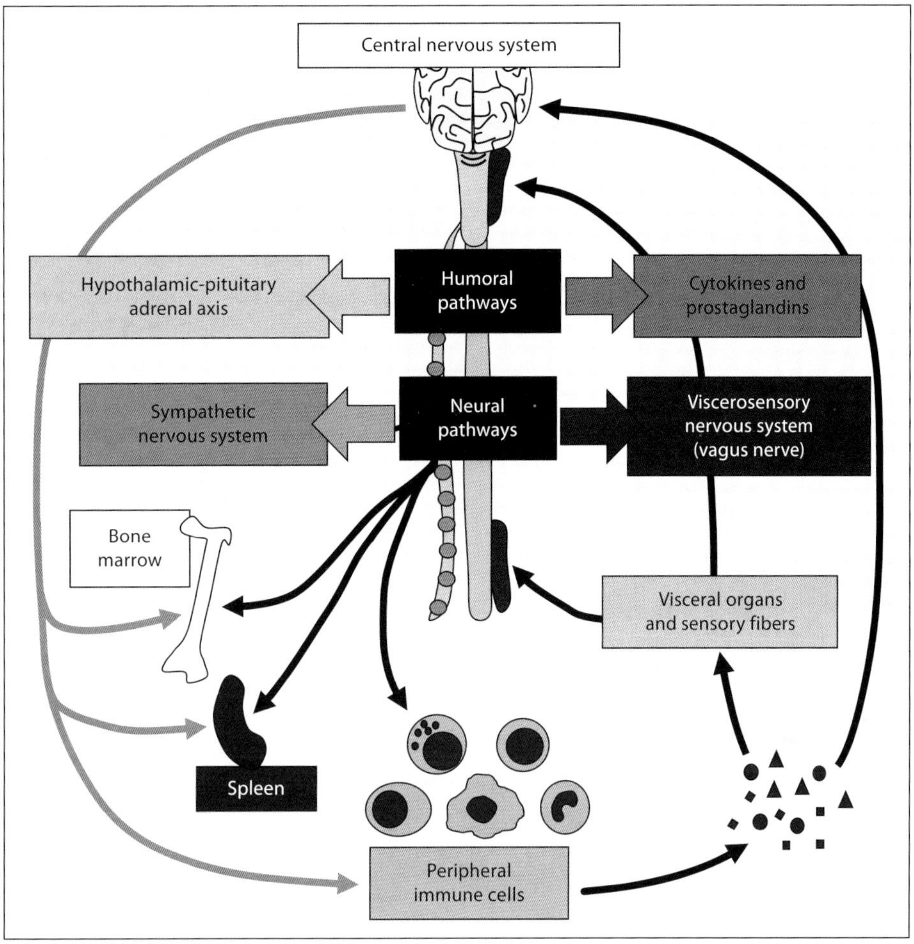

Fig. 4. Neuro-immune communication.

region. For example, the BBB near circumventricular organs and choroid plexus is more permissive to influx of peptides compared to other brain regions [17]. This is mediated by regional properties of endothelial cell junction proteins and the presence of the glia limitans. Astrocytes have processes that extend and entirely ensheath the vasculature, which produces a physiological barrier called the glia limitans [18]. Astrocytes and pericytes are proximal to endothelial cells and help provide BBB stability [19]. Pericytes are another cell type that is integral to the structure of the vasculature. For instance, ablation of pericytes can lead to vascular permeability and leave an individual susceptible to immunogenic responses in the CNS [20]. Between the endothelium and glia limitans is another compartment called the perivascular space. Immune cells including monocytes and lymphocytes have the capacity to take up residence in the perivascular space or traffic through the meninges to provide further immune surveillance [10, 21].

The BBB is an integral structure that has a dynamic role in CNS immunity. To infiltrate the CNS parenchyma immune cells must traverse through the endothelium to the perivascular space and then overcome the glia limitans [22]. Access to the endothelial cell layer is dependent on the expression of adhesion molecules. Important adhesion molecules on endothelial cells include intercellular adhesion molecule-1 (ICAM-1) and vascular cell adhesion molecule-1 (VCAM-1). Peripheral immune cells must express complementary integrins and selectins including leukocyte functional antigen-1 (LFA-1) and very late activation antigen-4 (VLA-4). Adhesion of peripheral immune cells to the endothelium results in transient tethering and rolling [23] (fig. 2). Because the CNS endothelium has relatively low expression of cellular adhesion molecules under basal conditions, inflammatory cytokines, such as IL-1β, are required to up-regulate their expression and promote adhesion [24]. Increased expression of integrins or selectins facilitates firm adhesion and extravasation of peripheral immune cells. Once immune cells have overcome the endothelium, they remain sequestered in the perivascular space sustained by the astroglial sheath. Although immune cell infiltration into the CNS parenchyma is limited, under certain circumstances increased proteinase production allows peripheral immune cells to enter the CNS parenchyma. For example, matrix metalloproteinase activation in trafficking immune cells degrade the astrocytic processes of the glia limitans to gain access to the CNS [25]. The utility of the BBB preserves CNS integrity through limited immune reactivity and selective access to the CNS parenchyma.

CNS Cells with Neuroinflammatory Potential

Microglia

The primary innate immune cells that reside in the CNS are microglia. Microglia are derived from primitive myeloid cells that migrate into the CNS from the yolk-sac early in embryogenesis and remain in the CNS throughout life [26, 27] (fig. 1). Under normal conditions, microglia maintain a low proliferative state and demonstrate limited, if any, turnover from bone marrow-derived monocytes. Microglia have many important non-immune functions including removal of apoptotic neurons, synapse clearance during development, and regular maintenance of synapses [28]. In terms of immune function, microglia continuously survey their micro-environment for damage or pathogens by active extension and retraction of cellular processes [29]. Thus, in homeostatic conditions microglia are described as ramified or 'surveying.' In addition, microglia make transient contact with the vascular interface and respond rapidly to injury by extending processes to reduce hemorrhage and infiltration of peripheral immune cells [5, 30]. Microglia also express an array of receptors for pathogen recognition (e.g. Toll-like receptors; TLRs), neurotransmitters, and hormones (e.g., glucocorticoids) [31, 32]. Although microglia are myeloid-derived cells, they express low levels of pathogen recognition receptors [33] and antigen presentation molecules [34]. Similar to activated peripheral macrophages, activated microglia upregulate antigen presentation molecules (MHC-II) and other pathogen recognition receptors (TLRs) following activation [35].

Many different stimuli promote microglia activation. This activation, however, is tightly regulated and without further perturbation microglia return to a 'surveying' phenotype. When activation signals are encountered microglia become de-ramified with retracted processes and hypertrophic cell body (fig. 1). Activated microglia exhibit macrophage-like functions, including proliferation, cytokine production, and phagocytosis. More specifically activated microglia increase production of inflammatory cytokines (e.g. IL-1β, TNFα, IL-6) and anti-inflammatory cytokines (e.g. IL-10, TGFβ) [36–38]. These cytokine cascades also lead to production of chemokines and secondary messengers (e.g. prostaglandins, nitric oxide, neurotransmitter metabolites). Intimate interactions with neurons and astrocytes help return activated and inflammatory microglia to a quiescent state. These interactions are important because impaired regulation of microglia activation causes prolonged expression of inflammatory cytokines [39]. In conditions such as aging, microglia have a 'primed' or 'sensitized' phenotype with a corresponding increased neuroinflammatory potential [40, 41]. Thus, primed microglia in aged rodents respond to inflammatory stimuli with amplified and prolonged neuroinflammation that causes extended behavioral consequences, including social withdrawal and depressive-like behavior [42–45]. Exaggerated neuroinflammation in aged mice is associated with reduced CX_3CR1 expression on microglia, which is an important receptor that maintains microglia in their homeostatic state [46]. Consistent with these findings CX_3CR1-deficient mice have prolonged neuroinflammation and activation of microglia that coincides with extended depressive-like behavior [47]. It is important to point out that aging is not the only context in which microglia dysregulation leads to behavioral deficits. Aberrant neuroinflammation and dysregulation of microglia is also associated with other psychiatric disorders related to anxiety including obsessive compulsive disorder and schizophrenia [48, 49]. In this context, dysregulation of microglia may diminish homeostatic functions, such as modulation of synapse formation. Taken together, microglia are the primary innate immune cells that reside in the CNS. They are immune competent cells that actively survey the CNS for pathogens and help regulate synaptogenesis. When microglia encounter an activation signal they produce inflammatory cytokines that induce secondary messengers. Microglia activation is tightly controlled through interactions with neurons and astrocytes. When these interactions are dysfunctional, however, prolonged microglia activation contributes to exaggerated neuroinflammation and behavioral complications.

CNS Macrophages

Another primary component of the innate immune system in the CNS is perivascular and meningeal macrophages. CNS macrophages are derived from circulating monocytes and reside in the perivascular space, meninges, or choroid plexus [50–52] (fig. 1). In addition, these cells are transiently associated with the CNS and steadily renew every 3–4 weeks [21, 50]. Although CNS macrophages make up a small portion of CNS immune cells, they have similar functions as other differentiated monocytes and express receptors for cytokines, pathogen recognition, hormones, and neurotransmitters. In addition, their proximity to vascular and parenchymal CNS cells makes them an important transducer of peripheral immune signals to the brain. For instance, in response to peripheral inflammatory cytokines, including IL-1β, perivascular macrophages produce prostaglandins (e.g. PGE_2) that regulate febrile and hypothalamic-pituitary-adrenal (HPA) axis responses [53] (fig. 3). Depending on the intensity and duration of the inflammatory stimulus, perivascular macrophages exert an inhibitory role on endothelial cells to reduce prostaglandin signaling in the CNS parenchyma. This relationship is important because depletion of CNS macrophages leads to exaggerated PGE_2 synthesis by endothelial cells. Amplified PGE_2 production caused increased and protracted febrile and HPA axis responses [54].

An additional function of monocytes is phagocytosis and antigen presentation to regulate adaptive immunity. Dendritic cells are differentiated monocytes that are primary antigen-presenting cells in the periphery [55]. The relative proportion of dendritic cells in the CNS under normal conditions is limited. Nonetheless, similar to CNS macrophages dendritic cells in the CNS can be located in the vasculature where they transiently sample CNS antigens [56]. Reports indicate that aging can increase the presence of dendritic cells in the brain. Dendritic cells, however, likely contribute significantly to neuroinflammation only during neurological disease or injury [57–59]. Peripheral monocytes contribute to the neuroinflammatory profile by differentiating into perivascular or meningeal macrophages that are integral to surveillance of the CNS. These monocytic cells interact with endothelial cells and produce factors that influence parenchymal CNS cells to shape neuroinflammatory and behavioral responses.

Astrocytes

Astrocytes are glia in the CNS that are derived from neural progenitor cells and reside throughout the CNS [60, 61]. The term astrocyte is descriptive of their stellate processes, however, their entire cell body encompasses a defined CNS domain [62]. Previous studies only considered astrocytes as a structural cell-type in the CNS, yet they have integral roles in modulation of the BBB and regulation of neurotransmission at synapses [62, 63]. As indicated, astrocytic processes ensheath CNS vasculature and develop the glia limitans of the BBB. In addition, astrocyte processes are engaged in synapses which allow them to influence neurotransmitters. The primary neurotransmitter that astrocytes regulate is glutamate. Astrocytes uptake excess glutamate at synapses and convert it into glutamine which is shuttled back to neurons [64]. Modulation of glutamate by astrocytes is relevant because activated astrocytes can release glutamate and glutamine that can exacerbate neuronal excitability and neuroinflammation [63, 65]. Several studies also revealed that astrocytes have the capacity to release inflammatory mediators including IL-1, TNFα, CCL2, and PGE_2 [66, 67]. In fact, chronic activation of astrocytes indicated by increased S100B levels, is implicated in prolonged neuroinflammation coinciding with schizophrenic symptomology [68, 69].

Aside from their role in the BBB and neurotransmission, astrocytes respond to immune signals and express a multitude of receptors for cytokines, neurotransmitters, and other small molecules [67]. One small molecule that has an impact on astrocytes is adenosine triphosphate (ATP). Because dystrophic neurons and activated immune cells release ATP, excess extracellular ATP may be indicative of CNS damage [70]. Moreover, purinergic receptor activation on astrocytes can initiate release of glutamate that regulates microglia activation and neuronal excitability [71]. Astrocytes also respond to activation signals following injury or damage in the CNS. For example, pathological injury in the CNS causes activated astrocytes to proliferate and develop glial scars that limit tissue damage and infiltration of peripheral immune cells [72]. Although astroglial scar formation can reduce inflammation following CNS injury, it also limits regeneration of tissue [73]. In this way, astrocytes play an integral role in modulating neurophysiology and CNS immunity. Because the immune capacity of astrocytes was only recently illuminated, the extent of their role in neuroinflammation will continue to be defined.

Cytokine and Secondary Messenger Production by CNS Immune Cells

Cytokines are small molecules that provide intercellular communication between immune cells and help direct immune responses. Activation of immune cells leads to production and secretion of inflammatory and anti-inflammatory cytokines including IL-1β, IL-6, TNFα, IL-10, IL-1 receptor antagonist (IL-1Ra) and TGFβ [74]. These cytokines bind to their corresponding receptor(s) to initiate cell signaling that dictate gene expression. The CNS maintains relatively low expression of cytokines, but inflammatory cytokines (IL-1, IL-6, and TNFα) are rapidly upregulated following CNS immune cell activation.

Inflammatory cytokine production by resident CNS immune cells is essential to propagate inflammatory signals and enact physiological responses to pathogens. In fact, peripheral immune activation cannot sufficiently signal the CNS without inflammatory cytokine production by resident CNS immune cells because cytokines are short-lived and do not diffuse easily through the CNS parenchyma. Cytokine production is mediated by several cellular signaling pathways depending on the origin and intensity of the stimulus. For example, IL-1 binding to IL-1 receptor activates mitogen-activated protein kinase (MAPK) [75] and nuclear factor-κβ (NF-κβ) [76] pathways that initiate inflammatory signaling cascades. In addition, cytokines promote production of secondary messengers including additional cytokines and synthesis of prostaglandins. Secondary mediators such as prostaglandins are essential for neuroimmune communication and directly target neuronal pathways that elicit physiological and behavioral responses [77]. For example, circulating IL-1 stimulates perivascular macrophages to produce PGE_2 that activates brainstem nuclei that in turn mediate HPA axis responses [78]. Coinciding with release of inflammatory cytokines and secondary messengers, there is production of acute phase proteins, neurotransmitter metabolites, and reactive oxygen species that contribute to neuroinflammation [79, 80]. In particular, neuroinflammation is associated with increased expression of inducible nitric oxide synthase (iNOS), which is an enzyme that generates nitric oxide [80]. Although nitric oxide can act as a neuromodulator, during neuroinflammation high concentrations can cause cellular damage [81]. To modulate neuroinflammation and reduce cytotoxicity, anti-inflammatory cytokines, including IL-10, IL-1Ra, and TGFβ are generated in conjunction with inflammatory cytokines. Anti-inflammatory cytokines can directly reduce transcription and translation of inflammatory cytokines through inhibition of MAPK and NF-κβ signaling [82]. In addition, anti-inflammatory cytokines induce suppressor of cytokine signaling pathways that reduce inflammatory cytokine production and promote cellular repair processes [83]. Production and secretion of anti-inflammatory cytokines is particularly important in the CNS because neurons have limited regenerative capacity.

Another key function of cytokine signaling is to promote recruitment of peripheral immune cells to sites of inflammation through expression of chemokines. Chemokines (e.g. CCL2) generate gradients that attract peripheral immune cells to sites of inflammation through binding of specific chemokine receptors (e.g., CCR2). In this way inflammatory cytokines (e.g. IL-1β, TNFα) can facilitate infiltration of peripheral immune cells into the CNS [84]. For example, in models of Alzheimer's disease CCL2 production at inflammatory foci near amyloid plaques recruit CCR2+ monocytes to help clear amyloid depositions [85]. During nonpathological neuroinflammation limited immune cell recruitment is detected. Nonetheless, persistent inflammatory responses can lead to CNS infiltration of peripheral immune cells [86]. In particular, recent evidence indicates that CCL2 and CX_3CL1 mediate the recruitment of monocytes into the CNS [87]. These are important chemokines because they can recruit different subsets of monocytes that express their respective receptors (CCR2, CX_3CR1). Expression of CCR2 is characteristic of a monocyte subset with an increased inflammatory potential, while high expression of CX_3CR1 is indicative of resident, homeostatic monocytes. For example, CX_3CL1-CX_3CR1 interactions enable vascular macrophages to traverse the endothelium patrolling for pathogens or injury. These patrolling macrophages are the first responders following local immune activation and have diminished efficacy in CX_3CR1-deficient mice [88]. This is important because constitutive expression of CX_3CL1 by neurons mediates microglia quiescence. Thus, high levels of CX_3CL1 in the brain can mediate rapid infiltration of patrolling macrophages into the inflamed CNS. The differential responses of these monocyte subsets are an emerging concept in immunity and their role in neuroinflammation will likely be further delineated.

Besides directly affecting neurophysiology and cellular infiltration, cytokines and their secondary messengers have downstream effects on neurotransmitter metabolism. Inflammatory cytokines are associated with activation of indoleamine 2,3-dioxygenase (IDO) and kynurenine 3-monooxygenase (KMO). IDO activation leads to breakdown of tryptophan, the precursor for serotonin production, into kynurenine. Thus, increased IDO activation following neuroinflammation reduces tryptophan availability and decreased serotonin levels [89]. Moreover, kynurenine can be converted by KMO into 3′hydroxy-kynurenine that can be further processed into quinolinic acid. In turn, quinolinic acid can initiate glutamatergic excitotoxicity and promote degeneration of susceptible dopaminergic neurons [90]. Therefore, extended neuroinflammation can directly diminish serotonergic, dopaminergic, and glutamatergic neurotransmission [91]. These inflammatory cytokine-induced pathways may underlie depressive-like behaviors because they reduce the availability of neurotransmitters implicated in depression. This is consistent with literature that indicates a strong association between prolonged neuroinflammation and induction of depression in humans and rodents [6, 92–94]. Moreover, recent studies demonstrate that IDO activation contributes to development of depression because inhibition of IDO by 1-methyl tryptophan reduces depressive-like symptoms [95]. Although kynurenine can also be metabolized into neuroprotective kynurenic acid, it is likely that neuroinflammation directs kynurenine away from this pathway because depressed patients have decreased levels of kynurenic acid [96]. In this way, activation of CNS immune cells can influence neurotransmitter metabolism that can contribute to the development of mental health disorders.

Communication between the Immune System and the CNS

The neuroimmune axis is a unique system of bidirectional communication that connects the CNS and immune system. There are afferent pathways that transmit the inflammatory state of peripheral organs to the CNS and initiate production of inflammatory cytokines. Afferent neuroimmune communication is mediated by humoral and neuronal routes that interact to coordinate appropriate physiological, immunological, and behavioral responses. Humoral pathways transmit the relative inflammatory status through the specialized BBB and circumventricular organs, while direct neuronal pathways can transmit alterations in the inflammatory state of specific peripheral tissues to the CNS [97]. In addition, neuronal and humoral efferent pathways exist that allow the CNS to modulate immune responses (fig. 3).

Afferent Neuroimmune Pathways → Immune to Brain

Afferent neuroimmune pathways communicate the inflammatory status of peripheral tissues to the brain. Inflammatory cytokine production following infections or various other stressors signal the CNS to elicit physiological and behavioral responses including immunomodulation, fever, lethargy, and decreased social interaction [98, 99]. Moreover, social withdrawal and other anti-social-like behaviors are an effective way to limit the spread of infection [100]. These responses are important to fight infections and disruption of normal sickness responses results in decreased host survival [101]. Circulating cytokines constitute the humoral arm of afferent neuroimmune communication. Inflammatory cytokines, including IL-1β, IL-6, and TNFα, are secreted by activated peripheral immune cells to help coordinate neuroimmune communication [97] (fig. 4). In order for cytokine signals initiated at the periphery to reach the brain, the signal must overcome the obstacle of the BBB. One way that cytokines can interact with the CNS is through activation of CNS immune cells or diffusion at specialized brain regions called circumventricular organs (CVOs). The CVOs are located around the third or fourth ventricles and exhibit specialized BBB properties that allow for diffusion of cytokines into the brain parenchyma [102]. These areas include the choroid plexus, subfornical organ, organum vasculosum of the lamina terminalis, median eminence, posterior pituitary, subcomissural organ, pineal gland, and area postrema. While these areas lack a definitive BBB, there is still a significant presence of microglia and astrocytes. This is important because local CNS immune cell activation in CVOs corresponds with activation of neurons that are integral for physiological responses to immune activation. For example, peripheral LPS injection increases IL-1β and iNOS production in microglia and CNS macrophages that coincides with neuronal activation in several CVOs including the area postrema [103]. In this manner, peripheral cytokine production can interact directly with CNS immune cells that target neuronal pathways.

Besides direct transmission at CVOs, cytokines can signal the CNS through endothelial cells of the BBB. Endothelial cells have receptors for many cytokines and can actively transport cytokines into the brain parenchyma. For example, endothelial cells have receptors for TNFα that when bound, will actively transport these cytokines into the brain [104]. Moreover, receptors on endothelial cells exhibit a dynamic responsiveness to inflammatory cytokines depending on their intensity and origin. Circulating IL-1β can be transported across the endothelium to propagate IL-1β signaling in the CNS, while IL-1β production in the brain parenchyma induces increased expression of adhesion molecules that promotes leukocyte infiltration into the parenchyma [105] (fig. 3). These studies highlight the functionality of the CNS endothelium depending on the signal and its source. Inflammatory cytokine signals are also transmitted across the BBB by de novo secretion of cytokines and secondary messengers by endothelial cells and CNS macrophages. In fact, inflammatory cytokines released following LPS administration cause endothelial cells and perivascular macrophages to produce prostaglandins and inflammatory cytokines including IL-1β, TNFα, and IL-6 [97]. Through direct binding or indirect activation these inflammatory mediators induce NF-κβ signaling pathways in microglia that promotes further production of IL-1β, TNFα, and IL-6 in the brain parenchyma [75, 106] (fig. 3). Ultimately, cytokines are integral components of neuroimmune communication that directly or indirectly through secondary messengers target neuronal substrates and initiate physiological responses following peripheral inflammation. Furthermore, cytokines influence activation of BBB and CNS immune cells to potentiate neuroinflammation.

The primary afferent neuronal pathway that communicates the inflammatory status of peripheral tissue and visceral organs to the CNS is the vagus nerve [103, 107–109] (fig. 4). The vagus nerve is a craniofacial nerve that innervates the nucleus of the solitary tract (NST). The NST is an important brain region for integration of sensory information. As a portion of the viscerosensory nervous system vagal nerve afferents are present in peripheral tissues and express cytokine receptors that transmit local levels of cytokine production [110]. Therefore, input from the vagus nerve is necessary to communicate peripheral immune activation and initiate cytokine and secondary messenger signaling in the CNS. This is relevant because without neuroimmune communication from the vagus nerve behavioral responses to peripheral inflammation are diminished. For example, resection of the vagus nerve below the diaphragm attenuated cytokine-induced social withdrawal [111]. Thus, activation of the vagus nerve is a direct neuronal afferent pathway that elicits adaptive behavioral responses in response to inflammatory cytokines induced by peripheral infection or other stressors.

In the past few decades, significant progress has been made in understanding afferent neuroimmune communication. Further studies directed at local cellular relationships can also reveal novel interactions between neuroimmune compartments. Recent studies are beginning to uncover the complex connections between activation of neuronal pathways and microglia following peripheral infection or stress. For instance, peripheral inflammation induced neuronal activation in specific amygdala nuclei that enhanced release of norepinephrine. Increased concentrations of norepinephrine coincided with a time-dependent increase of IL-1β, IL-6, and TNFα gene expression in the amygdala that resulted in lethargy [112]. This is relevant because localized activation of neuronal pathways likely enhanced CNS immune cell production of inflammatory cytokines that influenced the behavioral response to peripheral inflammation. Moreover, specific brain regions may be more susceptible to neuroinflammation depending on the stimulus. For instance, specific stress-responsive brain regions have corresponding activation of neurons and microglia that promote development of anxiety-like behavior and cognitive deficits [113–115]. Stress-induced activation of noradrenergic pathways is implicated because pretreatment with propranolol, a β-adrenergic receptor antagonist, reduced neuronal and microglial activation in these specific brain regions and prevented development of anxiety [115].

Region-specific alterations in inflammatory cytokine production and neurophysiology are relevant because many of these brain regions are linked to psychiatric disorders associated with stress responses [116, 117]. Furthermore, region-specific production of inflammatory cytokines may underlie behavioral and cognitive deficits noted in aging or following peripheral infection. Activation of CNS immune cells in the hippocampus has a detrimental impact on homeostatic CNS functions such as neurogenesis. This is relevant because models of aging and infection involve increased inflammatory cytokine production and reduced neurogenesis in the hippocampus that corresponds with depressive-like behavior and deficits in cognitive function [118–120]. Moreover, there is evidence that neurogenesis in the CNS is dependent on cross-talk between vascular cells (e.g. endothelial cells) and neural progenitors. For example, endothelial cells produce neurotrophic factors that are important to stimulate neural stem cell differentiation [121, 122]. Thus, it is possible in this context that inflammatory signals subjugate endothelial cells from their pro-neurogenic role to transmit inflammatory signals in the CNS. Although these findings help delineate the effects of region-specific neuroimmune signaling, there are many unanswered questions about the mechanisms that mediate cellular communication between neurons and CNS immune cells.

The fact that the CNS is organized to respond to region-specific immune signals may seem intuitive, but the specificity of neuroimmune signaling is not confined to the CNS. Recent studies indicate that inflammatory loci in the periphery are represented in the CNS with specific patterns of regional activation in brain nuclei. This relationship was revealed in a study that investigated the activation of neuronal pathways following localized peripheral inflammation induced by casein. Injection of casein in either the forelimb or hindlimb caused contralateral activation of neurons in the paraventricular nucleus of the hypothalamus. Moreover, forelimb inflammation tended to increase activation of neurons in the anterior part of the PVN, while hindlimb injections caused activation in the posterior regions of the PVN. This indicates that the location of peripheral inflammatory responses is represented in specific brain regions [123]. Local neuronal activation in the hypothalamus following peripheral inflammation is relevant because the hypothalamus can initiate anti-inflammatory reflexes mediated by the efferent parasympathetic neurons. Inflammatory-mediated activation of the hypothalamus causes local release of acetylcholine by parasympathetic neurons in target organs that reduces inflammation to prevent peripheral tissue damage [124]. The corresponding representation of peripheral inflammatory responses and responsiveness of the hypothalamus substantiates the intricate interactions between the CNS and immune system. Although these studies highlight some of the emerging concepts in neuroimmune communication, further research is essential to develop a better understanding of these integrated systems.

Efferent Neuroimmune Pathways → Brain to Immune

Reciprocal efferent neuroimmune pathways are present that influence immune reactivity and help restore homeostasis following peripheral inflammation. Similar to afferent neuroimmune pathways, there are humoral and neuronal efferent pathways. The hypothalamic-pituitary-adrenal (HPA) axis is the primary efferent humoral pathway that is activated in response to circulating inflammatory mediators and various stressors (fig. 4). Initial neuronal activation of the hypothalamus causes release of corticotrophin-releasing hormone (CRH) that instructs the pituitary gland to secrete adrenocorticotropic hormone (ACTH) into circulation. ACTH stimulates the adrenal gland to produce glucocorticoids that enter circulation. Because glucocorticoids are released directly into circulation, they exert widespread effects throughout the body [125]. Classical studies demonstrate that both peripheral and central immune cells express receptors for glucocorticoids, and receptor binding decreases immune cell viability and reactivity [31, 126]. More recent studies indicate that there is a dynamic response to glucocorticoids depending on temporal variables and length of exposure [127–129]. There is evidence that hyper-activation of the HPA axis and extended exposure to glucocorticoids reduces the sensitivity of immune cells to anti-inflammatory feedback [126]. For example, pretreatment of macrophages with glucocorticoids can increase the production of pro-inflammatory cytokines following immune activation with LPS [130]. Glucocorticoids can also influence production of peripheral immune cells from bone marrow progenitors. In fact, increased circulating levels of glucocorticoids shift hematopoiesis towards increased production of monocyte and granulocyte lineages [131–133]. This is relevant because an increased proportion of myeloid-derived cells that are glucocorticoid insensitive can increase basal levels of inflammation and cause amplified inflammation following peripheral immune activation [134]. This is important because enhanced peripheral inflammation following LPS challenge causes exaggerated inflammatory cytokine production in the CNS that can lead to extended sickness behavior [135]. In all, HPA axis activation and subsequent release of glucocorticoids are integral to regulate peripheral inflammatory responses; however, dysregulation of this humoral arm can lead to detrimental immune reactivity.

A primary efferent neuronal pathway stemming from the CNS is the sympathetic nervous system (SNS). The SNS is a component of the autonomic nervous system and innervates many organs throughout the body. Activation of the SNS causes release of norepinephrine at target tissues and systemic release of epinephrine and norepinephrine by the adrenal gland (fig. 4). Physiological effects of SNS activation involve constriction of blood vessels, increased heart rate, and mobilization of energy resources, which are associated with the 'fight or flight' response [126]. In addition to these classical autonomic responses, activation of the SNS has a profound influence on immune responses. Classical neuroanatomical studies demonstrate that SNS neurons directly innervate the bone marrow, spleen, and lymph nodes [136] (fig. 4). Because many immune cells express receptors for norepinephrine, SNS activation can regulate their activity [137]. For instance, peripheral immune activation and psychological stress exposure induced SNS activation that was associated with decreased TNFα and IL-1β production in splenic macrophages. An anti-inflammatory role for SNS activation was proposed because removal of the splenic nerve abrogated the suppressive effects of the stimuli [138].

Although acute SNS stimulation may be beneficial in reducing inflammatory cytokine production there is evidence that chronic activation of the SNS can have deleterious inflammatory and behavioral effects. Exposure to chronic psychological stress promotes extended SNS activation that is associated with inflammatory conditions. For example, peripheral blood monocytes ($CD11b^{+}$) obtained from caregivers exposed to extended stress have increased inflammatory signaling through the transcription factor NF-κβ [139, 140]. Moreover, gene transcription analyses indicate that increased inflammatory signaling is mediated by transcription factors that are activated following binding of noradrenergic receptors (e.g. GATA1) [141]. These alterations in gene expression result in reactive monocytes with an enhanced inflammatory potential. Furthermore, increased levels of epinephrine in the circulation leads to proliferation and motility of bone marrow hematopoietic cells [142]. When aversive stimuli initiate stress responses, this may be advantageous because the CNS is capable of arming and mobilizing immune cells through SNS activation. This is problematic if stress responses are generalized to non-aversive stimuli because chronic SNS activation can promote maladaptive inflammatory responses. As a consequence, increased peripheral inflammatory cytokine production can lead to stress-induced neuroinflammation. For example, rodent studies with inescapable foot shock and social defeat demonstrated that stress increased brain IL-1β, IL-6, and TNFα mRNA [115, 143–145]. Similar to other models of neuroinflammation, psychological stress promotes development of behavioral deficits, including anxiety, social withdrawal, and depressive-like behaviors [115, 146, 147]. This is consistent with clinical reports that indicate high levels of stress are associated with inflammatory conditions and a higher incidence of mental health disorders including depression and anxiety [139, 141, 148–153]. Thus, similar to the HPA axis, impaired regulation of SNS activation can have deleterious effects.

Conclusion

The purpose of this chapter was to introduce basic aspects that mediate neuroinflammation and CNS immunity. The BBB is comprised of endothelial cells, pericytes and glia limitans and maintains the immune-privileged nature of the CNS. To compensate for the lack of immune function, there are resident CNS immune cells that exhibit neuroinflammatory potential following activation. Microglia, CNS macrophages, and astrocytes produce cytokines and secondary messengers that potentiate neuroinflammation and directly influence neurophysiology. Because the CNS has limited regenerative capacity, negative feedback mechanisms including anti-inflammatory cytokines are integral to modulate inflammatory cytokine and secondary messenger production. Furthermore, neuroimmune communication facilitates CNS immune regulation through humoral and neuronal pathways that transmit the peripheral inflammatory status and direct appropriate behavioral responses. Each of these components is important because dysregulation in these systems results in amplified and prolonged inflammatory cytokine production that can contribute to mental health consequences. This is evident in models of aging, chronic infection, and psychological stress in which exaggerated neuroinflammation is associated with long-term behavioral and cognitive deficits, including prolonged sickness, depressive-like behavior, and memory impairment. Therefore, determining the mechanisms that mediate neuroimmune dysregulation associated with psychiatric disorders will provide therapeutic targets to alleviate behavioral consequences of neuroinflammation.

References

1 Lee YJ, Han SB, Nam SY, Oh KW, Hong JT: Inflammation and Alzheimer's disease. Arch Pharm Res 2010;33: 1539–1556.
2 Lee KS, Chung JH, Choi TK, Suh SY, Oh BH, et al: Peripheral cytokines and chemokines in Alzheimer's disease. Dement Geriatr Cogn Disord 2009;28:281–287.
3 Muir MT, Lovett-Racke AE, Racke MK: Novel therapeutic strategies targeting the pathogenic T-cells in multiple sclerosis. Expert Rev Clin Immunol 2005;1: 345–355.
4 Donnelly DJ, Popovich PG: Inflammation and its role in neuroprotection, axonal regeneration and functional recovery after spinal cord injury. Exp Neurol 2008;209:378–388.
5 McGavern DB, Kang SS: Illuminating viral infections in the nervous system. Nat Rev Immunol 2011;11:318–329.
6 Dantzer R, O'Connor JC, Freund GG, Johnson RW, Kelley KW: From inflammation to sickness and depression: when the immune system subjugates the brain. Nat Rev Neurosci 2008;9:46–56.
7 Dantzer R, Kelley KW: Twenty years of research on cytokine-induced sickness behavior. Brain Behav Immun 2007;21: 153–160.
8 Galea I, Bechmann I, Perry VH: What is immune privilege (not)? Trends Immunol 2007;28:12–18.
9 Fabry Z, Raine CS, Hart MN: Nervous tissue as an immune compartment: the dialect of the immune response in the CNS. Immunol Today 1994;15:218–224.
10 Hickey WF: Basic principles of immunological surveillance of the normal central nervous system. Glia 2001;36:118–124.
11 Stuve O, Marra CM, Bar-Or A, Niino M, Cravens PD, et al: Altered CD4+/CD8+ T-cell ratios in cerebrospinal fluid of natalizumab-treated patients with multiple sclerosis. Arch Neurol 2006;63: 1383–1387.
12 Hellwig K, Gold R: Progressive multifocal leukoencephalopathy and natalizumab. J Neurol 2011;258:1920–1928.
13 Stuve O, Gold R, Chan A, Mix E, Zettl U, et al: Alpha4-Integrin antagonism with natalizumab: effects and adverse effects. J Neurol 2008;255(suppl 6):58–65.
14 Carson MJ, Doose JM, Melchior B, Schmid CD, Ploix CC: CNS immune privilege: hiding in plain sight. Immunol Rev 2006;213:48–65.
15 Abbott NJ: Astrocyte-endothelial interactions and blood-brain barrier permeability. J Anat 2002;200:629–638.
16 Zlokovic BV: The blood-brain barrier in health and chronic neurodegenerative disorders. Neuron 2008;57:178–201.
17 Ermisch A, Ruhle HJ, Landgraf R, Hess J: Blood-brain barrier and peptides. J Cereb Blood Flow Metab 1985;5:350–357.
18 Mathiisen TM, Lehre KP, Danbolt NC, Ottersen OP: The perivascular astroglial sheath provides a complete covering of the brain microvessels: an electron microscopic 3D reconstruction. Glia 2010;58:1094–1103.
19 Abbott NJ, Ronnback L, Hansson E: Astrocyte-endothelial interactions at the blood-brain barrier. Nat Rev Neurosci 2006;7:41–53.
20 Armulik A, Genove G, Mae M, Nisancioglu MH, Wallgard E, et al: Pericytes regulate the blood-brain barrier. Nature 2010;468:557–561.
21 Hickey WF, Kimura H: Perivascular microglial cells of the CNS are bone marrow-derived and present antigen in vivo. Science 1988;239:290–292.
22 Bechmann I, Galea I, Perry VH: What is the blood-brain barrier (not)? Trends Immunol 2007;28:5–11.
23 Wilson EH, Weninger W, Hunter CA: Trafficking of immune cells in the central nervous system. J Clin Invest 2010;120:1368–1379.
24 Greenwood J, Wang Y, Calder VL: Lymphocyte adhesion and transendothelial migration in the central nervous system: the role of LFA-1, ICAM-1, VLA-4 and VCAM-1. Immunology 1995;86:408–415.
25 Yong VW, Zabad RK, Agrawal S, Goncalves Dasilva A, Metz LM: Elevation of matrix metalloproteinases (MMPs) in multiple sclerosis and impact of immunomodulators. J Neurol Sci 2007;259:79–84.
26 Ransohoff RM, Cardona AE: The myeloid cells of the central nervous system parenchyma. Nature 2010;468:253–262.
27 Ginhoux F, Greter M, Leboeuf M, Nandi S, See P, et al: Fate mapping analysis reveals that adult microglia derive from primitive macrophages. Science 2010;330:841–845.
28 Tremblay ME, Stevens B, Sierra A, Wake H, Bessis A, et al: The role of microglia in the healthy brain. J Neurosci 2011;31:16064–16069.
29 Nimmerjahn A, Kirchhoff F, Helmchen F: Resting microglial cells are highly dynamic surveillants of brain parenchyma in vivo. Science 2005;308: 1314–1318.
30 Grossmann R, Stence N, Carr J, Fuller L, Waite M, et al: Juxtavascular microglia migrate along brain microvessels following activation during early postnatal development. Glia 2002;37:229–240.
31 Rivest S: Regulation of innate immune responses in the brain. Nat Rev Immunol 2009;9:429–439.
32 Pocock JM, Kettenmann H: Neurotransmitter receptors on microglia. Trends Neurosci 2007;30:527–535.
33 Olson JK, Miller SD: Microglia initiate central nervous system innate and adaptive immune responses through multiple TLRs. J Immunol 2004;173:3916–3924.
34 de Haas AH, Boddeke HW, Biber K: Region-specific expression of immunoregulatory proteins on microglia in the healthy CNS. Glia 2008;56:888–894.

35 Aloisi F: Immune function of microglia. Glia 2001;36:165–179.
36 Davalos D, Grutzendler J, Yang G, Kim JV, Zuo Y, et al: ATP mediates rapid microglial response to local brain injury in vivo. Nat Neurosci 2005;8:752–758.
37 Garden GA, Moller T: Microglia biology in health and disease. J Neuroimmune Pharmacol 2006;1:127–137.
38 Ransohoff RM, Perry VH: Microglial physiology: unique stimuli, specialized responses. Annu Rev Immunol 2009;27: 119–145.
39 Jurgens HA, Johnson RW: Dysregulated neuronal-microglial cross-talk during aging, stress and inflammation. Exp Neurol 2012;233:40–48.
40 Dilger RN, Johnson RW: Aging, microglial cell priming, and the discordant central inflammatory response to signals from the peripheral immune system. J Leukoc Biol 2008;84:932–939.
41 Corona AW, Fenn AM, Godbout JP: Cognitive and behavioral consequences of impaired immunoregulation in aging. J Neuroimmune Pharmacol 2012;7: 7–23.
42 Henry CJ, Huang Y, Wynne AM, Godbout JP: Peripheral lipopolysaccharide (LPS) challenge promotes microglial hyperactivity in aged mice that is associated with exaggerated induction of both pro-inflammatory IL-1beta and anti-inflammatory IL-10 cytokines. Brain Behav Immun 2009;23:309–317.
43 Wynne AM, Henry CJ, Huang Y, Cleland A, Godbout JP: Protracted downregulation of CX(3)CR1 on microglia of aged mice after lipopolysaccharide challenge. Brain Behav Immun 2010;24:1190–1201.
44 Godbout JP, Moreau M, Lestage J, Chen J, Sparkman NL, et al: Aging exacerbates depressive-like behavior in mice in response to activation of the peripheral innate immune system. Neuropsychopharmacology 2008;33:2341–2351.
45 Huang Y, Henry CJ, Dantzer R, Johnson RW, Godbout JP: Exaggerated sickness behavior and brain proinflammatory cytokine expression in aged mice in response to intracerebroventricular lipopolysaccharide. Neurobiol Aging 2008; 29:1744–1753.
46 Cardona AE, Pioro EP, Sasse ME, Kostenko V, Cardona SM, et al: Control of microglial neurotoxicity by the fractalkine receptor. Nat Neurosci 2006;9: 917–924.
47 Corona AW, Huang Y, O'Connor JC, Dantzer R, Kelley KW, et al: Fractalkine receptor (CX3CR1) deficiency sensitizes mice to the behavioral changes induced by lipopolysaccharide. J Neuroinflammation 2010;7:93.
48 Bechter K, Reiber H, Herzog S, Fuchs D, Tumani H, et al: Cerebrospinal fluid analysis in affective and schizophrenic spectrum disorders: identification of subgroups with immune responses and blood-CSF barrier dysfunction. J Psychiatr Res 2010;44:321–330.
49 Chen SK, Tvrdik P, Peden E, Cho S, Wu S, et al: Hematopoietic origin of pathological grooming in Hoxb8 mutant mice. Cell 2010;141:775–785.
50 Bechmann I, Priller J, Kovac A, Bontert M, Wehner T, et al: Immune surveillance of mouse brain perivascular spaces by blood-borne macrophages. Eur J Neurosci 2001;14:1651–1658.
51 Vallieres L, Sawchenko PE: Bone marrow-derived cells that populate the adult mouse brain preserve their hematopoietic identity. J Neurosci 2003;23:5197–5207.
52 Prinz M, Priller J, Sisodia SS, Ransohoff RM: Heterogeneity of CNS myeloid cells and their roles in neurodegeneration. Nat Neurosci 2011;14:1227–1235.
53 Schiltz JC, Sawchenko PE: Signaling the brain in systemic inflammation: the role of perivascular cells. Front Biosci 2003;8: s1321–s1329.
54 Serrats J, Schiltz JC, Garcia-Bueno B, van Rooijen N, Reyes TM, et al: Dual roles for perivascular macrophages in immune-to-brain signaling. Neuron 2010;65:94–106.
55 McMahon EJ, Bailey SL, Miller SD: CNS dendritic cells: critical participants in CNS inflammation? Neurochem Int 2006;49:195–203.
56 Prodinger C, Bunse J, Kruger M, Schiefenhovel F, Brandt C, et al: CD11c-expressing cells reside in the juxtavascular parenchyma and extend processes into the glia limitans of the mouse nervous system. Acta Neuropathol 2011; 121:445–458.
57 Bulloch K, Miller MM, Gal-Toth J, Milner TA, Gottfried-Blackmore A, et al: CD11c/EYFP transgene illuminates a discrete network of dendritic cells within the embryonic, neonatal, adult, and injured mouse brain. J Comp Neurol 2008;508:687–710.
58 Felger JC, Abe T, Kaunzner UW, Gottfried-Blackmore A, Gal-Toth J, et al: Brain dendritic cells in ischemic stroke: time course, activation state, and origin. Brain Behav Immun 2010;24:724–737.
59 Kaunzner UW, Miller MM, Gottfried-Blackmore A, Gal-Toth J, Felger JC, et al: Accumulation of resident and peripheral dendritic cells in the aging CNS. Neurobiol Aging 2012;33:681e1–693e1.
60 He F, Ge W, Martinowich K, Becker-Catania S, Coskun V, et al: A positive autoregulatory loop of Jak-STAT signaling controls the onset of astrogliogenesis. Nat Neurosci 2005;8:616–625.
61 Bonni A, Sun Y, Nadal-Vicens M, Bhatt A, Frank DA, et al: Regulation of gliogenesis in the central nervous system by the JAK-STAT signaling pathway. Science 1997;278:477–483.
62 Allen NJ, Barres BA: Neuroscience: Glia – more than just brain glue. Nature 2009;457:675–677.
63 Parpura V, Heneka MT, Montana V, Oliet SH, Schousboe A, et al: Glial cells in (patho)physiology. J Neurochem 2012; 121:4–27.
64 Anderson CM, Swanson RA: Astrocyte glutamate transport: review of properties, regulation, and physiological functions. Glia 2000;32:1–14.
65 Galic MA, Riazi K, Pittman QJ: Cytokines and brain excitability. Front Neuroendocrinol 2012;33:116–125.
66 Dong Y, Benveniste EN: Immune function of astrocytes. Glia 2001;36:180–190.
67 Sofroniew MV, Vinters HV: Astrocytes: biology and pathology. Acta Neuropathol 2010;119:7–35.
68 Pedersen A, Diedrich M, Kaestner F, Koelkebeck K, Ohrmann P, et al: Memory impairment correlates with increased S100B serum concentrations in patients with chronic schizophrenia. Prog Neuropsychopharmacol Biol Psychiatry 2008;32:1789–1792.
69 Rothermundt M, Falkai P, Ponath G, Abel S, Burkle H, et al: Glial cell dysfunction in schizophrenia indicated by increased S100B in the CSF. Mol Psychiatry 2004;9:897–899.
70 Haydon PG, Carmignoto G: Astrocyte control of synaptic transmission and neurovascular coupling. Physiol Rev 2006;86:1009–1031.
71 Pascual O, Ben Achour S, Rostaing P, Triller A, Bessis A: Microglia activation triggers astrocyte-mediated modulation of excitatory neurotransmission. Proc Natl Acad Sci USA 2011;109:E197–E205.

72 Herrmann JE, Imura T, Song B, Qi J, Ao Y, et al: STAT3 is a critical regulator of astrogliosis and scar formation after spinal cord injury. J Neurosci 2008;28: 7231–7243.
73 Silver J, Miller JH: Regeneration beyond the glial scar. Nat Rev Neurosci 2004;5: 146–156.
74 Janeway CA Jr, Medzhitov R: Innate immune recognition. Annu Rev Immunol 2002;20:197–216.
75 Parker LC, Luheshi GN, Rothwell NJ, Pinteaux E: IL-1 beta signalling in glial cells in wildtype and IL-1RI deficient mice. Br J Pharmacol 2002;136:312–320.
76 Lawrence T: The nuclear factor NF-kappaB pathway in inflammation. Cold Spring Harb Perspect Biol 2009;1: a001651.
77 Lazarus M, Yoshida K, Coppari R, Bass CE, Mochizuki T, et al: EP3 prostaglandin receptors in the median preoptic nucleus are critical for fever responses. Nat Neurosci 2007;10:1131–1133.
78 Ericsson A, Arias C, Sawchenko PE: Evidence for an intramedullary prostaglandin-dependent mechanism in the activation of stress-related neuroendocrine circuitry by intravenous interleukin-1. J Neurosci 1997;17:7166–7179.
79 Anthony DC, Couch Y, Losey P, Evans MC: The systemic response to brain injury and disease. Brain Behav Immun 2012;26:534–540.
80 Saha RN, Pahan K: Regulation of inducible nitric oxide synthase gene in glial cells. Antioxid Redox Signal 2006;8: 929–947.
81 Paakkari I, Lindsberg P: Nitric oxide in the central nervous system. Ann Med 1995;27:369–377.
82 Strle K, Zhou JH, Shen WH, Broussard SR, Johnson RW, et al: Interleukin-10 in the brain. Crit Rev Immunol 2001;21: 427–449.
83 Baker BJ, Akhtar LN, Benveniste EN: SOCS1 and SOCS3 in the control of CNS immunity. Trends Immunol 2009;30: 392–400.
84 Cardona AE, Li M, Liu L, Savarin C, Ransohoff RM: Chemokines in and out of the central nervous system: much more than chemotaxis and inflammation. J Leukoc Biol 2008;84:587–594.
85 El Khoury J, Toft M, Hickman SE, Means TK, Terada K, et al: Ccr2 deficiency impairs microglial accumulation and accelerates progression of Alzheimer-like disease. Nat Med 2007;13:432–438.
86 D'Mello C, Le T, Swain MG: Cerebral microglia recruit monocytes into the brain in response to tumor necrosis factor-alpha signaling during peripheral organ inflammation. J Neurosci 2009;29: 2089–2102.
87 Prinz M, Priller J: Tickets to the brain: role of CCR2 and CX3CR1 in myeloid cell entry in the CNS. J Neuroimmunol 2010;224:80–84.
88 Auffray C, Fogg D, Garfa M, Elain G, Join-Lambert O, et al: Monitoring of blood vessels and tissues by a population of monocytes with patrolling behavior. Science 2007;317:666–670.
89 Myint AM, Kim YK: Cytokine-serotonin interaction through IDO: a neurodegeneration hypothesis of depression. Med Hypoth 2003;61:519–525.
90 Stone TW, Darlington LG: Endogenous kynurenines as targets for drug discovery and development. Nat Rev Drug Discov 2002;1:609–620.
91 Muller N, Schwarz MJ: The immune-mediated alteration of serotonin and glutamate: towards an integrated view of depression. Mol Psychiatry 2007;12: 988–1000.
92 Leonard BE, Myint A: The psychoneuroimmunology of depression. Hum Psychopharmacol 2009;24:165–175.
93 Raison CL, Capuron L, Miller AH: Cytokines sing the blues: inflammation and the pathogenesis of depression. Trends Immunol 2006;27:24–31.
94 Haroon E, Raison CL, Miller AH: Psychoneuroimmunology meets neuropsychopharmacology: translational implications of the impact of inflammation on behavior. Neuropsychopharmacology 2012;37:137–162.
95 O'Connor JC, Lawson MA, Andre C, Moreau M, Lestage J, et al: Lipopolysaccharide-induced depressive-like behavior is mediated by indoleamine 2,3-dioxygenase activation in mice. Mol Psychiatry 2009;14:511–522.
96 Myint AM, Kim YK, Verkerk R, Scharpe S, Steinbusch H, et al: Kynurenine pathway in major depression: evidence of impaired neuroprotection. J Affect Disord 2007;98:143–151.
97 Quan N, Banks WA: Brain-immune communication pathways. Brain Behav Immun 2007;21:727–735.
98 Kent S, Bret-Dibat JL, Kelley KW, Dantzer R: Mechanisms of sickness-induced decreases in food-motivated behavior. Neurosci Biobehav Rev 1996;20:171–175.
99 Konsman JP, Parnet P, Dantzer R: Cytokine-induced sickness behaviour: mechanisms and implications. Trends Neurosci 2002;25:154–159.
100 Maier SF, Watkins LR: Cytokines for psychologists: implications of bidirectional immune-to-brain communication for understanding behavior, mood, and cognition. Psychol Rev 1998;105:83–107.
101 Kluger MJ, Rothenburg BA: Fever and reduced iron: their interaction as a host defense response to bacterial infection. Science 1979;203:374–376.
102 Benarroch EE: Circumventricular organs: receptive and homeostatic functions and clinical implications. Neurology 2011;77:1198–1204.
103 Goehler LE, Erisir A, Gaykema RP: Neural-immune interface in the rat area postrema. Neuroscience 2006;140:1415–1434.
104 Quan N: Immune-to-brain signaling: how important are the blood-brain barrier-independent pathways? Mol Neurobiol 2008;37:142–152.
105 Ching S, He L, Lai W, Quan N: IL-1 type I receptor plays a key role in mediating the recruitment of leukocytes into the central nervous system. Brain Behav Immun 2005;19:127–137.
106 Pinteaux E, Parker LC, Rothwell NJ, Luheshi GN: Expression of interleukin-1 receptors and their role in interleukin-1 actions in murine microglial cells. J Neurochem 2002;83:754–763.
107 Marvel FA, Chen CC, Badr N, Gaykema RP, Goehler LE: Reversible inactivation of the dorsal vagal complex blocks lipopolysaccharide-induced social withdrawal and c-Fos expression in central autonomic nuclei. Brain Behav Immun 2004;18:123–134.
108 Konsman JP, Luheshi GN, Bluthe RM, Dantzer R: The vagus nerve mediates behavioural depression, but not fever, in response to peripheral immune signals; a functional anatomical analysis. Eur J Neurosci 2000;12:4434–4446.
109 Goehler LE, Gaykema RP, Hammack SE, Maier SF, Watkins LR: Interleukin-1 induces c-Fos immunoreactivity in primary afferent neurons of the vagus nerve. Brain Res 1998;804:306–310.
110 Goehler LE, Relton JK, Dripps D, Kiechle R, Tartaglia N, et al: Vagal paraganglia bind biotinylated interleukin-1 receptor antagonist: a possible mechanism for immune-to-brain communication. Brain Res Bull 1997;43:357–364.

111 Luheshi GN, Bluthe RM, Rushforth D, Mulcahy N, Konsman JP, et al: Vagotomy attenuates the behavioural but not the pyrogenic effects of interleukin-1 in rats. Auton Neurosci 2000;85:127–132.
112 Engler H, Doenlen R, Engler A, Riether C, Prager G, et al: Acute amygdaloid response to systemic inflammation. Brain Behav Immun 2011;25:1384–1392.
113 Hinwood M, Morandini J, Day TA, Walker FR: Evidence that microglia mediate the neurobiological effects of chronic psychological stress on the medial prefrontal cortex. Cereb Cortex 2012;22:1442–1454.
114 Tynan RJ, Naicker S, Hinwood M, Nalivaiko E, Buller KM, et al: Chronic stress alters the density and morphology of microglia in a subset of stress-responsive brain regions. Brain Behav Immun 2010;24:1058–1068.
115 Wohleb ES, Hanke ML, Corona AW, Powell ND, Stiner LM, et al: Beta-adrenergic receptor antagonism prevents anxiety-like behavior and microglial reactivity induced by repeated social defeat. J Neurosci 2011;31:6277–6288.
116 Morilak DA, Barrera G, Echevarria DJ, Garcia AS, Hernandez A, et al: Role of brain norepinephrine in the behavioral response to stress. Prog Neuropsychopharmacol Biol Psychiatry 2005;29: 1214–1224.
117 Shin LM, Liberzon I: The neurocircuitry of fear, stress, and anxiety disorders. Neuropsychopharmacology 2010;35: 169–191.
118 Yirmiya R, Goshen I: Immune modulation of learning, memory, neural plasticity and neurogenesis. Brain Behav Immun 2011;25:181–213.
119 Lucin KM, Wyss-Coray T: Immune activation in brain aging and neurodegeneration: too much or too little? Neuron 2009;64:110–122.
120 Villeda SA, Luo J, Mosher KI, Zou B, Britschgi M, et al: The ageing systemic milieu negatively regulates neurogenesis and cognitive function. Nature 2011;477: 90–94.
121 Palmer TD, Willhoite AR, Gage FH: Vascular niche for adult hippocampal neurogenesis. J Comp Neurol 2000;425: 479–494.
122 Carmeliet P: Blood vessels and nerves: common signals, pathways and diseases. Nat Rev Genet 2003;4:710–720.
123 Belevych N, Buchanan K, Chen Q, Bailey M, Quan N: Location-specific activation of the paraventricular nucleus of the hypothalamus by localized inflammation. Brain Behav Immun 2010;24: 1137–1147.
124 Rosas-Ballina M, Tracey KJ: The neurology of the immune system: neural reflexes regulate immunity. Neuron 2009;64:28–32.
125 Webster JI, Tonelli L, Sternberg EM: Neuroendocrine regulation of immunity. Annu Rev Immunol 2002;20:125–163.
126 Sternberg EM: Neural regulation of innate immunity: a coordinated nonspecific host response to pathogens. Nat Rev Immunol 2006;6:318–328.
127 Munhoz CD, Sorrells SF, Caso JR, Scavone C, Sapolsky RM: Glucocorticoids exacerbate lipopolysaccharide-induced signaling in the frontal cortex and hippocampus in a dose-dependent manner. J Neurosci 2010;30:13690–13698.
128 Sorrells SF, Sapolsky RM: An inflammatory review of glucocorticoid actions in the CNS. Brain Behav Immun 2007;21: 259–272.
129 Frank MG, Baratta MV, Sprunger DB, Watkins LR, Maier SF: Microglia serve as a neuroimmune substrate for stress-induced potentiation of CNS pro-inflammatory cytokine responses. Brain Behav Immun 2007;21:47–59.
130 Sorrells SF, Sapolsky RM: Glucocorticoids can arm macrophages for innate immune battle. Brain Behav Immun 2010;24:17–18.
131 Trottier MD, Newsted MM, King LE, Fraker PJ: Natural glucocorticoids induce expansion of all developmental stages of murine bone marrow granulocytes without inhibiting function. Proc Natl Acad Sci USA 2008;105:2028–2033.
132 Engler H, Engler A, Bailey MT, Sheridan JF: Tissue-specific alterations in the glucocorticoid sensitivity of immune cells following repeated social defeat in mice. J Neuroimmunol 2005;163:110–119.
133 Engler H, Bailey MT, Engler A, Sheridan JF: Effects of repeated social stress on leukocyte distribution in bone marrow, peripheral blood and spleen. J Neuroimmunol 2004;148:106–115.
134 Kinsey SG, Bailey MT, Sheridan JF, Padgett DA: The inflammatory response to social defeat is increased in older mice. Physiol Behav 2008;93:628–636.
135 Wohleb ES, Fenn AM, Pacenta AM, Powell ND, Sheridan JF, et al: Peripheral innate immune challenge exaggerated microglia activation, increased the number of inflammatory CNS macrophages, and prolonged social withdrawal in socially defeated mice. Psychoneuroendocrinology 2012;37:1491–1505.
136 Felten DL, Felten SY, Carlson SL, Olschowka JA, Livnat S: Noradrenergic and peptidergic innervation of lymphoid tissue. J Immunol 1985;135:755s–765s.
137 Nance DM, Sanders VM: Autonomic innervation and regulation of the immune system (1987–2007). Brain Behav Immun 2007;21:736–745.
138 Meltzer JC, MacNeil BJ, Sanders V, Pylypas S, Jansen AH, et al: Stress-induced suppression of in vivo splenic cytokine production in the rat by neural and hormonal mechanisms. Brain Behav Immun 2004;18:262–273.
139 Miller GE, Chen E, Sze J, Marin T, Arevalo JM, et al: A functional genomic fingerprint of chronic stress in humans: blunted glucocorticoid and increased NF-kappaB signaling. Biol Psychiatry 2008;64:266–272.
140 Miller GE, Chen E, Fok AK, Walker H, Lim A, et al: Low early-life social class leaves a biological residue manifested by decreased glucocorticoid and increased proinflammatory signaling. Proc Natl Acad Sci USA 2009;106:14716–14721.
141 Cole SW, Arevalo JM, Takahashi R, Sloan EK, Lutgendorf SK, et al: Computational identification of gene-social environment interaction at the human IL-6 locus. Proc Natl Acad Sci USA 2010;107: 5681–5686.
142 Spiegel A, Shivtiel S, Kalinkovich A, Ludin A, Netzer N, et al: Catecholaminergic neurotransmitters regulate migration and repopulation of immature human CD34+ cells through Wnt signaling. Nat Immunol 2007;8:1123–1131.
143 Johnson JD, Cortez V, Kennedy SL, Foley TE, Hanson H 3rd, et al: Role of central beta-adrenergic receptors in regulating proinflammatory cytokine responses to a peripheral bacterial challenge. Brain Behav Immun 2008;22:1078–1086.
144 Johnson JD, Campisi J, Sharkey CM, Kennedy SL, Nickerson M, et al: Catecholamines mediate stress-induced increases in peripheral and central inflammatory cytokines. Neuroscience 2005;135:1295–1307.

145 Audet MC, Jacobson-Pick S, Wann BP, Anisman H: Social defeat promotes specific cytokine variations within the prefrontal cortex upon subsequent aggressive or endotoxin challenges. Brain Behav Immun 2011;25:1197–1205.
146 Anisman H, Merali Z: Anhedonic and anxiogenic effects of cytokine exposure. Adv Exp Med Biol 1999;461:199–233.
147 Krishnan V, Han MH, Graham DL, Berton O, Renthal W, et al: Molecular adaptations underlying susceptibility and resistance to social defeat in brain reward regions. Cell 2007;131:391–404.
148 O'Donovan A, Neylan TC, Metzler T, Cohen BE: Lifetime exposure to traumatic psychological stress is associated with elevated inflammation in the Heart and Soul Study. Brain Behav Immun 2012;26:642–649.
149 Kiecolt-Glaser JK, Glaser R: Depression and immune function: central pathways to morbidity and mortality. J Psychosom Res 2002;53:873–876.
150 Kiecolt-Glaser JK, Preacher KJ, MacCallum RC, Atkinson C, Malarkey WB, et al: Chronic stress and age-related increases in the proinflammatory cytokine IL-6. Proc Natl Acad Sci USA 2003; 100:9090–9095.
151 Glaser R, Kiecolt-Glaser JK: Stress-induced immune dysfunction: implications for health. Nat Rev Immunol 2005;5:243–251.
152 Suarez EC, Krishnan RR, Lewis JG: The relation of severity of depressive symptoms to monocyte-associated proinflammatory cytokines and chemokines in apparently healthy men. Psychosom Med 2003;65:362–368.
153 Suarez EC, Lewis JG, Krishnan RR, Young KH: Enhanced expression of cytokines and chemokines by blood monocytes to in vitro lipopolysaccharide stimulation are associated with hostility and severity of depressive symptoms in healthy women. Psychoneuroendocrinology 2004;29:1119–1128.

Dr. Jonathan P. Godbout
Department of Neuroscience, Institute for Behavioral Medicine Research
Center for Brain and Spinal Cord Repair, The Ohio State University
259 IBMR Building, 460 Medical Center Drive, Columbus, OH 43210 (USA)
E-Mail jonathan.godbout@osumc.edu

Halaris A, Leonard BE (eds): Inflammation in Psychiatry.
Mod Trends Pharmacopsychiatry. Basel, Karger, 2013, vol 28, pp 20–32 (DOI: 10.1159/000343965)

Stress and Neuroinflammation

Angela J. Grippo[a] · Melissa-Ann L. Scotti[a,b]

[a]Department of Psychology, Northern Illinois University, DeKalb, Ill., and [b]Department of Psychiatry and Brain-Body Center, University of Illinois at Chicago, Chicago, Ill., USA

Abstract

It has been well established that there is bidirectional communication between the immune and central nervous systems. One context in which this interaction has been extensively studied is that of the stress response. Stress, whether physical or psychological, induces alterations in immune function. Often exposure to a stressor results in pro-inflammatory responses in the brain and periphery. These responses are mediated by a variety of inflammatory molecules including neuropeptides, cytokines, and stress hormones among others. Here, we will discuss several of the more comprehensively studied of these inflammatory mediators and their role(s) in stress-induced neurogenic inflammation.

Stress is defined as a set of responses to an instigating stimulus (e.g. a stressor). This definition implies that stress is a state of the body (and mind) in which the demands of the situation surpass one's resources available to cope with those demands [1–3]. These responses may be physiological, psychological, or behavioral; the measurement of which have been discussed for several decades [4]. Stressors, which elicit a *stress response*, are often distinguished by severity (e.g. mild vs. moderate vs. severe), duration (e.g. short-term vs. chronic), time course (intermittent vs. sustained), and predictability (predictable vs. unpredictable) [1–3]. The precise effects of various types of stressors on the body have been debated; however, in many contexts chronic, sustained, unpredictable stressors are the most deleterious, contributing to immune and endocrine dysfunction, altered mood, and several neurobiological and psychological diseases [1, 5]. In contrast, brief, predictable stressors may sometimes be beneficial in terms of enhancing cognition, emotion, and neurobiological systems such as the immune system [6, 7].

Pioneering work by Cannon [8], Selye [9], Mason [10], and Lazarus [11] helped define the concept of the stress response. Selye's concepts of the *general adaptation syndrome* and *diseases of adaptation* [9, 12] have provided a foundation for our current understanding of the body's responses to stressors. The concept of stress is linked to functions of the hypothalamic-pituitary-adrenal (HPA) axis, renin-angiotensin-aldosterone system, and autonomic nervous system, all of which have inter-communication with the immune system. The stress response is an adaptive response involving physiological, psychological, and behavioral changes to promote survival and reduce the consequences of a threatening situation. For example, the stress response involves changes in blood pressure, heart rate, oxygen delivery, and tissue breakdown, along with down-regulation of neurobiological systems that hinder the adaptive reaction such as digestion, reproduction and growth [13, 14]. Affective and behavioral responses mirror these physiological changes, including exhibiting defensive behaviors, fear, and distress.

Parvocellular neurons in the hypothalamic paraventricular nucleus (PVN) contain corticotropin-releasing factor (CRF), and project to the median eminence to control pituitary adrenocorticotropic hormone (ACTH) release in response to a stressor. ACTH and the sympathetic nervous system signal the adrenal glands to secrete cortisol and catecholamines, respectively. Adrenal glucocorticoids (GCs) can modulate activity of the HPA axis by providing feedback to the pituitary, hippocampus and hypothalamus [5, 14]. An extensive network of CRF-containing neurons is present in other central nervous system (CNS) structures, and under conditions of stress there is an increase in activity in neurons where CRF is serving as a classic hypothalamic releasing factor as well as in others where it is released in the brain as a neurotransmitter. CRF-containing axons from cell bodies in the hypothalamic PVN project to the ventrolateral medulla and the intermediolateral column of the spinal cord. Extrahypothalamic CRF projections, originating in the central nucleus of the amygdala and the bed nucleus of the stria terminalis, also communicate with the parabrachial and dorsal motor nucleus of the vagus (10th cranial nerve) [15]. Intracerebroventricular (i.c.v.) injections of CRF induce behaviors resembling those of the stress response, including decreased food intake and sexual activity, disturbed sleep, altered motor behavior and impaired learning [16–18]. A heightened arousal to various stimuli may reflect an action of CRF on the noradrenergic neurons of the locus coeruleus or serotonin neurons in the raphe nuclei [15, 19].

Through this communication, the HPA axis interacts with the autonomic nervous system, renin-angiotensin-aldosterone system, and other visceral functions. CRF activation of the locus coeruleus stimulates the release of norepinephrine (NE) [14, 20]. For instance, CRF produces an increase in blood pressure and heart rate via sympathetic activation [21, 22]. The locus coeruleus also inhibits the parasympathetic nervous system as well as certain neurovegetative functions such as sleep [23]. When the balance of the autonomic nervous system shifts to increased sympathetic drive, this stimulates the kidneys to release renin. This in turn initiates the renin-angiotensin-aldosterone system cascade whereby angiotensin II is produced in part to cause vasoconstriction and a further increase in heart rate.

The CNS not only controls the stress response via the mechanisms mentioned above, but it also has the capacity to induce an inflammatory response to both physiological and psychological stressors [24, 25]. It is widely accepted that nerves containing inflammatory neuropeptides participate in both local inflammatory reactions in response to tissue trauma or infection, and also in response to stress [24]; the relationship between stress and the development of inflammatory diseases (e.g. cardiovascular diseases, inflammatory bowel disorders) has been well-documented [26, 27]. The following sections will focus on several neuropeptides, steroid hormones, cytokines, and other inflammatory factors that play important roles in the inflammatory stress response.

Neurogenic Inflammation

Both somatic and autonomic nerves are associated with inflammation; however, the majority of data providing evidence for neurogenic inflammation comes from studies of the unmyelinated sensory nerve fibers, or 'C' fibers [24]. Electrical, mechanical, heat, and chemical stimulation of somatic nerves containing C fibers results in an inflammatory response [28]. The response includes vasodilatation of arterioles (flare), and swelling (wheal) caused by the extravasation of plasma from postcapillary venules [24]. This inflammatory reaction is brought about by neuropeptides (e.g. substance P; SP) that are released from the peripheral sensory nerve terminals [24, 27]. This response requires an intact peripheral nervous system, as cutting the nerves or blockade by a local anesthetic will prevent this inflammatory response [24].

Substance P

One of the more potent neuropeptides involved in neurogenic inflammation is SP [24]. SP is a member of the tachykinin family of bioactive peptides and functions as a neurotransmitter/neuromodulator in the CNS [29, 30]. This 11 amino acid peptide is not only widely distributed in the CNS but in the peripheral and enteric nervous systems as well [24, 30, 31]. Both SP and its primary binding site, the neurokinin-1 receptor, are found in areas of the brain that are important for processing and responding to physiological and psychological stressors (i.e. amygdala, prefrontal cortex) [24, 30]. SP is involved in sensory pathways, including nociceptive signaling, within the dorsal horn of the spinal cord. In the peripheral nervous system and throughout most of the body, SP has been found in both autonomic afferents and in C-type (i.e. unmyelinated) sensory fibers [24, 30, 31]. Based on where this peptide is located it is, perhaps, not surprising that SP has been shown to play an important role in the stress response as well as to contribute to stress-induced inflammation.

SP is elevated in the brain in response to a variety of psychological stressors including restraint stress, and as a result of anxiety [24, 30]. The regions in which SP shows the greatest increase are in the limbic system, specifically the amygdala [24, 30]. Additionally, SP interacts with the HPA axis, leading to increases in CRF and ACTH during psychological stressors and SP is essential for the activation of the HPA in response to somatosensory stress [24]. Typical stress responses, such as isolation-induced vocalizations in guinea pigs, can be triggered by treatment with SP and abolished by treatment with a SP receptor antagonist [24, 32].

SP has repeatedly been shown to play an important role in both the neurogenic and non-neurogenic inflammatory response. For example, SP receptors have been found in a variety of immune cell types (e.g. leukocytes, lymphocytes and mast cells), and SP can trigger the release of cytokines from these cells [24]. Generally, however, SP is released to a site of tissue damage from primary afferent fibers that have been activated by the injury; SP then acts on mast cells to release serotonin and histamine, promoting vasodilatation [31]. Inoculation of the skin with SP alone, however, can also induce many of the signs of an acute inflammatory response including vasodilatation, contributing to the production of swelling and uticaria [33]. Treatment with an SP antagonist can block this response [33].

SP has also been identified as a possible contributor to tissue damage in chronic inflammatory diseases including arthritis [31]. For example, treatment of a peripheral nerve with capsaicin, which depletes SP, also reduces swelling in an experimentally induced arthritic limb [34]. Additionally, in animals that are injected with Freund's adjuvant – an antigen that induces a syndrome in rats that is similar to human rheumatoid arthritis – the ankles are more severely affected than the knee joints [31]. The synovial fluid from the ankle joints has significantly higher levels of SP, and denervation significantly reduced the amount of SP in the ankle [31, 35]. Recent data suggest that psychological stress, such as social isolation, can trigger mast cell degranulation, which releases several inflammatory mediators

[36]. The mast cell degranulation appears to be mediated, in part, by SP released from sensory nerve fibers [27, 36].

Neuropeptide Y

Neuropeptide Y (NPY) is a peptide secreted from postganglionic sympathetic nerve fibers [37, 38]. NPY is considered a stress hormone and is released from sympathetic nerves and the adrenal medulla during various stressful conditions in animals (e.g. acute cold, immobilization, foot shock) [24, 37, 39] and humans (e.g. psychological/uncontrollable stress) [40]. NPY has been found to have anxiolytic effects in animal models of anxiety [41], and human studies suggest that post-stress levels of NPY are negatively correlated with psychological distress, suggesting that NPY may buffer against psychological stress [40]. NPY is a nonadrenergic co-mediator of catecholamine action [39]. Specifically, NPY amplifies NE's actions, including regulation of vasoconstriction [37, 39]. NPY also participates in other physiological and behavioral processes such as energy balance, feeding and anxiety [38]. NPY receptors are widely distributed in the CNS and are also found on immune cells including T cells, B cells, dendritic cells, mast cells and NK cells [38]. NPY helps to modulate some T cell responses, chemotaxis, and adhesion of leukocytes, platelet aggregation, and macrophage activation [38]. The latter effect on the immune system, as well as NPY's role in vasoconstriction and as a vascular mitogen, have led to the hypothesis that it may be involved in the inflammatory processes that are involved in atherosclerosis [39, 42]. For example, chronic cold water stress accelerated rat carotid stenosis after a vascular injury via angioplasty [43], and these stress-induced changes could be mimicked by the implantation of a NPY pellet. Administration of NPY receptor antagonists prevented the development of stress-induced or NPY-induced occlusions [43].

Glucocorticoids

GCs are steroid hormones that are produced by the zona fasciculata of the adrenal cortex [44]. The synthesis and secretion of these hormones is regulated by pituitary ACTH [44]. GCs diffuse into the brain and can then bind to two types of intracellular receptors, mineralocorticoid receptors (MR; type I) and glucocorticoid receptors (GR; type II). The MR has a much higher affinity for GCs than does the GR [45]. As such, at basal GC levels GCs bind predominantly to MRs [46, 47]. It is only when GC levels rise in response to a stressor, i.e. when MRs become saturated with GCs, that GRs become occupied [47]. GRs become heavily occupied only after moderate-to-severe stress exposure and during the GC circadian peak [45, 46]. GRs and MRs are expressed differently across different brain regions, and overall the concentration of GR in the brain is greater than that of MR [25]. The distribution of these receptors can be altered, however, by stress and early experience [48]. The differences in GC affinity and distribution have led to the suggestion that MR and GR may have independent functions [46, 47]. For example, the MR is responsible for much of the effects GCs have at low-stress or basal levels (permissive effects) whereas the GR is responsible for the effects of GCs at high-stress levels [47].

The actions of GCs (i.e. cortisol, corticosterone) are numerous and varied, however, one of their primary roles is to control and modulate the stress response in the period following exposure to a stressor, specifically providing feedback inhibition of the HPA axis [47]. GCs, along with catecholamines, serve to mobilize energy stores to provide a ready source to the skeletal muscles, presumably to allow an organism to effect an escape from or face a stressor [47, 49]. At this point, GCs are also reducing the activity of less immediately necessary functions [49]. One of the functions in which activity is reduced is the immune system; as such GCs have been shown to actively suppress inflammation [49].

In the periphery, GCs inhibit lymphocyte proliferation and dendritic cell maturation as well as trigger apoptosis in basophils, eosinophils and T cells [5, 46, 47]. Further, GCs inhibit or reduce the expression of several pro-inflammatory cytokines (e.g. TNFα, IL-6) and enhance the expression of anti-inflammatory cytokines (e.g. IL-10, TNFβ) [5, 46, 47]. As such, GCs are generally thought to be anti-inflammatory. More recent data, however, provide a more complex view of the effects GCs have on immune function, including inflammatory responses. GCs may have both permissive and suppressive effects on immune function following a stressor [5, 46, 47]. This is especially true within the CNS, although evidence of pro-inflammatory effects has been found in the periphery as well. In the periphery, for example, during the initial stress response when GC levels have not yet begun to increase, GCs initially help activate the immune system, and then later when GC levels are high suppress the immune system possibly to prevent autoimmune damage and inflammatory disorders [47].

The effect GCs have in the periphery, whether enhancive or suppressive, often depends on the type of stressor. Acute stressors seem to enhance immune function, whereas chronic stressors are suppressive. An acute stressor prior to an immune challenge significantly enhances cell-mediated immune response against the antigen; chronic stress suppresses this response [6, 7]. Additionally, acute low-dose and chronic treatment with corticosterone mimic the effects of acute and chronic stress on the immune responses measured [50]. Such experiments altered the initial belief that the depletion of leukocytes in response to the increase in GCs following an acute stressor was the result of immunosuppression by GCs; it is now recognized that these cells leave the circulation and relocate to the area of injury [7, 51].

GCs can also be effective at reducing certain types of inflammation in the CNS; however there are also circumstances during which they fail to provide anti-inflammatory action, and even enhance inflammation [46, 47]. The duration and severity of the stressor also play a large role in the effect that the GCs have on the CNS, although the mechanisms by which the GCs have their effects in the CNS differ from those in the periphery [25, 46, 47, 52]. In the healthy CNS acute stress can activate and increase the activity or levels of several inflammatory mediators (e.g. NFκβ, TNFα, prostaglandins) [46, 53–55]. Some of these stress-induced changes are dependent on GCs [46, 56]. Acute GC exposure prior to a challenge, as is the case for the periphery, also primes the inflammatory response in the CNS [46]. In the injured nervous system the pattern, however, is opposite to that observed in healthy tissue and in the periphery. In the injured CNS, acute stressors reduce the inflammatory actions of GCs; whereas chronic stressors increase them [47].

An important mechanism by which GCs promote inflammation in injured brain tissue may be via the inhibition of glucose uptake [47]. This exacerbates the injury-induced reductions in adenosine triphosphate (ATP) and mitochondrial potential, which diminishes the ability of neurons to perform essential functions, leading to cell death. The dead neurons release a variety of chemicals including cytokines and prostaglandins, which activate microglia and signal the migration of leukocytes and other white blood cells to the area of damage [47]. These cells release inflammatory cytokines and oxygen radicals, thus enhancing inflammation [47]. This response may be initially beneficial to surviving neurons if the inflammation is moderate; however, more extensive and prolonged inflammation is likely detrimental to surviving neurons [47]. Similarly, chronic unpredictable stress results in significantly increased neuronal damage after direct cortical lipopolysaccharide (LPS) infusion, and a GR antagonist is capable of preventing this effect by preventing the migration of inflammatory cells to the injury site [57].

Catecholamines

Like GCs, catecholamines, i.e. epinephrine (E), NE and dopamine (DA), are released in response to a stressor and act, together with GCs, to affect an organism's response to stress [25, 44]. Here we will focus on the two members of this chemical family that play a primary role in the stress response, E and NE. Both of these molecules are produced in and secreted from the adrenal medulla, however, the primary product of this tissue is E, whereas NE is also secreted from sympathetic nerve fibers [44]. As is the case for the hypothalamic-adrenal arm of the stress response, the hormones released by the sympatho-adrenal arm of this response also have effects on the immune system.

Similar to GCs, catecholamines appear to have both pro- and anti-inflammatory effects; however, the majority of the effects of these molecules in both the CNS and periphery seems to be pro-inflammatory [25]. Specifically, it has been shown that these molecules, especially NE, stimulate the release of cytokines (e.g. TNF-α, IL-1β, IL-6, IL-8) from microglia and macrophages [24, 25, 58]. The effect of these catecholamines on cytokine secretion is regulated through both adrenoreceptor types (i.e. α and β); however, the mechanisms by which they do so differs between the CNS and periphery [58]. In rats pretreated with an α_1-receptor antagonist (prazosin), and then subjected to a tail shock, stress-induced circulating IL-1β and IL-6 levels were attenuated in the periphery, whereas the nonspecific β-receptor antagonist (propranolol) completely prevented stress-induced circulating IL-1β in the hippocampus, hypothalamus and pituitary [58]. In addition, nonspecific β-receptor activation (via isoproterenol) was sufficient to increase plasma levels of IL-1β and IL-6 within 60 min. This treatment also significantly increased levels of IL-1β in the hypothalamus, hippocampus and pituitary within 30 min [58]. These results are consistent with the early stages of the stress response being pro-inflammatory as the actions of catecholamines occur on a shorter timescale than those of GCs [25, 47].

Pro-Inflammatory Cytokines

Cytokines are large soluble proteins that are secreted by a variety of accessory immune cells [59]. One of their primary functions is to coordinate local and systemic inflammatory response to pathogens. In the periphery cytokines are primarily produced by cells such as macrophages and lymphocytes, and in the CNS are produced by cells such as microglia and astrocytes [25, 60]. Interestingly, although large proteins, cytokines produced in the periphery are able to affect the CNS and even cross the blood brain barrier via several proposed mechanisms including: active transport, via areas of the brain with a weak blood-brain barrier, activation of vagal afferents that transmit cytokine signals to various brain regions, or binding to receptors in the blood vessels of the brain which activate second messengers that can enter the brain directly [61, 62].

Cytokines are divided into four broad groups: (1) pro-inflammatory cytokines (e.g. IL-1β, IL-6, TNF-α) play a major role in non-specific innate immunity; (2) type I T helper (Th1) cytokines help induce cytotoxic immunity; (3) type II T helper cell (Th2) cytokines are anti-inflammatory, and (4) type III T helper cell (Th3) cytokines are immunosuppressive [60, 63]. It is believed that the balance between the pro-inflammatory and anti-inflammatory cytokines plays an important role in neuro-inflammatory responses, and that an imbalance in this system can lead to physiological and psychological disorders [63]. For example, stress-related disorders such as major depression and cardiovascular disease are accompanied by enhanced expression of pro-inflammatory cytokines [64].

Interleukin-1

IL-1, the most extensively studied forms of which are IL-1β and IL-1α, is produced by a wide variety of cell types (e.g. synovial cells, osteoblasts, glial cells, and neutrophils), but is primarily produced by cells from the mononuclear phagocytic lineage [60]. IL-1 has diverse functions including: the induction of growth factors, activation of T cells via IL-2 upregulation, and increasing the action of inducible nitric oxide synthase (iNOS) [25, 60]. IL-1 has been implicated as an important mediator of neurodegeneration following cerebral injury (e.g. ischemia, excitatory or traumatic brain damage) and is believed to be neurotoxic [65–68]. It is rapidly released from microglia following CNS injury [65, 69, 70]. Specifically, studies suggest that the IL-1β form might be a major mediator of injury following CNS trauma [70–72]. For example, i.c.v. administration of an anti-IL-1β antibody reduced ischemic brain injury in rats, whereas i.c.v. administration of IL-1β enhances ischemic and other brain injury [71, 72].

Exposure to a stressor also affects IL-1 expression in the brain as well as the periphery [63]. IL-1β expression is often elevated in the CNS of animals exposed to an acute psychological stressor (e.g. immobilization, inescapable tail shock) and in the plasma of humans that undergo a speech task stressor [63, 73, 74]. The effects of chronic stress on IL-1β expression are less clear. Some studies report increases in IL-1β expression, whereas others show suppression [63, 73, 75].

Recent data suggest that IL-1β is an essential mediator of the antineurogenic and anhedonic effects of stress. Animals that are exposed to an acute stressor or receive an acute administration of IL-1β showed decreased hippocampal cell proliferation, and an IL-1 receptor antagonist (IL-1Ra) inhibited this change for both treatments [76]. Similarly, IL-1Ra prevents the expression of anhedonia, which typically occurs following exposure to chronic mild stressors [76]. This is not surprising given the volume of research that indicates the relationship between depression and inflammatory processes, specifically elevated levels of pro-inflammatory cytokines.

Tumor Necrosis Factor-α

TNFα is a central mediator of tissue inflammation that is secreted from a variety of cell types (e.g. lymphocytes, mast cells, natural killer cells), but primarily mononuclear phagocytes [60, 77]. TNFα exerts an array of biological activities including the activation of the acute-phase response (APR), inducing vascular permeability, and the activation of inflammatory cells (e.g. neutrophils, lymphocytes, monocytes) [77, 78]. This cytokine has long been known to play a pivotal role in the periphery and has been found to be involved in antitumor effects, septic shock, and hematopoiesis [60, 68, 77]. TNFα and its receptors (TNF receptor I (TNFR-1; p55) and TNF receptor II (TNFR-2; p75)) have been identified in the CNS [25, 79], and TNF-α has been implicated in the pathogenesis of a variety of neurological conditions (e.g. Alzheimer's diseases, Parkinson's disease, ischemia) because its production is increased in the CNS after tissue damage or degenerative processes occur [25, 52, 67, 68]. Site-directed administration of TNFα increases tissue damage due to ischemia, whereas blocking TNFα reduces injury [68, 80]. Interestingly, treatment with TNFα prior to experimentally induced ischemia results in a significantly reduced infarct size, suggesting that TNFα has both deleterious and protective properties [68]. These opposing effects could potentially be explained by the fact that TNFα has two receptors which may mediate different effects of TNFα [25, 81].

The role of TNFα in stress-induced inflammation is not completely clear. GCs and catecholamines, which are released in response to a stressor, work to increase the expression of pro-inflammatory cytokines and cytokine receptors (see above), including TNFα and its receptors [1, 5, 58]. Immobilization stress increases soluble TNFα in the cortex after 1 h [55]. In humans, a mock interview, parachuting, or other psychological stressors resulted in increased circulating levels of TNFα, and social stress (e.g. social rejection)

resulted in upregulation of the TNFR-2 receptor [63, 82]. Additionally, TNFα release is responsible for stress-induced NFκβ and iNOS expression, both of which are pro-inflammatory mediators (see below) [52, 55]. However, other studies have found (1) no effect on TNFα expression [63], or (2) lower levels of TNFα expression in response to inescapable shock or other psychogenic stressors [25, 83]. Clearly more research is required to better understand the role TNFα plays in stress-induced inflammation.

Interleukin-6

The main sources of IL-6 are mononuclear phagocytic cells; however, T and B lymphocytes, fibroblasts, and endothelial cells, among others, also produce this cytokine [60]. IL-6 has several important biological functions including inducing B lymphocytes to mature into plasma cells and the mediation of T cell activation, growth and differentiation [60]. Further, IL-6 is the most important inducer of acute-phase proteins (APP) including C-reactive protein [3, 27, 60, 78]. In general, brain levels of IL-6 are low, but increase markedly in response to trauma (e.g. stroke, traumatic brain injury), and IL-6 is elevated in a variety of human CNS disorders (e.g. epilepsy, depression, Alzheimer's disease, Parkinson's disease, schizophrenia) [68]. These increases in IL-6 are likely induced by other molecules including cytokines such as IL-1β, TNFα and TGFβ [68]. IL-6 is also an endogenous pyrogen, which exerts multiple effects and, like TNFα, can have both beneficial and deleterious effects on CNS cells [25, 68]. For example, i.c.v. IL-6 reduces the amount of brain damage that occurs in experimentally induced ischemia and IL-6 knockout mice have significant deficits in the inflammatory response to experimentally induced brain injury, such that there is more cell death and impaired cell repair [25, 68, 84]. In contrast, overexpression of IL-6 leads to neurological disease states and neurodegeneration [66], suggesting that IL-6 may be neurotoxic at chronically elevated levels [25, 66]. Further, IL-6 has been shown to have both pro- and anti-inflammatory properties [67] and brain levels of IL-6 are likely to be critical in determining the manner in which it acts [67].

IL-6 secretion can be induced by psychological stress in animals. For example, IL-6 levels are increased, along with IL-1 and TNFα, after restraint stress, and exposure to an open-field stressor significantly increased plasma IL-6 activity in rats [63]. It is possible that the increase in IL-6 may be triggered by catecholamines, as it is known that NE stimulates the release of IL-6 and other cytokines [58]. In humans, psychosocial stress, specifically social rejection, elevated IL-6 in the saliva of adults [82]. Further, plasma levels of IL-6 and its soluble receptor, sIL-6R, were elevated in individuals with post-traumatic stress disorder [85].

Nuclear Factor-κβ

The transcription factor NFκβ is widely expressed in the CNS [52]. Once activated via the sequential activation of different enzyme complexes (e.g. NFκβ-induced kinase, Iκβ kinase), NFκβ can enter the nucleus of cells where it binds to specific DNA sequences in the promoter regions of target genes [25, 52]. Among these genes are those that code for proteins involved in oxidative stress/damage and inflammation (e.g. iNOS and cyclo-oxygenase (COX)-2) [25, 52]. NFκβ can be activated by many stimuli including infection, UV light, and cytokines [52]. Further, NFκβ activation is one of the earliest events in the stress-inflammation response and has been observed as soon as 4 h after the onset of stress (restraint) in rats [53, 86]. Similarly in humans, psychological stressor (forced speech, mental arithmetic tasks) increase NFκβ in cell nuclei [25, 86, 87]. Interestingly, these changes co-occur with increases in GCs and catecholamines and it seems that both the pro- and anti-inflammatory actions of GCs may be due to their effect on NFκβ [88]. For example, the activated GC receptor may have the ability to bind directly to NFκβ, preventing its migration to the nucleus and thus preventing cytokine gene transcription [46, 88], but GCs also mediate increases in LPS-induced NFκβ in the frontal cortex and hippocampus upon exposure to chronic unpredictable stress [25, 89].

Inducible Nitric Oxide Synthase and Nitric Oxide

Nitric oxide (NO) plays many important roles in the immune system. Many cell types can generate NO, including but not limited to: NK cells, macrophages, microglia, dendritic cells, phagocytes, smooth muscles cells, epithelial cells, neurons, and neutrophils [90]. NO, as a diffusible gas, is not restricted to its site of production [90]. Rather, NO radicals can, with the aid of other molecules, travel long distances [90]. Additionally, NO interactions, unlike those of cytokines, are not limited to specific receptors; NO can react directly with proteins, inorganic molecules (e.g. oxygen, metals), DNA, and a variety of other molecules [90, 91]. There are three forms of NO synthase (NOS): (1) neuronal (nNOS); (2) endothelial (eNOS), and (3) inducible (iNOS), which is expressed during inflammatory reactions [25, 90]. iNOS expression is regulated by cytokines, immune complexes, and can be induced in many cell types by these and other noxious stimuli (e.g. LPS, microbial or viral products) [92]. iNOS mediates cytotoxicity due to the oxidative/nitrosative stress produced by extensive and prolonged release of NO and the production of other oxidant species such as peroxynitrate ($ONOO^-$). $ONOO^-$ damages targets such as DNA, proteins and mitochondria [25, 93] and is able to participate in the upregulation of pro-inflammatory signal transduction pathways [25, 93]. In this manner, a vicious cycle of inflammatory injury occurs, whereby increased iNOS expression is induced by the upregulated pro-inflammatory signals (e.g. cytokines) leading to increases in NO production. This mechanism is an important part of the pathophysiology of inflammation [25, 93]. As such, NO and iNOS expression are thought to contribute to many CNS diseases including multiple sclerosis, stroke, Parkinson's disease and Alzheimer's disease [68].

iNOS and NO expression are also upregulated by stress exposure. For example, acute and chronic restraint stress increases brain iNOS activity and $ONOO^-$ production and this can occur as early as 6 h after the stressor [53]. Chronic stress produces even higher levels of iNOS activity [94]. The result of chronically elevated iNOS can include mitochondrial impairment and degradation of the blood-brain barrier [25, 95]. These changes are likely due to the production of $ONOO^-$ at tissue-damaging levels.

Cyclo-Oxygenase-2

COX exists in two distinct isoforms (i.e. COX-1, COX-2), with an additional form, COX-3, that is a splicing variant of COX-2 [54]. COX-1 is expressed constitutively in nearly all tissue types. COX-2 is constitutively expressed in only a few tissues, including the brain [52, 54]; however, COX-2 can be rapidly and transitively up regulated in the brain and in cells that are involved in inflammation (e.g. macrophages, monocytes) via a variety of molecules including cytokines, mitogens and LPS [52, 54]. COX-2 is induced in several negative neurobiological processes such as seizures, Alzheimer's disease, and stroke-induced ischemia [25, 68, 96, 97]. Cox-2 is primarily responsible for the production of prostanoids involved in pathological inflammatory conditions and the deleterious effects of this enzyme are thought to involve the synthesis of peroxides and prostanoids, including prostaglandin E_2 (PGE_2), in toxic amounts [25]. These molecules cause damage by (1) generating free radicals (i.e. oxidative/nitrosative damage), and (2) inducing apoptosis by the induction of glutamate release from astrocytes [25, 98, 99].

Stress can induce increases in COX-2. For example, COX-2 mRNA levels are increased in mice that undergo a forced swim stressor [54]. Further, significant increases in COX-2 and PGE_2 occur in the cerebral cortex of rats that are subjected to immobilization stress as soon as 2 h (PGE_2) and 4 h (COX-2) after the onset of the stressor [54]. Additionally, significant increases in COX-2 immunoreactivity in the hippocampus and cortex also are found within 4 h of a restraint stressor [54].

Conclusions

Animal and human studies have made it clear that stress can have direct effects on immune function, and that these effects have important health implications. Specifically, stress is a potent trigger of inflammation. Stress can induce or modulate inflammation via the alteration of various pathways in the CNS and periphery that influence the inflammatory response. In this review, we have discussed data about some of the main pro-inflammatory mediators of stress-induced inflammation and their relationship to the etiology and progress of various disease states.

References

1 Glaser R, Kiecolt-Glaser JK: Stress-induced immune dysfunction: implications for health. Nat Rev Immunol 2005;5:243–251.
2 Johnson AK, Anderson EA: Stress and arousal; in Cacioppo JT, Tassinary LG, (eds): Principles of Psychophysiology: Physical, Social, and Inferential Elements. New York, Cambridge University Press, 1990.
3 Maier SF, Watkins LR: Cytokines for psychologists: implications of bidirectional immune-to-brain communication for understanding behavior, mood, and cognition. Psychol Rev 1998;105:83–107.
4 McGrath JE: Settings, measures, and themes: an integrative review of some research on social-psychological factors in stress; in McGrath JE (ed): Social and Psychological Factors in Stress. New York, Holt, Rinehart & Winston, 1970, pp 76–83.
5 Sapolsky RM, Romero LM, Munck AU: How do glucocorticoids influence stress responses? Integrating permissive, suppressive, stimulatory, and preparative actions. Endocr Rev 2000;21:55–89.
6 Dhabhar FS, McEwen BS: Stress-induced enhancement of antigen-specific cell-mediated immunity. J Immunol 1996; 156:2608–2615.
7 Dhabhar FS, McEwen BS: Acute stress enhances while chronic stress suppresses cell-mediated immunity in vivo: a potential role for leukocyte trafficking. Brain Behav Immun 1997; 11:286–306.
8 Cannon WB: Bodily Changes in Pain, Hunger, Fear and Rage. Oxford, Appleton, 1929.
9 Selye H: A syndrome produced by diverse nocuous agents. Nature 1936; 138:32–33.
10 Mason JW: A review of psychoendocrine research on the pituitary-adrenal cortical system. Psychosom Med 1968;30: 576–607.
11 Lazarus RS: From psychological stress to the emotions: a history of changing outlooks. Annu Rev Psychol 1993;44:1–21.
12 Selye H: In Vivo: The Case for Supramolecular Biology. New York, Liveright, 1967.
13 Chrousos GP: The stress response and immune function: clinical implications: the 1999 Novera H. Spector Lecture. Ann NY Acad Sci 2000;917:38–67.
14 Gold P, Chrousos G: Organization of the stress system and its dysregulation in melancholic and atypical depression: high vs. low CRH/NE states. Mol Psychiatry 2002;7:254–275.
15 De Souza EB, Grigoriadis DE: Physiology, pharmacology, and role in central nervous system and immune disorders; in Bloom FE, Kupfer DJ (eds): Psychopharmacology: The Fourth Generation of Progress. New York, Raven Press, 1995, pp 505–517.
16 Sirinathsinghji DJS, Rees LH, Rivier J, Vale W: Corticotropin-releasing factor is a potent inhibitor of sexual receptivity in the female rat. Nature 1983;305: 232–235.
17 Glowa J, Gold P: Corticotropin releasing hormone produces profound anorexigenic effects in the rhesus monkey. Neuropeptides 1991;18:55–61.
18 Krahn DD, Gosnell BA, Levine AS, Morley JE: Behavioral effects of corticotropin-releasing factor: localization and characterization of central effects. Brain Res 1988;443:63–69.
19 Kirby LG, Rice KC, Valentino RJ: Effects of corticotropin-releasing factor on neuronal activity in the serotonergic dorsal raphe nucleus. Neuropsychopharmacology 2000;22:148–162.
20 Lehnert H, Schulz C, Dieterich K: Physiological and neurochemical aspects of corticotropin-releasing factor actions in the brain: the role of the locus coeruleus. Neurochem Res 1998;23:1039–1052.

21 Brown M, Fisher L, Spiess J, River C, River J, Vale W: Corticotropin-releasing factor: actions on the sympathetic nervous system and metabolism. Endocrinology 1982;111:928–931.
22 Fisher LA: Corticotropin-releasing factor: endocrine and autonomic integration of responses to stress. Trends Pharmacol Sci 1989;10:189–193.
23 Aston-Jones G, Rajkowski J, Kubiak P, Valentino RJ, Shipley MT: Role of the locus coeruleus in emotional activation. Prog Brain Res 1996;107:379–402.
24 Black PH: Stress and the inflammatory response: a review of neurogenic inflammation. Brain Behav Immun 2002;16:622–653.
25 García-Bueno B, Caso JR, Leza JC: Stress as a neuroinflammatory condition in brain: damaging and protective mechanisms. Neurosci Biobehav Rev 2008;32:1136–1151.
26 Collins SM: Stress and the gastrointestinal tract. IV. Modulation of intestinal inflammation by stress: basic mechanisms and clinical relevance. Am J Physiol 2001;280:G315–G318.
27 Black PH, Garbutt LD: Stress, inflammation and cardiovascular disease. J Psychosom Res 2002;52:1–23.
28 Daemen M, Kurvers H, Kitslaar PJ, Slaaf DW, Bullens P, Van den Wildenberg F: Neurogenic inflammation in an animal model of neuropathic pain. Neurol Res 1998;20:41.
29 Payan DG: Neuropeptides and inflammation: the role of substance P. Annu Rev Med. 1989;40:341–352.
30 Rosenkranz MA: Substance P at the nexus of mind and body in chronic inflammation and affective disorders. Psychol Bull 2007;133:1007–1037.
31 Basbaum AI, Levine JD: The contribution of the nervous system to inflammation and inflammatory disease. Can J Physiol Pharmacol 1991;69:647–651.
32 Kramer MS, Cutler N, Feighner J, Shrivastava R, Carman J, Sramek JJ, Reines SA, Liu G, Snavely D, Wyatt-Knowles E: Distinct mechanism for antidepressant activity by blockade of central substance P receptors. Science 1998;281:1640–1645.
33 Saria A: Substance P in sensory nerve fibres contributes to the development of oedema in the rat hind paw after thermal injury. Br J Pharmacol 1984;82:217–222.
34 Colpaert F, Donnerer J, Lembeck F: Effects of capsaicin on inflammation and on the substance P content of nervous tissues in rats with adjuvant arthritis. Life Sci. 1983;32:1827–1834.
35 Levine J, Dardick S, Basbaum A, Scipio E: Reflex neurogenic inflammation. I. Contribution of the peripheral nervous system to spatially remote inflammatory responses that follow injury. J Neurosci 1985;5:1380–1386.
36 Singh LK, Pang X, Alexacos N, Letourneau R, Theoharides TC: Acute immobilization stress triggers skin mast cell degranulation via corticotropin releasing hormone, neurotensin, and substance P: a link to neurogenic skin disorders. Brain Behav Immun 1999;13:225–239.
37 Zukowska-Grojec Z, Konarska M, McCarty R: Differential plasma catecholamine and neuropeptide Y responses to acute stress in rats. Life Sci 1988;42:1615–1624.
38 Prod'homme T, Weber MS, Steinman L, Zamvil SS: A neuropeptide in immune-mediated inflammation, Y? Trends Immunol 2006;27:164–167.
39 Li L, Lee EW, Ji H, Zukowska Z: Neuropeptide Y-induced acceleration of post-angioplasty occlusion of rat carotid artery. Arterioscler Thromb Vasc Biol 2003;23:1204–1210.
40 Morgan CA, Rasmusson AM, Wang S, Hoyt G, Hauger RL, Hazlett G: Neuropeptide-Y, cortisol, and subjective distress in humans exposed to acute stress: replication and extension of previous report. Biol Psychiatry 2002;52:136–142.
41 Heilig M, Söderpalm B, Engel JA, Widerlöv E: Centrally administered neuropeptide Y (NPY) produces anxiolytic-like effects in animal anxiety models. Psychopharmacology 1989;98:524–529.
42 Zukowska Z, Pons J, Lee EW, Li L: Neuropeptide Y: a new mediator linking sympathetic nerves, blood vessels and immune system? Can J Physiol Pharmacol 2003;81:89–94.
43 Li L, Jönsson-Rylander AC, Abe K, Zukowska Z: Chronic stress induces rapid occlusion of angioplasty-injured rat carotid artery by activating neuropeptide Y and its Y1 receptors. Arterioscler Thromb Vasc Biol 2005;25:2075–2080.
44 Hadley ME: Endocrinology, ed 5. Upper Saddle River, Prentice-Hall, 2000.
45 De Kloet ER, Vreugdenhil E, Oitzl MS, Joëls M: Brain corticosteroid receptor balance in health and disease. Endocr Rev 1998;19:269–301.
46 Sorrells SF, Caso JR, Munhoz CD, Sapolsky RM: The stressed CNS: when glucocorticoids aggravate inflammation. Neuron 2009;64:33–39.
47 Sorrells SF, Sapolsky RM: An inflammatory review of glucocorticoid actions in the CNS. Brain Behav Immun 2007;21:259–272.
48 Patel PD, Katz M, Karssen AM, Lyons DM: Stress-induced changes in corticosteroid receptor expression in primate hippocampus and prefrontal cortex. Psychoneuroendocrinology 2008;33:360–367.
49 Sapolsky RM: Endocrinology of the stress response; in Becker JB, Breedlove SB, Crews D, McCarthy MM, (eds): Behavioral Endocrinology. Cambridge, MIT Press, 2002, pp 409–450.
50 Dhabhar FS, McEwen BS: Enhancing versus suppressive effects of stress hormones on skin immune function. Proc Natl Acad Sci USA 1999;96:1059–1064.
51 Dhabhar FS, Miller AH, McEwen BS, Spencer RL: Stress-induced changes in blood leukocyte distribution. Role of adrenal steroid hormones. J Immunol 1996;157:1638–1644.
52 Munhoz C, García-Bueno B, Madrigal J, Lepsch L, Scavone C, Leza J: Stress-induced neuroinflammation: mechanisms and new pharmacological targets. Braz J Med Biol Res 2008;41:1037–1046.
53 Madrigal JLM, Moro MA, Lizasoain I, Lorenzo P, Castrillo A, Boscá L, Leza JC: Inducible nitric oxide synthase expression in brain cortex after acute restraint stress is regulated by nuclear factor κβ-mediated mechanisms. J Neurochem 2001;76:532–538.
54 Madrigal J, Moro MA, Lizasoain I, Lorenzo P, Fernandez AP, Rodrigo J, Bosca L, Leza JC: Induction of cyclooxygenase-2 accounts for restraint stress-induced oxidative status in rat brain. Neuropsychopharmacology 2003;28:1579–1588.
55 Madrigal JLM, Hurtado O, Moro MA, Lizasoain I, Lorenzo P, Castrillo A, Bosca L, Leza JC: The increase in TNF-[alpha] levels is implicated in NF-[kappa] B activation and inducible nitric oxide synthase expression in brain cortex after immobilization stress. Neuropsychopharmacology 2002;26:155–163.

56 García-Bueno B, Madrigal JLM, Pérez-Nievas BG, Leza JC: Stress mediators regulate brain prostaglandin synthesis and peroxisome proliferator-activated receptor-γ activation after stress in rats. Endocrinology 2008;149:1969–1978.
57 De Pablos R, Villaran R, Argüelles S, Herrera A, Venero J, Ayala A, Cano J, Machado A: Stress increases vulnerability to inflammation in the rat prefrontal cortex. J Neurosci 2006;26:5709–5719.
58 Johnson J, Campisi J, Sharkey C, Kennedy S, Nickerson M, Greenwood B, Fleshner M: Catecholamines mediate stress-induced increases in peripheral and central inflammatory cytokines. Neuroscience 2005;135:1295–1307.
59 Dantzer R: Cytokine, sickness behavior, and depression. Neurol Clin 2006;24: 441–460.
60 Commins SP, Borish L, Steinke JW: Immunologic messenger molecules: cytokines, interferons and chemokines. J Allergy Clin Immunol 2010; 125:S53–S72.
61 Licinio J, Wong ML: Pathways and mechanisms for cytokine signaling of the central nervous system. J Clin Invest 1997;100:2941–2947.
62 Dantzer R, O'Connor JC, Freund GG, Johnson RW, Kelley KW: From inflammation to sickness and depression: when the immune system subjugates the brain. Nat Rev Neurosci 2008;9:46–56.
63 Kim YK, Maes M: The role of the cytokine network in psychological stress. Acta Neuropsychiatr 2003;15:148–155.
64 Kop WJ, Gottdiener JS: The role of immune system parameters in the relationship between depression and coronary artery disease. Psychosom Med 2005;67:S37–S41.
65 Boutin H, LeFeuvre R, Horai R, Asano M, Iwakura Y, Rothwell N: Role of IL-1α and IL-1β in ischemic brain damage. J Neurosci 2001;21:5528–5534.
66 Allan SM, Rothwell NJ: Cytokines and acute neurodegeneration. Nat Rev Neurosci 2001;2:734–744.
67 Arvin B, Neville LF, Barone FC, Feuerstein GZ: The role of inflammation and cytokines in brain injury. Neurosci Biobehav Rev 1996;20:445–452.
68 Lucas SM, Rothwell NJ, Gibson RM: The role of inflammation in CNS injury and disease. Br J Pharmacol 2006;147: S232–S240.
69 Hillhouse E, Kida S, Iannotti F: Middle cerebral artery occlusion in the rat causes a biphasic production of immunoreactive interleukin-1[beta] in the cerebral cortex. Neurosci Lett 1998;249: 177–179.
70 Toulmond S, Rothwell N: Interleukin-1 receptor antagonist inhibits neuronal damage caused by fluid percussion injury in the rat. Brain Res 1995;671: 261–266.
71 Yamasaki Y, Matsuura N, Shozuhara H, Onodera H, Itoyama Y, Kogure K: Interleukin-1 as a pathogenetic mediator of ischemic brain damage in rats. Stroke 1995;26:676–681.
72 Lawrence CB, Allan SM, Rothwell NJ: Interleukin-1β and the interleukin-1 receptor antagonist act in the striatum to modify excitotoxic brain damage in the rat. Eur J Neurosci 1998;10:1188–1195.
73 Nguyen KT, Deak T, Owens SM, Kohno T, Fleshner M, Watkins LR, Maier SF: Exposure to acute stress induces brain interleukin-1β protein in the rat. J Neurosci 1998;18:2239–2246.
74 Ackerman KD, Martino M, Heyman R, Moyna NM, Rabin BS: Stressor-induced alteration of cytokine production in multiple sclerosis patients and controls. Psychosom Med 1998;60:484–491.
75 McCarthy DO, Ouimet ME, Daun JM: The effects of noise stress on leukocyte function in rats. Res Nurs Health 1992; 15:131–137.
76 Koo JW, Duman RS: IL-1β is an essential mediator of the antineurogenic and anhedonic effects of stress. Proc Natl Acad Sci USA 2008;105:751–756.
77 Tracey KJ, Cerami A: Tumor necrosis factor: a pleiotropic cytokine and therapeutic target. Annu Rev Med 1994;45: 491–503.
78 Baumann H, Gauldie J: The acute phase response. Immunol Today 1994;15:74–80.
79 Kinouchi K, Brown G, Pasternak G, Donner DB: Identification and characterization of receptors for tumor necrosis factor-[alpha] in the brain. Biochem Biophys Res Commun 1991;181:1532–1538.
80 Hallenbeck JM: The many faces of tumor necrosis factor in stroke. Nat Med 2002;8:1363–1368.
81 Gary DS, Bruce-Keller AJ, Kindy MS, Mattson MP: Ischemic and excitotoxic brain injury is enhanced in mice lacking the p55 tumor necrosis factor receptor. J Cereb Flow Metab 1998;18:1283–1287.
82 Slavich GM, Way BM, Eisenberger NI, Taylor SE: Neural sensitivity to social rejection is associated with inflammatory responses to social stress. Proc Natl Acad Sci USA 2010;107:14817–14822.
83 O'Connor KA, Johnson JD, Hansen MK, Wieseler Frank JL, Maksimova E, Watkins LR, Maier SF: Peripheral and central proinflammatory cytokine response to a severe acute stressor. Brain Res 2003;991:123–132.
84 Penkowa M, Giralt M, Carrasco J, Hadberg H, Hidalgo J: Impaired inflammatory response and increased oxidative stress and neurodegeneration after brain injury in interleukin-6-deficient mice. Glia 2000;32:271–285.
85 Maes M, Lin A, Delmeire L, Van Gastel A, Kenis G, De Jongh R, Bosmans E: Elevated serum interleukin-6 (IL-6) and IL-6 receptor concentrations in posttraumatic stress disorder following accidental man-made traumatic events. Biol Psychiatry 1999;45:833–839.
86 Black PH: The inflammatory consequences of psychologic stress: relationship to insulin resistance, obesity, atherosclerosis and diabetes mellitus, type II. Med Hypotheses 2006;67:879–891.
87 Bierhaus A, Wolf J, Andrassy M, Rohleder N, Humpert PM, Petrov D, Ferstl R, Von Eynatten M, Wendt T, Rudofsky G: A mechanism converting psychosocial stress into mononuclear cell activation. Proc Natl Acad Sci USA 2003;100:1920–1925.
88 Padgett DA, Glaser R: How stress influences the immune response. Trends Immunol 2003;24:444–448.
89 Munhoz CD, Lepsch LB, Kawamoto EM, Malta MB, de Sá Lima L, Avellar MCW, Sapolsky RM, Scavone C: Chronic unpredictable stress exacerbates lipopolysaccharide-induced activation of nuclear factor-κB in the frontal cortex and hippocampus via glucocorticoid secretion. J Neurosci 2006;26:3813–3820.
90 Bogdan C: Nitric oxide and the immune response. Nat Immunol 2001;2:907–916.
91 Gross SS, Wolin M: Nitric oxide: pathophysiological mechanisms. Annu Rev Physiol 1995;57:737–769.
92 Moncada S, Palmer R, Higgs E: Nitric oxide: physiology, pathophysiology, and pharmacology. Pharmacol Rev 1991;43:109–142.

93 Pacher P, Beckman JS, Liaudet L: Nitric oxide and peroxynitrite in health and disease. Physiol Rev 2007;87:315–424.
94 Olivenza R, Moro MA, Lizasoain I, Lorenzo P, Fernández AP, Rodrigo J, Boscá L, Leza JC: Chronic stress induces the expression of inducible nitric oxide synthase in rat brain cortex. J Neurochem 2000;74:785–791.
95 Madrigal JLM, Moro MA, Lizasoain I, Lorenzo P, Leza JC: Stress-induced increase in extracellular sucrose space in rats is mediated by nitric oxide. Brain Res 2002;938:87–91.
96 Pasinetti GM: Cyclooxygenase and inflammation in Alzheimer's disease: experimental approaches and clinical interventions. J Neurosci Res 1998;54: 1–6.
97 Nogawa S, Zhang F, Ross ME, Iadecola C: Cyclo-oxygenase-2 gene expression in neurons contributes to ischemic brain damage. J Neurosci 1997;17:2746–2755.
98 Takadera T, Yumoto H, Tozuka Y, Ohyashiki T: Prostaglandin E_2 induces caspase-dependent apoptosis in rat cortical cells. Neurosci Lett 2002;317:61–64.
99 Phillis JW, Horrocks LA, Farooqui AA: Cyclooxygenases, lipoxygenases, and epoxygenases in CNS: their role and involvement in neurological disorders. Brain Res Rev 2006;52:201–243.

Angela J. Grippo, PhD
Department of Psychology
Northern Illinois University, PM 357
DeKalb, IL 60115 (USA)
E-Mail angelagrippo@niu.edu

Halaris A, Leonard BE (eds): Inflammation in Psychiatry.
Mod Trends Pharmacopsychiatry. Basel, Karger, 2013, vol 28, pp 33–48 (DOI: 10.1159/000343966)

Role of Inflammation in Depression: Implications for Phenomenology, Pathophysiology and Treatment

Charles L. Raison[a] · Andrew H. Miller[b]

[a]Department of Psychiatry, College of Medicine, John and Doris Norton School of Family and Consumer Sciences, University of Arizona, Tucson, Ariz., and [b]Department of Psychiatry and Behavioral Sciences, Emory University School of Medicine, Atlanta, Ga., USA

Abstract

Like all psychiatric conditions, depression is a complex phenomenon that is unlikely to yield to simple monolithic explanatory paradigms. Nonetheless, increasing evidence suggests that the immune system in general and inflammatory processes in particular, may contribute to depressive pathogenesis in a significant proportion of otherwise medically healthy individuals struggling with the disorder. In this chapter, we review the best current evidence suggesting that inflammatory processes contribute to the development of depression, both via direct actions on the brain as well as by effects on secondary pathways that marry brain to body. We review epidemiological evidence linking inflammation to depression before reviewing findings that exposure to inflammatory stimuli produce depressive symptoms in concert with brain-body changes known to be common in depression. Following this review of the role of inflammation in depressive causation, we consider emerging evidence that immunomodulatory interventions may hold promise as antidepressants, especially in individuals with elevations in peripheral inflammatory biomarkers. Interventions discussed include cytokine and cyclo-oxygenase antagonists, as well as agents that impact inflammatory transcription factors/signaling cascades. We conclude with a brief discussion of the potential role of various behavioral strategies in reducing inflammation and thereby enhancing emotional well-being.

Because the brain and body are so complex, nothing in psychiatry is certain, and physiological findings reliably associated with a given diagnosis can be remarkably absent in any given individual with the same condition. Consistent with this frustrating truth, despite a half century of concerted searching no single physiological process or abnormality has been shown to be either necessary or sufficient to cause depression. We should not expect the rapidly expanding body of evidence linking inflammatory processes to depression to be any different, and it isn't. While from a '10,000 foot' vantage point it is increasingly clear that bidirectional interactions between the brain and bodily immune processes are far more relevant to the development of depression in response to both stress and disease than anyone could have guessed a quarter century ago, when one descends from the heights of generality down into the mess of details, the picture becomes far more complicated, and more interesting.

With this perspective in mind, this chapter aims to provide an overview of current knowledge regarding the association of depression with inflammation. With the understanding that inflammatory processes are almost certainly not going to be shown to be of relevance to all cases of depression as a given, we propose to discuss findings that are now more or less universally accepted as valid for groups of depressed people before briefly enumerating important areas related to depression and inflammation that are less well established. Said differently, we'll first review what we believe are established findings related to inflammation and depression, and then examine areas that most researchers in the field would consider unsettled and/or controversial.

Established Findings Linking Inflammation to Depression

As a Group People with Major Depression and/or Significant Depressive Symptoms Exhibit Increased Biomarkers of Inflammation

Findings that depression is associated with increases in inflammatory biomarkers were initially met with surprise and more than a little skepticism [1], because they occurred in the context of a pre-existing literature on the immune system in depression which was dominated by reports of decreased cellular (lymphocyte) responses and reduced natural killer cell activity [2]. More recently, multiple lines of evidence suggest that the inflammatory response itself may contribute to the various measures of suppressed immunity observed in major depression, including direct effects of inflammatory cytokines on signaling through the T cell receptor [2–4, 41, 43].

Since the initial reports of immune activation in major depression, a vast literature has reproduced these findings, and meta-analyses confirm that peripheral blood elevations in the cytokines, IL-6, tumor necrosis factor (TNF)-α, and the acute phase reactant, C-reactive protein (CRP) are among the most reliable biomarkers of increased inflammation in depressed patients [5–7]. Increases in peripheral blood chemokines and cellular adhesion molecules as well as increased stress-induced nuclear factor-κβ (NFκβ), a central signaling molecule in the inflammatory response have also been reported in patients with major depression [8]. Importantly, not only have mean differences in inflammatory markers between depressives and controls been reported, but many studies have found significant associations between blood concentrations of inflammatory factors and the severity of depressive symptoms [9–12].

Although not entirely consistent, increased inflammation in depression appears to be a state dependent phenomenon, with a number of studies demonstrating that increased inflammatory markers in depressed patients return to control levels following successful antidepressant treatment [13]. These findings suggest that failure to respond to antidepressant treatment should be associated with ongoing elevations in inflammatory biomarkers, and indeed several studies support this [14, 15]. Moreover, a corollary prediction would be that elevations in inflammation prior to treatment might predict poor response. Indeed, patients with increased inflammatory markers at baseline have been repeatedly reported to be less likely to show a response to a variety of treatments, both pharmacologic and behavioral, suggesting a relationship between inflammation and treatment resistance [14–17].

Cytokine Administration Induces Depressive Symptoms

If inflammation plays a causative role in depression, as opposed to being merely a downstream physiological association, one would expect that exposure to inflammatory cytokines should induce the disorder. In fact, data incontrovertibly

demonstrate that this is the case. For example, acute administration of endotoxin and typhoid vaccination to humans leads to a host of behavioral changes including depressed mood, fatigue, cognitive dysfunction and feelings of social isolation [18–21]. Chronic administration of the inflammatory cytokine, interferon (IFN)-α, has also been found to induce depressive symptoms, with as many as 30–50% of IFNα-treated patients meeting symptom criteria for major depression [22]. Moreover, a comparison of symptoms in patients with IFNα-induced depression versus medically healthy depressed patients revealed a large degree of overlap in both symptom expression and severity [23]. Further supporting the similarity between the depression associated with IFNα and depression in other populations is the fact that IFNα-induced depression can be prevented and/or treated by conventional antidepressants [24–27].

Stress Increases Inflammation

Psychosocial stressors are among the most replicated and reliable risk factors for major depression [28]. Thus, the data that acute and chronic psychosocial stress activates the inflammatory response represent a major breakthrough in the link between stress and depression as it relates to the immune system. One of the important early findings in this area was that exposure to the Trier Social Stress Test (TSST), a public speaking and mental arithmetic stressor, was associated with a significant increase in DNA binding of the inflammatory transcription factor, NFκβ, in peripheral blood mononuclear cells (PBMCs), whereas merely watching others undergoing the TSST had no such effect [29]. Subsequent work has suggested that psychosocial stressors may result in enhanced inflammatory responses in individuals with early life adversity, whether such people are currently depressed [30] or nondepressed [31]. These data suggest that early life stress (and/or childhood maltreatment) may represent so powerful a risk factor for adult depression as a result of inducing long-term increases in inflammatory signaling, a vulnerability heightened when these people are depressed, but that exists even in the nondepressed state [32].

Of the various stress-related brain-to-body pathways by which stress might activate the inflammatory response, data most strongly support a central role for the autonomic nervous system [33, 34]. Much of the early work linking the autonomic nervous system to immune activity focused on the sympathetic nervous system. However, an emerging literature indicates that the parasympathetic nervous system is also involved in immune regulation and has potent anti-inflammatory effects [35]. These inhibitory effects on the inflammatory response (often referred to as the 'cholinergic anti-inflammatory pathway') have been shown to be mediated by vagal release of acetylcholine that in turn activates the alpha7 subunit of the nicotinic acetylcholine receptor which can regulate both cytokine transcription and translation [35]. In addition, recent data suggest that T cells may contribute to this inhibitory cholinergic reflex [36]. Finally, reduced heart rate variability, a reflection of reduced parasympathetic tone, has been associated with increased inflammatory biomarkers including IL-6 and CRP in medically healthy individuals as well as patients with major depression and heart disease [37, 38].

Mechanisms of Cytokine Effects on the Brain

Given the association of depression with increased inflammatory biomarkers and the capacity of inflammatory stimuli and cytokines to induce depressive symptoms, there has been considerable attention paid to the mechanisms by which cytokines influence behavior.

Cytokines Alter Neurotransmitter Metabolism

Probably best studied of the mechanisms by which cytokines influence behavior is their effects on neurotransmitter metabolism. Numerous human and laboratory animal studies have demonstrated that multiple neurotransmitter systems are affected by acute and chronic administration of cytokines including the monoamines, serotonin and dopamine, as well as glutamate [13].

Effects on Monoamine Metabolism. The repeatedly observed fact that serotonin reuptake inhibitors are able to prevent and/or treat depressive symptoms during chronic exposure to IFNα in patients with infectious diseases and cancer provides strong evidence that serotonin pathways are involved in cytokine effects on behavior [24, 39]. Genetic studies have complemented this work by suggesting that polymorphisms in the promoter region of the serotonin transporter gene (5-HTTLPR) that have been associated with risk of developing depression in response to stress may be associated with IFNα-induced behavioral changes in patients with hepatitis C [40, 41]. Further implicating alterations in serotonin metabolism in cytokine-induced behavioral changes are data showing that IFNα-associated increases in cerebrospinal fluid (CSF) concentrations of IL-6 are inversely correlated with the serotonin metabolite, 5-hydroxyindoleacetic acid (5-HIAA), which, in turn, negatively correlate with IFNα-induced depression severity [42].

Cytokines also have a significant impact on dopamine pathways. Studies have found altered blood flow and metabolic activity in basal ganglia nuclei during exposure to inflammatory stimuli [19, 43, 44], and in studies in non-human primates, reduced CSF concentrations of the dopamine metabolite, homovanillic acid (HVA), were associated with depressive-like huddling behavior secondary to chronic IFNα administration [45]. That such huddling behavior reflects, at least in part, an effect of cytokines on CNS monoamine signaling is supported by the fact that similar huddling behavior is observed following chronic administration of the monoamine depleting drug, reserpine [46]. Rodent studies have also indicated that cytokines target the basal ganglia and dopamine pathways [47], and chronic administration of high dose endotoxin into the brain was associated with a 70% reduction of nigral dopaminergic neurons within 10 weeks [48].

Mechanisms by Which Cytokines Affect Monoamine Metabolism. The enzyme indoleamine 2,3-dioxygenase (IDO) is activated by various cytokines alone or in combination [49, 50]. IDO catabolizes tryptophan, the primary amino acid precursor of serotonin, into kynurenine. Depletion of tryptophan provides protection against a variety of pathogens, but also inhibits effector T cell responses and thereby contributes to immune tolerance [51]. Evidence of a role of IDO in cytokine-induced depression comes from a number of studies which have demonstrated correlations between IFNα-induced depression and decreases in tryptophan and increases in kynurenine and/or the kynurenine to tryptophan ratio [52, 53]. Although much of the attention regarding IDO was initially focused on the depletion of tryptophan and in turn of serotonin, multiple lines of evidence have increasingly pointed to the relevance of increased production of kynurenine and its metabolites for immune-induced depression [54–56].

Another pathway that can influence monoamine metabolism is the cytokine signaling pathway mitogen-activated protein kinase (MAPK). Both in vitro and in vivo data demonstrate that stimulation of p38 MAPK pathways increases the expression and function of the serotonin transporter (SERT), which in animal studies has been shown to increase depressive-like behavior in response to inflammatory stimuli [57]. In humans, increased phosphorylation of p38 MAPK following the first injection of IFNα was associated with the subsequent development of IFNα-induced depression and fatigue [58]. In addition to effects on serotonin metabolism, MAPK pathways have also been found to influence the dopamine transporter (DAT) [59].

Cytokine Effects on Glutamate Metabolism. In addition to affecting monoamines, inflammatory mediators are known to profoundly impact glutamatergic functioning in the CNS. For example, cytokines have been shown to decrease the expression of glutamate transporters on glial cells and to increase the release of glutamate from astrocytes [60–64]. This astrocyte-derived glutamate has preferential access to extrasynaptic NMDA receptors, which can mediate excitotoxicity and lead to decreased production of trophic factors including brain-derived neurotrophic factor (BDNF) [65, 66]. As noted above, inflammation-induced quinolinic acid can compound glutamatergic excitotoxicity through direct activation of the NMDA receptor, while also contributing to oxidative stress that further impacts relevant cell types including oligodendrocytes, which are especially vulnerable to oxidative damage and overactivation of calcium-permeable glutamate receptors [64, 67]. Interestingly, loss of glial elements such as oligodendrocytes in multiple mood relevant brain regions including the subgenual PFC and amygdala has emerged as a fundamental morphologic abnormality in major depression [68].

Cytokine Effects on Neurogenesis and Neurotrophic Factors

Neurogenesis has emerged as an important process in the development of depression and the activity of antidepressant medications [69]. In terms of the role of cytokines in these findings, data have shown that stress-induced decreases in neurogenesis as well as the expression of relevant nerve growth factors, including BDNF, which support neurogenesis, can be reversed by administration of the IL-1 receptor antagonist (IL-1Ra) or transplantation of IL-1Ra secreting neural precursor cells into the hippocampus or the use of IL-1 receptor knock-out (KO) mice [70–72]. In vitro studies indicate that the inhibitory effect of IL-1 on neurogenesis is mediated by activation of NFκβ [72]. In humans, treatment with IFNα has been associated with reduced levels of peripheral BDNF, which are known to correlate well with BDNF availability in the CNS [73]. Of note, reduced BDNF during IFNα treatment correlated with increased depression [73].

Cytokine Effects on Neuroendocrine Function

Alterations in the hypothalamic-pituitary-adrenal (HPA) axis are some of the most reproducible findings in patients with major depression. Patients with major depression have been found to exhibit increased concentrations of the HPA axis hormones, ACTH and cortisol, as well as increases in CSF measures of the HPA axis regulatory neuropeptide, corticotrophin-releasing hormone (CRH) [74, 75]. Of relevance to inflammation, administration of inflammatory cytokines to laboratory animals has been shown to profoundly stimulate not only ACTH and cortisol but also the expression and release of CRH [76]. Human studies also have demonstrated that the acute ACTH and cortisol response to the first injection of IFNα (presumably due to activation of CRH pathways) is robust and correlates with the subsequent development of depressive symptoms during IFNα therapy in patients with cancer [77]. In contrast to acute administration of IFNα, chronic IFNα administration is associated with flattening of the diurnal curve and increased evening cortisol concentrations, both of which correlate with the development of depression and fatigue [78]. Flattening of the diurnal cortisol rhythm has been seen in a number of medical disorders associated with inflammation including cardiovascular disease and cancer, where it has been associated with a worse outcome in these diseases [78]. Indeed, patients with metastatic breast cancer with flattened cortisol slope were found to exhibit decreased survival as well as decreased natural killer cell number and activity [79]. Of note, previous studies have shown that the flattened cortisol rhythm is associated with nonsuppression of cortisol by dexamethasone (DEX) in the DEX suppression test (DST) [80]. DST nonsuppression is a common finding in major depression both in vivo and in vitro in peripheral blood mononuclear cells (PBMCs) [74, 75, 81]. DEX nonsuppression, also referred to as glucocorticoid resistance, has additionally been correlated with stimulated production of IL-1β by PBMCs of patients with major depression [82]. Moreover, increased TNFα has been associated with glucocorticoid resistance in the skin as measured by reduced cutaneous blanching in response to topically applied glucocorticoids [83].

It is generally believed that glucocorticoid resistance is a result of decreased expression and/or function of the receptor for glucocorticoids. Of relevance in this regard, there is a rich literature indicating that inflammatory cytokines can disrupt glucocorticoid receptor (GR) function while decreasing GR expression [84]. For example, IFNα has been shown to inhibit GR function by activating STAT5 which in turns binds to the activated GR in the nucleus, thus disrupting GR-DNA binding [85]. A similar protein-protein interaction between the GR and NFκβ in the nucleus has also been described [86]. On the other hand, IL-1α and IL-1β have been shown both in vitro and in vivo to inhibit GR translocation from the cytoplasm to the nucleus through activation of p38 MAPK [87–89]. Of note, stress-induced alterations in GR translocation leading to glucocorticoid resistance in laboratory mice has been found to be mediated by IL-1 through the use of IL-1 KO mice, who fail to exhibit impaired GR translocation following social disruption stress [88]. Given the potent anti-inflammatory effects of glucocorticoids, it is not surprising that glucocorticoid resistance secondary to stress in mice is associated with increased lethality in response to endotoxin administration [90]. Thus, the effects of cytokines on GR function may lead to a feed-forward cascade whereby increased inflammation through its effects of the GR undercuts the well-known ability of glucocorticoids to restrain inflammatory responses [91], leading to further increases in inflammation and reduced GR function. Finally, it should be noted that chronic exposure to inflammatory glucocorticoids can lead to decreased GR expression as well as increased expression of GRβ, a GR isoform which has a distinct hormone binding domain (that is unable to bind known glucocorticoids) and a unique pattern of gene regulation [84, 92]. Of note, GRβ expression has been shown to be increased in patients with inflammatory disorders including patients with bronchial asthma and rheumatoid arthritis [92].

Impact of Cytokines on Neurocircuitry
As interest in the impact of inflammation on the brain has grown, there has been an emerging neuroimaging literature examining the brain regions that are most affected by administration of inflammatory stimuli. Brain regions that have been most reliably identified include the basal ganglia and the dorsal anterior cingulate cortex (dACC).

Basal Ganglia
Using positron emission tomography (PET), early studies in patients undergoing IFNα therapy for cancer or hepatitis C revealed marked increases in glucose metabolic activity in the basal ganglia which correlated with symptoms of fatigue [93, 94]. These increases are consistent with metabolic changes in the basal ganglia found in patients with Parkinson's disease, and are believed to reflect increased oscillatory burst activity in relevant basal ganglia nuclei secondary to DA depletion [95–98]. Increased metabolic activity in the basal ganglia of Parkinson's disease patients can be reversed by the administration of levodopa [99]. Of note, IFNα has been associated with development of Parkinson-like symptoms that were relieved by levodopa administration [100]. Using functional magnetic resonance imaging (fMRI), typhoid vaccination was also found to alter activation in the basal ganglia. In particular, vaccinated volunteers (compared to a control condition) exhibited increased evoked activity in the substantia nigra that was associated with both prolonged reaction times and increased peripheral blood concentrations of IL-6 [19]. Finally, administration of endotoxin to healthy volunteers led to reduced activation in the ventral striatum during a monetary reward task [101]. Between-group differences in striatal activation were in turn associated with increases in depressed mood.

The impact of cytokines on the basal ganglia, and ultimately motor activity and motivation, has been suggested to be an evolutionarily derived process, which may serve to reduce exploratory behavior and motor activity in order to facilitate the reallocation of energy resources to fighting infection, mounting a fever and wound healing.

Anterior Cingulate Cortex

Another brain region that has been found to be influenced by cytokine administration and inflammatory stimuli is the dACC. The dACC plays an important role in error detection and conflict monitoring [102]. Patients with hepatitis C treated with IFNα were found to have significantly greater activation of the dACC (Broadman's area (BA) [24]) using fMRI and a task of visuospatial attention [103]. Interestingly a strong correlation was found in this study between activation of the dACC in IFNα-treated patients and the number of errors made during the task. The error rate was quite low for the task, and no such correlation was found in control subjects. Of note, patients with high-trait anxiety have also been shown to exhibit increased dACC activation during fMRI in the context of low error rates [104]. In addition, increased activation of the dACC has been observed in subjects with neuroticism and obsessive compulsive disorder, both of which are associated with increases in anxiety as well as arousal [105, 106]. Aside from IFNα, increased activation of the dACC has been found using fMRI after a high demand color word Stroop task in patients administered typhoid vaccination [107]. Moreover, increased activation of the dACC after a psychosocial stressor (TSST) was correlated with the degree of activation of peripheral blood IL-6, suggesting a role for stress-induced inflammation in the findings [108].

Based on fMRI studies and a social rejection task, it has been suggested by Eisenberger and Lieberman [109] that the dACC not only plays a role in error detection and conflict monitoring, but also processes social pain. Given the connection of the dACC with downstream autonomic nervous system arousal pathways, these investigators have further hypothesized that the dACC may serve as a 'neural alarm system' that can both detect and respond (with arousal and distress) to threatening environmental stimuli in the social domain [109]. Thus, cytokines may sensitize the responsivity of the dACC, thereby contributing to the anxiety, arousal and alarm that often accompany chronic exposure to inflammatory stimuli such as IFNα. From an evolutionary perspective, this heightened dACC activity (and heightened sensitivity to social threat) may subserve the survival priority of vigilance against attack in an animal that is otherwise vulnerable due to infection and/or wounding [110]. Taken together with the effects of inflammation on the basal ganglia, the effects of inflammation on neurocircuitry in the brain appear to subserve two competing evolutionary survival priorities; to lay low to conserve energy resources for fighting infection and wound healing and to be vigilant for protection against future attack [110]. Finally, it should be noted that administration of an inflammatory stimulus (typhoid vaccination) has been associated with activation of the subgenual ACC, the target of deep brain stimulation strategies in patients with treatment-resistant depression [111].

Potential of Immunomodulatory Strategies in the Treatment of Depression

Because immune pathways interact with virtually every system in the body and the brain, an array of strategies that impact one aspect or other of immune function have shown promise for the treatment of depression, either directly, or more hypothetically, via animal models, or by effects on disease states that share key pathophysiologic mechanisms with depression. Given the extent of these data, only a sampling of potential immune-related antidepressant strategies can be discussed here.

Anti-Cytokine Interventions

Cytokines are probably the most obvious targets for treating the impact of peripheral inflammation on the brain. Early case reports indicated that several of the currently available biological response modifiers (or simply 'biologics') that target cytokines were effective in reducing symptoms of depression and fatigue as well as quality of life [112]. Biologics are compounds derived from biological processes versus medicinal chemistry. Most of the biologics approved by the US Food and Drug Administration (FDA) for treatment for autoimmune or inflammatory diseases use monoclonal antibodies, soluble cytokine receptors or other antagonists which combine cytokine receptors with other fusion proteins. Therefore, there are limited drug-drug interactions between biologics and psychotropic medications, although the side effect profile for these agents, especially as it relates to the risk of infection, is cause for concern. Currently FDA-approved drugs target both TNFα and IL-1.

Recent work by our group suggests that biologic cytokine antagonists may indeed hold promise for the treatment of depression, but only in individuals with evidence of elevated peripheral inflammatory activity. Specifically, in a double-blind, placebo-controlled trial of the TNFα antagonist infliximab in sixty medically-healthy adults with treatment resistant major depression, increasing baseline plasma concentrations of either CRP or TNFα strongly predicted response to infliximab. For individuals with baseline CRP levels >5 mg/dl, the difference in response on the Hamilton Depression Rating Scale of 3 points is consistent with the effect size observed for standard antidepressants.

But the results of this trial were complex, and therefore instructive regarding any urge one might have to jump to simple conclusions about the usefulness of anti-cytokine therapy for MDD. First, no between-group differences were observed, indicating that TNFα blockade does not appear to hold promise for the treatment of MDD in general. Second, the relationship between baseline inflammatory status and clinical response was a true one – meaning that individuals who entered the trial with low levels of peripheral inflammatory biomarkers actually had poorer responses to infliximab than to placebo. These findings, while unexpected, are consistent with several lines of evidence suggesting that at lower levels cytokines may have neurotrophic and other effects that might actually protect against depression.

Two additional points emerged from this study that are worthy of mention. First, because infliximab is too large to cross the blood-brain barrier, this study provides a first demonstration that targeting peripheral inflammation specifically may be sufficient to induce antidepressant effects in individuals with elevated peripheral inflammation. Many of us have tended to view peripheral inflammatory markers as inexact surrogates for the inflammatory state of the CNS that we assumed was the true source of the association between the immune system and depression. Findings from the infliximab study challenge this dominant paradigm and suggest that increased peripheral inflammation may itself be a sufficient physiologic source for depression as a result of its ability to drive CNS functioning in directions that promote depression. A second striking finding from the infliximab study is that individuals with elevated baseline inflammatory biomarkers had very low rates of placebo response, suggesting that prior observations of an association between increased inflammation and antidepressant nonresponse may be accounted for – at least in part – by the fact that inflammatory mediators disrupt social perceptions/affective responses in ways that rob antidepressants of their placebo component.

Cyclo-Oxygenase

Aside from the study noted above that directly blocked cytokine activity, probably the best studied anti-inflammatory strategy for the treatment of depression is the targeting of inflammatory signaling pathways. The most common approach in this regard has been the administration of pharmacologic agents that inhibit cyclo-oxygenase (COX), the enzyme that converts arachidonic acid into prostaglandin. As noted above, prostaglandins have been shown to be increased in depression and are known to play an important role in the inflammatory response including the mediation of fever and sensitivity to pain. Moreover, inhibition of COX-1 and COX-2 selectively and in combination have been shown in laboratory animals to inhibit depressive-like behavior following administration of the inflammatory stimulus, LPS, without influencing cytokine responses [113, 114]. Of note, there has been some controversy as to whether COX-1 versus COX-2 is the most relevant target, with some recent animal data suggesting that COX-1 but not COX-2 selective inhibitors were able to inhibit low dose LPS-induced behavioral responses. As might be expected, indomethacin and ibuprofen, nonselective COX inhibitors, have reliably shown efficacy in animal models [114].

In human studies, the addition of acetylsalicylic acid (which blocks COX-1 and COX-2) to fluoxetine led to increased remission rates in an open label study of 24 depressed patients previously nonresponsive to fluoxetine alone [115]. Similarly, in a study of 20 subjects per group, medically healthy depressed patients who received the COX-2 inhibitor, celecoxib, in combination with reboxetine showed greater symptomatic improvement compared to patients randomized to reboxetine plus placebo [116]. Nevertheless, both studies involved small sample sizes, and in the celecoxib study, 50% or more of the sample in each group dropped out before conclusion of the study [116]. Interestingly, the largest trial to date examining the antidepressant efficacy of a COX-2 inhibitor (cimicoxib) in combination with sertraline versus sertraline plus placebo (conducted in Austria, the Czech Republic and Germany) has completed enrollment, and results are pending (ClinicalTrials.gov: NCT00510822).

p38 MAPK

In laboratory animals, inhibition of the p38 mitogen-activated protein kinase (MAPK) pathway has been shown to reverse the development of LPS-induced behavioral changes, in part, through reversing the effects of p38 MAPK on the serotonin transporter [117]. A number of p38 MAPK inhibitors are in development for the treatment of a range of diseases including autoimmune and inflammatory disorders, cardiovascular disease, pulmonary disorders including asthma and COPD as well as pain [118]. According to a recent market analysis, about ten, next generation p38 MAPK inhibitors (which exhibit fewer side effects such as liver toxicity) are in phase I or II clinical trials. One drug, GW856553, manufactured by GlaxoSmithKline, has just completed phase II testing in patients with major depression (ClinicalTrials.gov: NCT00976560). However, no results are available at this time. Of note, minocycline, a tetracycline derivative, which inhibits activation of microglia as well as microglial production of IL-1 and nitric oxide, has been shown to inhibit p38 MAPK as well as poly(ADP-ribose) polymerase-1 (PARP-1), which along with inhibition of NFκβ (see below), may serve as the basis for its anti-inflammatory efficacy [119–122]. Interestingly, minocycline has also been found to exhibit antidepressant properties in laboratory animals using the FST both alone and in combination with subthreshold doses of the tricyclic antidepressant, desipramine and several glutamate antagonists [123]. Minocycline has also been shown to inhibit the development of neuroinflammation and behavioral changes after LPS administration [124], and has been found to attenuate the development of opioid tolerance, which may in part be related to the capacity of opioids to induce an inflammatory response [125–127].

NFκβ

NFκβ plays a central role in the inflammatory response, given its primacy in transmitting inflammatory signals from the periphery to the brain, where it has been shown to mediate inflammation-induced inhibition of neurogenesis in vitro [128, 129]. Moreover, intracerebroventricular administration of NFκβ antagonists to rats subjected to acute or chronic stress has been found to reverse stress-induced suppression of hippocampal neurogenesis as well as the effects of chronic unpredictable stress on behavior including sucrose consumption [128]. There are hundreds of compounds that have been shown to inhibit NFκβ; however, special interest has been recently paid to natural compounds such as curcumin (a derivative of the curry spice, turmeric), which has been shown to decrease NFκβ activation in PBMCs of patients with pancreatic cancer [130]; α-tocopherol (a form of vitamin E) which has been shown to block NFκβ and the associated neuroinflammatory and behavioral response to LPS [131], and resveratrol (found in the skin of red grapes) which has been shown to inhibit LPS-induced induction of inflammatory cytokines, chemokines and inducible nitric oxide synthase (iNOS) in murine microglia and astrocytes in association with inhibition of NFκβ [132]. These natural compounds are currently in a number of phase I–III clinical trials for multiple indications including independent examination of effects on inflammation and behavior.

Indoleamine 2,3-Dioxygenase

The importance of IDO as a potential target for the unique contributions of the immune system to depression is evidenced by studies that have administered an IDO antagonist or used IDO KO mice in the context of immune activation or infection. For example, treatment of mice with the IDO antagonist, 1-methyl tryptophan (1-MT), abrogates the impact of LPS as well as infection with bacillus Calmette-Guerin (BCG), an attenuated form of *Mycobacterium bovis*, on depressive-like behavior including increased immobility in the FST and TST [54, 133]. In addition, IDO-deficient mice are resistant to BCG-induced behavioral changes, while exhibiting a normal inflammatory cytokine response to BCG administration [133]. Targeted deletion of KAT II (the enzyme that converts KYN to KA) has also been shown to increase cognitive performance in association with an increase in the amplitude of long-term potentiation in vitro, while reducing extracellular KA as measured by hippocampal in vivo microdialysis [134].

In terms of pharmacologic agents in clinical trials, 1-MT is currently being tested in patients with advanced stage cancers, given alone or in combination with vaccine therapies (e.g. ClinicalTrials.gov: NCT00617422 and NCT01302821). These studies are based on the inhibitory effect of IDO-mediated TRP depletion on T cell function [135]. No studies are currently underway examining the impact of 1-MT on behavioral alterations in cancer patients or other patient populations, although efficacy in regard to behavior would be an important proof of concept for clinical studies of IDO inhibitors in depressive disorders. Until recently, the majority of IDO inhibitors displayed enzyme affinities in the micromolar range and were derived from metabolites of tryptophan (e.g. 1-MT) and β-carbolines [136]. Nevertheless, more potent (nanomolar range) IDO inhibitors have been identified in extracts from marine invertebrates and screening of the NCI Diversity Set Library [136–138]. In addition, elaboration of the crystal structure of human IDO has paved the way for the in silico design of new IDO inhibitors [136]. Thus, there is great promise regarding the development of more potent IDO inhibitors, although these compounds are still somewhat removed from clinical trials [138]. In similar fashion, given the elaboration of the kynurenine pathway as well as data indicating the role of kynurenic acid and quinolinic acid in neuropsychiatric disease, especially neurodegenerative disorders, a number of compounds which interfere with multiple steps in the kynurenine pathway have been identified including inhibitors of 3-hydroxyanthranilic acid oxygenase which converts 3-hydroxyanthranilic acid to quinolinic acid. Nevertheless, these compounds have yet to advance to phase I–III testing for neuropsychiatric disorders [55, 139].

Behavioral Interventions with Immunomodulatory Potential

Interestingly, most behavioral interventions that have shown antidepressant properties, including psychotherapy, exercise and weight loss, also tend to reduce peripheral inflammatory biomarkers. More recently, meditation practices have also shown evidence of anti-inflammatory activity. For example, the practice of a secularized compassion meditation technique has been reported to reduce inflammatory responses (i.e. plasma IL-6) to a standardized laboratory stressor in healthy young adults and to reduce resting state levels of CRP in highly traumatized adolescents in state custody [140–142]. These studies also observed behavioral effects of the training that might be expected to either reduce or prevent depressive pathology. More recently, mindfulness-based stress reduction (MBSR) has been reported to reduce feelings of loneliness in older individuals and to down-regulate NFκβ-associated gene expression profiles [143]. Similar changes in the balance of pro- to anti-inflammatory transcription activity as assessed by gene expression analyses have been observed in caregivers of dementia patients trained in yogic meditation [144].

Conclusions

While many questions remain, increasing data demonstrate that immune processes, both within the central nervous system and the periphery of the body, may play previously underappreciated roles in the development of major depression, and by extension, may hold promise as novel treatment modalities for the disorder. These data highlight the urgent need to both test novel pharmacologic and behavioral strategies that target the immune system to treat neuropsychiatric disorders and to develop better biomarkers to measure the relative status of the immunologic response as this response relates to depressive pathogenesis. In some ways, the immune hypothesis of depression has some advantages in regard to both these areas of need, because of the general recognition that inflammation is a central pathophysiologic mechanism in a host of disorders. Thus, there are numerous partners in the search for new anti-inflammatory treatments as well as new biomarkers relevant to both diagnosis and treatment response. Finally, the fact that inflammation may contribute to the development of depression in a significant proportion of depressed patients while also contributing to the development and progression of multiple other illnesses, places psychiatry squarely in line with other medical disciplines in the search for new strategies to treat and prevent immunologic contributions to disease development.

References

1 Maes M: Major depression and activation of the inflammatory response system. Adv Exp Med Biol 1999;461:25–46.

2 Irwin MR, Miller AH: Depressive disorders and immunity: 20 years of progress and discovery. Brain Behav Immun 2007;21:374–383.

3 Blume J, Douglas SD, Evans DL: Immune suppression and immune activation in depression. Brain Behav Immun 2011;25:221–229.

4 Clark J, Vagenas P, Panesar M, Cope AP: What does tumour necrosis factor excess do to the immune system long term? Ann Rheum Dis 2005;64(suppl 4):iv70–iv76.

5 Zorilla E, Luborsky L, McKay J, Roesnthal R, et al: The relationship of depression and stressors to immunological assays: a meta-analytic review. Brain Behav Immun 2001;15:199–226.

6 Dowlati Y, Herrmann N, Swardfager W, et al: A meta-analysis of cytokines in major depression. Biol Psychiatry 2010;67:446–457.

7 Howren MB, Lamkin DM, Suls J: Associations of depression with C-reactive protein, IL-1, and IL-6: a meta-analysis. Psychosom Med 2009;71:171–186.

8 Haroon E, Raison CL, Miller AH: Psychoneuroimmunology meets neuropsychopharmacology: translational implications of the impact of inflammation on behavior. Neuropsychopharmacology 2012;37:137–162.
9 Bower JE, Ganz PA, Aziz N, Fahey JL: Fatigue and proinflammatory cytokine activity in breast cancer survivors. Psychosom Med 2002;64:604–611.
10 Miller GE, Stetler CA, Carney RM, Freedland KE, Banks WA: Clinical depression and inflammatory risk markers for coronary heart disease. Am J Cardiol 2002;90:1279–1283.
11 Alesci S, Martinez PE, Kelkar S, et al: Major depression is associated wwith significant diurnal elevations in plasma interleukin-6 levels, a shift of its circadian rhythm, and loss of physiological complexity in its secretion: clinical implications. J Clin Endocrinol Metab 2005;90:2522–2530.
12 Motivala SJ, Sarfatti A, Olmos L, Irwin MR: Inflammatory markers and sleep disturbance in major depression. Psychosom Med 2005;67:187–194.
13 Miller AH, Maletic V, Raison CL: Inflammation and its discontents: the role of cytokines in the pathophysiology of major depression. Biol Psychiatry 2009;65:732–741.
14 Lanquillon S, Krieg JC, Bening-Abu-Shach U, Vedder H: Cytokine production and treatment response in major depressive disorder. Neuropsychopharmacology 2000;22:370–379.
15 Fitzgerald P, O'Brien SM, Scully P, Rijkers K, Scott LV, Dinan TG: Cutaneous glucocorticoid receptor sensitivity and pro-inflammatory cytokine levels in antidepressant-resistant depression. Psychol Med 2006;36:37–43.
16 Benedetti F, Lucca A, Brambilla F, Colombo C, Smeraldi E: Interleukine-6 serum levels correlate with response to antidepressant sleep deprivation and sleep phase advance. Prog Neuro-Psychopharmacol Biol Psychiatry 2002;26:1167–1170.
17 Harley J, Luty S, Carter J, Mulder R, Joyce P: Elevated C-reactive protein in depression: a predictor of good long-term outcome with antidepressants and poor outcome with psychotherapy. J Psychopharmacol 2010;24:625–626.
18 Reichenberg A, Yirmiya R, Schuld A, et al: Cytokine-associated emotional and cognitive disturbances in humans. Arch Gen Psychiatry 2001;58:445–452.
19 Brydon L, Harrison NA, Walker C, Steptoe A, Critchley HD: Peripheral inflammation is associated with altered substantia nigra activity and psychomotor slowing in humans. Biol Psychiatry 2008;63:1022–1029.
20 Harrison NA, Brydon L, Walker C, Gray MA, Steptoe A, Critchley HD: Inflammation causes mood changes through alterations in subgenual cingulate activity and mesolimbic connectivity. Biol Psychiatry 2009;66:407–414.
21 Eisenberger NI, Inagaki TK, Mashal NM, Irwin MR: Inflammation and social experience: an inflammatory challenge induces feelings of social disconnection in addition to depressed mood. Brain Behav Immun 2010;24:558–563.
22 Raison CL, Demetrashvili M, Capuron L, Miller AH: Neuropsychiatric side effects of interferon-alpha: recognition and management. CNS Drugs 2005;19:1–19.
23 Capuron L, Fornwalt FB, Knight BT, Harvey PD, Ninan PT, Miller AH: Does cytokine-induced depression differ from idiopathic major depression in medically healthy individuals? J Affect Disord 2009;119:181–185.
24 Musselman DL, Lawson DH, Gumnick JF, et al: Paroxetine for the prevention of depression induced by high-dose interferon alfa. N Engl J Med 2001;344: 961–966.
25 Raison CL, Woolwine BJ, Demetrashvili MF, et al: Paroxetine for prevention of depressive symptoms induced by interferon-alpha and ribavirin for hepatitis C. Aliment Pharmacol Ther 2007;25:1163–1174.
26 Kraus MR, Schafer A, Schottker K, et al: Therapy of interferon-induced depression in chronic hepatitis C with citalopram: a randomised, double-blind, placebo-controlled study. Gut 2008;57:531–536.
27 Hauser P, Khosla J, Aurora H, et al: A prospective study of the incidence and open-label treatment of interferon-induced major depressive disorder in patients with hepatitis C. Mol Psychiatry 2002;7:942–947.
28 Kendler KS, Gardner CO: Dependent stressful life events and prior depressive episodes in the prediction of major depression: the problem of causal inference in psychiatric epidemiology. Arch Gen Psychiatry 2010;67:1120–1127.
29 Bierhaus A, Wolf J, Andrassy M, et al: A mechanism converting psychosocial stress into mononuclear cell activation. Proc Natl Acad Sci USA 2003;100:1920–1925.
30 Pace TW, Mletzko TC, Alagbe O, et al: Increased stress-induced inflammatory responses in male patients with major depression and increased early life stress. Am J Psychiatry 2006;163:1630–1633.
31 Carpenter LL, Gawuga CE, Tyrka AR, Lee JK, Anderson GM, Price LH: Association between plasma IL-6 response to acute stress and early-life adversity in healthy adults. Neuropsychopharmacology 2010;35:2617–2623.
32 Brown DW, Anda RF, Tiemeier H, et al: Adverse childhood experiences and the risk of premature mortality. Am J Prev Med 2009;37:389–396.
33 Johnson JD, Campisi J, Sharkey CM, et al: Catecholamines mediate stress-induced increases in peripheral and central inflammatory cytokines. Neuroscience 2005;135:1295–1307.
34 Mazzeo RS, Donovan D, Fleshner M, et al: Interleukin-6 response to exercise and high-altitude exposure: influence of alpha-adrenergic blockade. J Appl Physiol 2001;91:2143–2149.
35 Tracey KJ: Reflex control of immunity. Nat Rev Immunol 2009;9:418–428.
36 O'Mahony C, van der Kleij H, Bienenstock J, Shanahan F, O'Mahony L: Loss of vagal anti-inflammatory effect: in vivo visualization and adoptive transfer. Am J Physiol Regul Integr Comp Physiol 2009;297:R1118–R1126.
37 Frasure-Smith N, Lesperance F, Irwin MR, Talajic M, Pollock BG: The relationships among heart rate variability, inflammatory markers and depression in coronary heart disease patients. Brain Behav Immun 2009;23:1140–1147.
38 Thayer JF, Fischer JE: Heart rate variability, overnight urinary norepinephrine and C-reactive protein: evidence for the cholinergic anti-inflammatory pathway in healthy human adults. J Intern Med 2009;265:439–447.
39 Raison CL, Woolwine BJ, Demetrashvili MF, et al: Paroxetine for prevention of depressive symptoms induced by interferon-alpha and ribavirin for hepatitis C. Aliment Pharm Ther 2007;25:1163–1174.

40 Bull SJ, Huezo-Diaz P, Binder EB, et al: Functional polymorphisms in the interleukin-6 and serotonin transporter genes, and depression and fatigue induced by interferon-alpha and ribavirin treatment. Mol Psychiatry 2009;14:1095–1104.

41 Lotrich FE, Ferrell RE, Rabinovitz M, Pollock BG: Risk for depression during interferon-alpha treatment is affected by the serotonin transporter polymorphism. Biol Psychiatry 2009;65:344–348.

42 Raison CL, Borisov AS, Majer M, et al: Activation of central nervous system inflammatory pathways by interferon-alpha: relationship to monoamines and depression. Biol Psychiatry 2009;65: 296–303.

43 Juengling FD, Ebert D, Gut O, et al: Prefrontal cortical hypometabolism during low-dose interferon alpha treatment. Psychopharmacology 2000;152:383–389.

44 Capuron L, Pagnoni G, Demetrashvili MF, et al: Basal ganglia hypermetabolism and symptoms of fatigue during interferon-alpha therapy. Neuropsychopharmacology 2007;32:2384–2392.

45 Felger JC, Alagbe O, Hu F, et al: Effects of interferon-alpha on rhesus monkeys: a nonhuman primate model of cytokine-induced depression. Biol Psychiatry 2007;62:1324–1333.

46 McKinney WT Jr, Eising RG, Moran EC, Suomi SJ, Harlow HF: Effects of reserpine on the social behavior of rhesus monkeys. Dis Nerv Syst 1971;32:735–741.

47 Qin L, Wu X, Block ML, et al: Systemic LPS causes chronic neuroinflammation and progressive neurodegeneration. Glia 2007;55:453–462.

48 Gao HM, Jiang J, Wilson B, Zhang W, Hong JS, Liu B: Microglial activation-mediated delayed and progressive degeneration of rat nigral dopaminergic neurons: relevance to Parkinson's disease. J Neurochem 2002;81:1285–1297.

49 Pemberton LA, Kerr SJ, Smythe G, Brew BJ: Quinolinic acid production by macrophages stimulated with IFN-gamma, TNF-alpha, and IFN-alpha. J Interferon Cytokine Res 1997;17:589–595.

50 Fujigaki H, Saito K, Fujigaki S, et al: The signal transducer and activator of transcription 1alpha and interferon regulatory factor 1 are not essential for the induction of indoleamine 2,3-dioxygenase by lipopolysaccharide: involvement of p38 mitogen-activated protein kinase and nuclear factor-kappaB pathways, and synergistic effect of several proinflammatory cytokines. J Biochem 2006;139:655–662.

51 Huang L, Baban B, Johnson BA 3rd, Mellor AL: Dendritic cells, indoleamine 2,3 dioxygenase and acquired immune privilege. Int Rev Immunol 2010;29:133–155.

52 Bonaccorso S, Marino V, Puzella A, et al: Increased depressive ratings in patients with hepatitis C receiving interferon-alpha-based immunotherapy are related to interferon-alpha-induced changes in the serotonergic system. J Clin Psychopharmacol 2002;22:86–90.

53 Capuron L, Neurauter G, Musselman DL, et al: Interferon-alpha-induced changes in tryptophan metabolism: relationship to depression and paroxetine treatment. Biol Psychiatry 2003;54:906–914.

54 O'Connor JC, Lawson MA, Andre C, Moreau M, Lestage J, Castanon N, Kelley KW, Dantzer R: Lipopolysaccharide-induced depressive-like behavior is mediated by indoleamine 2,3-dioxygenase activation in mice. Mol Psychiatry 2009;14:511–522.

55 Schwarcz R, Pellicciari R: Manipulation of brain kynurenines: glial targets, neuronal effects, and clinical opportunities. J Pharmacol Exp Ther 2002;303:1–10.

56 Raison CL, Dantzer R, Kelley KW, et al: CSF concentrations of brain tryptophan and kynurenines during immune stimulation with IFN-alpha: relationship to CNS immune responses and depression. Mol Psychiatry 2010;15:393–403.

57 Zhu CB, Blakely RD, Hewlett WA: The proinflammatory cytokines interleukin-1beta and tumor necrosis factor-alpha activate serotonin transporters. Neuropsychopharmacology 2006;31:2121–2131.

58 Felger JC, Alagbe O, Pace TW, et al: Early activation of p38 mitogen activated protein kinase is associated with interferon-alpha-induced depression and fatigue. Brain Behav Immun 2011;25:1094–1098.

59 Moron JA, Zakharova I, Ferrer JV, et al: Mitogen-activated protein kinase regulates dopamine transporter surface expression and dopamine transport capacity. J Neurosci 2003;23:8480–8488.

60 Bezzi P, Domercq M, Brambilla L, et al: CXCR4-activated astrocyte glutamate release via TNFalpha: amplification by microglia triggers neurotoxicity. Nat Neurosci 2001;4:702–710.

61 Pitt D, Nagelmeier IE, Wilson HC, Raine CS: Glutamate uptake by oligodendrocytes: implications for excitotoxicity in multiple sclerosis. Neurology 2003;61:1113–1120.

62 Volterra A, Meldolesi J: Astrocytes, from brain glue to communication elements: the revolution continues. Nat Rev Neurosci 2005;6:626–640.

63 Tilleux S, Hermans E: Neuroinflammation and regulation of glial glutamate uptake in neurological disorders. J Neurosci Res 2007;85:2059–2070.

64 Ida T, Hara M, Nakamura Y, Kozaki S, Tsunoda S, Ihara H: Cytokine-induced enhancement of calcium-dependent glutamate release from astrocytes mediated by nitric oxide. Neurosci Lett 2008;432:232–236.

65 Hardingham GE, Fukunaga Y, Bading H: Extrasynaptic NMDARs oppose synaptic NMDARs by triggering CREB shut-off and cell death pathways. Nat Neurosci 2002;5:405–414.

66 Haydon PG, Carmignoto G: Astrocyte control of synaptic transmission and neurovascular coupling. Physiol Rev 2006;86:1009–1031.

67 Matute C, Domercq M, Sanchez-Gomez MV: Glutamate-mediated glial injury: mechanisms and clinical importance. Glia 2006;53:212–224.

68 Rajkowska G, Miguel-Hidalgo JJ: Gliogenesis and glial pathology in depression. CNS Neurol Disord Drug Targets 2007;6:219–233.

69 Duman RS, Monteggia LM: A neurotrophic model for stress-related mood disorders. Biol Psychiatry 2006;59: 1116–1127.

70 Barrientos RM, Sprunger DB, Campeau S, et al: Brain-derived neurotrophic factor mRNA downregulation produced by social isolation is blocked by intrahippocampal interleukin-1 receptor antagonist. Neuroscience 2003;121:847–853.

71 Ben Menachem-Zidon O, Goshen I, Kreisel T, et al: Intrahippocampal transplantation of transgenic neural precursor cells overexpressing interleukin-1 receptor antagonist blocks chronic isolation-induced impairment in memory and neurogenesis. Neuropsychopharmacology 2008;33:2251–2262.

72 Koo JW, Duman RS: IL-1beta is an essential mediator of the antineurogenic and anhedonic effects of stress. Proc Natl Acad Sci USA 2008;105:751–756.
73 Kenis G, Prickaerts J, van Os J, et al: Depressive symptoms following interferon-alpha therapy: mediated by immune-induced reductions in brain-derived neurotrophic factor? Int J Neuropsychopharmacol 2011;14:247–253.
74 Pariante CM, Miller AH: Glucocorticoid receptors in major depression: relevance to pathophysiology and treatment. Biol Psychiatry 2001;49:391–404.
75 Pariante CM, Lightman SL: The HPA axis in major depression: classical theories and new developments. Trends Neurosci 2008;31:464–468.
76 Besedovsky HO, del Rey A: Immune-neuro-endocrine interactions: facts and hypotheses. Endocr Rev 1996;17:64–102.
77 Capuron L, Raison CL, Musselman DL, Lawson DH, Nemeroff CB, Miller AH: Association of exaggerated HPA axis response to the initial injection of interferon-alpha with development of depression during interferon-alpha therapy. Am J Psychiatry 2003;160: 1342–1345.
78 Raison CL, Borisov AS, Woolwine BJ, Massung B, Vogt G, Miller AH: Interferon-alpha effects on diurnal hypothalamic-pituitary-adrenal axis activity: relationship with proinflammatory cytokines and behavior. Mol Psychiatry 2010;15:535–547.
79 Sephton SE, Sapolsky RM, Kraemer HC, Spiegel D: Diurnal cortisol rhythm as a predictor of breast cancer survival. J Natl Cancer Inst 2000;92:994–1000.
80 Spiegel D, Giese-Davis J, Taylor CB, Kraemer H: Stress sensitivity in metastatic breast cancer: analysis of hypothalamic-pituitary-adrenal axis function. Psychoneuroendocrinology 2006;31:1231–1244.
81 Pariante CM: Glucocorticoid receptor function in vitro in patients with major depression. Stress 2004;7:209–219.
82 Maes M, Bosmans E, Meltzer HY, Scharpe S, Suy E: Interleukin-1 beta: a putative mediator of HPA axis hyperactivity in major depression? Am J Psychiatry 1993;150:1189–1193.
83 Fitzgerald P, O'Brien SM, Scully P, Rijkers K, Scott LV, Dinan TG: Cutaneous glucocorticoid receptor sensitivity and pro-inflammatory cytokine levels in antidepressant-resistant depression. Psychol Med 2006;36:37–43.
84 Pace TW, Hu F, Miller AH: Cytokine-effects on glucocorticoid receptor function: relevance to glucocorticoid resistance and the pathophysiology and treatment of major depression. Brain Behav Immun 2007;21:9–19.
85 Hu F, Pace TW, Miller AH: Interferon-alpha inhibits glucocorticoid receptor-mediated gene transcription via STAT5 activation in mouse HT22 cells. Brain Behav Immun 2009;23:455–463.
86 Smoak KA, Cidlowski JA: Mechanisms of glucocorticoid receptor signaling during inflammation. Mech Age Dev 2004;125:697–706.
87 Wang X, Wu H, Miller AH: Interleukin 1alpha (IL-1alpha) induced activation of p38 mitogen-activated protein kinase inhibits glucocorticoid receptor function. Mol Psychiatry 2004;9:65–75.
88 Engler H, Bailey MT, Engler A, Stiner-Jones LM, Quan N, Sheridan JF: Interleukin-1 receptor type 1-deficient mice fail to develop social stress-associated glucocorticoid resistance in the spleen. Psychoneuroendocrinology 2008;33: 108–117.
89 Pariante CM, Pearce BD, Pisell TL, et al: The proinflammatory cytokine, interleukin-1alpha, reduces glucocorticoid receptor translocation and function. Endocrinology 1999;140:4359–4366.
90 Quan N, Avitsur R, Stark JL, et al: Social stress increases the susceptibility to endotoxic shock. J Neuroimmunol 2001;115:36–45.
91 Rhen T, Cidlowski JA: Anti-inflammatory action of glucocorticoids – new mechanisms for old drugs. N Engl J Med 2005;353:1711–1723.
92 Kino T, Manoli I, Kelkar S, Wang Y, Su YA, Chrousos GP: Glucocorticoid receptor (GR) beta has intrinsic, GRalpha-independent transcriptional activity. Biochem Biophys Res Commun 2009;381:671–675.
93 Capuron L, Pagnoni G, Demetrashvili MF, et al: Basal ganglia hypermetabolism and symptoms of fatigue during interferon-alpha therapy. Neuropsychopharmacology 2007;32:2384–2392.
94 Juengling FD, Ebert D, Gut O, et al: Prefrontal cortical hypometabolism during low-dose interferon alpha treatment. Psychopharmacology (Berl) 2000;152:383–389.
95 Mentis MJ, McIntosh AR, Perrine K, et al: Relationships among the metabolic patterns that correlate with mnemonic, visuospatial, and mood symptoms in Parkinson's disease. Am J Psychiatry 2002;159:746–754.
96 Spetsieris PG, Moeller JR, Dhawan V, Ishikawa T, Eidelberg D: Visualizing the evolution of abnormal metabolic networks in the brain using PET. Comput Med Imaging Graph 1995;19:295–306.
97 Eidelberg D, Moeller JR, Dhawan V, et al: The metabolic topography of parkinsonism. J Cereb Blood Flow Metab 1994;14: 783–801.
98 Wichmann T, DeLong MR: Oscillations in the basal ganglia. Nature 1999;400: 621–622.
99 Feigin A, Fukuda M, Dhawan V, et al: Metabolic correlates of levodopa response in Parkinson's disease. Neurology 2001;57:2083–2088.
100 Bersano A, Aghemo A, Rumi MG, Ballabio E, Candelise L, Colombo M: Recovery after L-DOPA treatment in peginterferon and ribavirin induced parkinsonism. Eur J Intern Med 2008;19:370–371.
101 Eisenberger NI, Berkman ET, Inagaki TK, Rameson LT, Mashal NM, Irwin MR: Inflammation-induced anhedonia: endotoxin reduces ventral striatum responses to reward. Biol Psychiatry 2010;68:748–754.
102 Carter CS, Braver TS, Barch DM, Botvinick MM, Noll D, Cohen JD: Anterior cingulate cortex, error detection, and the online monitoring of performance. Science 1998;280:747–749.
103 Capuron L, Pagnoni G, Demetrashvili M, et al: Anterior cingulate activation and error processing during interferon-alpha treatment. Biol Psychiatry 2005;58:190–196.
104 Paulus MP, Feinstein JS, Simmons A, Stein MB: Anterior cingulate activation in high trait anxious subjects is related to altered error processing during decision making. Biol Psychiatry 2004;55:1179–1187.
105 Eisenberger NI, Lieberman MD, Satpute AB: Personality from a controlled processing perspective: an fMRI study of neuroticism, extraversion, and self-consciousness. Cogn Affect Behav Neurosci 2005;5:169–181.

106 Ursu S, Stenger VA, Shear MK, Jones MR, Carter CS: Overactive action monitoring in obsessive-compulsive disorder: evidence from functional magnetic resonance imaging. Psychol Sci 2003;14: 347–353.
107 Harrison NA, Brydon L, Walker C, et al: Neural origins of human sickness in interoceptive responses to inflammation. Biol Psychiatry 2009;66:415–422.
108 Slavich GM, Way BM, Eisenberger NI, Taylor SE: Neural sensitivity to social rejection is associated with inflammatory responses to social stress. Proc Natl Acad Sci USA 2010;107:14817–14822.
109 Eisenberger NI, Lieberman MD: Why rejection hurts: a common neural alarm system for physical and social pain. Trends Cogn Sci 2004;8:294–300.
110 Miller AH: Mechanisms of cytokine-induced behavioral changes: psychoneuroimmunology at the translational interface. Brain Behav Immun 2009;23: 149–158.
111 Lozano AM, Mayberg HS, Giacobbe P, Hamani C, Craddock RC, Kennedy SH: Subcallosal cingulate gyrus deep brain stimulation for treatment-resistant depression. Biol Psychiatry 2008;64: 461–467.
112 Soczynska JK, Kennedy SH, Goldstein BI, Lachowski A, Woldeyohannes HO, McIntyre RS: The effect of tumor necrosis factor antagonists on mood and mental health-associated quality of life: novel hypothesis-driven treatments for bipolar depression? Neurotoxicology 2009;30: 497–521.
113 de Paiva VN, Lima SN, Fernandes MM, Soncini R, Andrade CA, Giusti-Paiva A: Prostaglandins mediate depressive-like behaviour induced by endotoxin in mice. Behav Brain Res 2010;215:146–151.
114 Teeling JL, Cunningham C, Newman TA, Perry VH: The effect of non-steroidal anti-inflammatory agents on behavioural changes and cytokine production following systemic inflammation: implications for a role of COX-1. Brain Behav Immun 2010;24:409–419.
115 Mendlewicz J, Kriwin P, Oswald P, Souery D, Alboni S, Brunello N: Shortened onset of action of antidepressants in major depression using acetylsalicylic acid augmentation: a pilot open-label study. Int Clin Psychopharmacol 2006;21:227–231.
116 Muller N, Schwarz MJ, Dehning S, et al: The cyclooxygenase-2 inhibitor celecoxib has therapeutic effects in major depression: results of a double-blind, randomized, placebo controlled, add-on pilot study to reboxetine. Mol Psychiatry 2006;11:680–684.
117 Zhu CB, Lindler KM, Owens AW, Daws LC, Blakely RD, Hewlett WA: Interleukin-1 receptor activation by systemic lipopolysaccharide induces behavioral despair linked to MAPK regulation of CNS serotonin transporters. Neuropsychopharmacology 2010;35:2510–2520.
118 Kumar S, Boehm J, Lee JC: p38 MAP kinases: key signalling molecules as therapeutic targets for inflammatory diseases. Nat Rev Drug Discov 2003;2: 717–726.
119 Tikka T, Fiebich BL, Goldsteins G, Keinanen R, Koistinaho J: Minocycline, a tetracycline derivative, is neuroprotective against excitotoxicity by inhibiting activation and proliferation of microglia. J Neurosci 2001;21:2580–2588.
120 Tikka TM, Koistinaho JE: Minocycline provides neuroprotection against N-methyl-D-aspartate neurotoxicity by inhibiting microglia. J Immunol 2001;166:7527–7533.
121 Si Q, Cosenza M, Kim MO, et al: A novel action of minocycline: inhibition of human immunodeficiency virus type 1 infection in microglia. J Neurovirol 2004;10:284–292.
122 Alano CC, Kauppinen TM, Valls AV, Swanson RA: Minocycline inhibits poly(ADP-ribose) polymerase-1 at nanomolar concentrations. Proc Natl Acad Sci USA 2006;103:9685–9690.
123 Molina-Hernandez M, Tellez-Alcantara NP, Perez-Garcia J, Olivera-Lopez JI, Jaramillo-Jaimes MT: Antidepressant-like actions of minocycline combined with several glutamate antagonists. Prog Neuropsychopharmacol Biol Psychiatry 2008;32:380–386.
124 Henry CJ, Huang Y, Wynne A, et al: Minocycline attenuates lipopolysaccharide (LPS)-induced neuroinflammation, sickness behavior, and anhedonia. J Neuroinflammation 2008;5:15.
125 Hutchinson MR, Coats BD, Lewis SS, et al: Proinflammatory cytokines oppose opioid-induced acute and chronic analgesia. Brain Behav Immun 2008;22:1178–1189.
126 Hutchinson MR, Northcutt AL, Chao LW, et al: Minocycline suppresses morphine-induced respiratory depression, suppresses morphine-induced reward, and enhances systemic morphine-induced analgesia. Brain Behav Immun 2008;22:1248–1256.
127 Habibi-Asl B, Hassanzadeh K, Charkhpour M: Central administration of minocycline and riluzole prevents morphine-induced tolerance in rats. Anesth Analg 2009;109:936–942.
128 Koo JW, Russo SJ, Ferguson D, Nestler EJ, Duman RS: Nuclear factor-kappaB is a critical mediator of stress-impaired neurogenesis and depressive behavior. Proc Natl Acad Sci USA 2010;107: 2669–2674.
129 Nadjar A, Bluthe RM, May MJ, Dantzer R, Parnet P: Inactivation of the cerebral NFkappaB pathway inhibits interleukin-1beta-induced sickness behavior and c-Fos expression in various brain nuclei. Neuropsychopharmacology 2005;30: 1492–1499.
130 Dhillon N, Aggarwal BB, Newman RA, et al: Phase II trial of curcumin in patients with advanced pancreatic cancer. Clin Cancer Res 2008;14: 4491–4499.
131 Godbout JP, Berg BM, Krzyszton C, Johnson RW: Alpha-tocopherol attenuates NFkappaB activation and pro-inflammatory cytokine production in brain and improves recovery from lipopolysaccharide-induced sickness behavior. J Neuroimmunol 2005;169: 97–105.
132 Lu X, Ma L, Ruan L, et al: Resveratrol differentially modulates inflammatory responses of microglia and astrocytes. J Neuroinflammation 2010;7:46.
133 O'Connor JC, Lawson MA, Andre C, et al: Induction of IDO by bacille Calmette-Guerin is responsible for development of murine depressive-like behavior. J Immunol 2009;182:3202–3212.
134 Potter MC, Elmer GI, Bergeron R, et al: Reduction of endogenous kynurenic acid formation enhances extracellular glutamate, hippocampal plasticity, and cognitive behavior. Neuropsychopharmacology 2010;35:1734–1742.
135 Katz JB, Muller AJ, Prendergast GC: Indoleamine 2,3-dioxygenase in T-cell tolerance and tumoral immune escape. Immunol Rev 2008;222:206–221.

136 Rohrig UF, Awad L, Grosdidier A, et al: Rational design of indoleamine 2,3-dioxygenase inhibitors. J Med Chem 2010;53:1172–1189.
137 Volgraf M, Lumb JP, Brastianos HC, et al: Biomimetic synthesis of the IDO inhibitors exiguamine A and B. Nat Chem Biol 2008;4:535–537.
138 Di Pucchio T, Danese S, De Cristofaro R, Rutella S: Inhibitors of indoleamine 2,3-dioxygenase: a review of novel patented lead compounds. Expert Opin Ther Pat 2010;20:229–250.
139 Stone TW: Kynurenic acid antagonists and kynurenine pathway inhibitors. Expert Opin Investig Drugs 2001;10: 633–645.
140 Pace TW, Negi LT, Adame DD, et al: Effect of compassion meditation on neuroendocrine, innate immune and behavioral responses to psychosocial stress. Psychoneuroendocrinology 2009;34:87–98.
141 Pace TW, Negi LT, Sivilli TI, et al: Innate immune, neuroendocrine and behavioral responses to psychosocial stress do not predict subsequent compassion meditation practice time. Psychoneuroendocrinology 2010;35:310–315.
142 Pace TW, Negi LT, Dodson-Lavelle B, et al: Engagement with cognitively-based compassion training is associated with reduced salivary C-reactive protein from before to after training in foster care program adolescents. Psychoneuroendocrinology 2012, E-pub ahead of print.
143 Creswell JD, Irwin MR, Burklund LJ, et al: Mindfulness-based stress reduction training reduces loneliness and pro-inflammatory gene expression in older adults: a small randomized controlled trial. Brain Behav Immun 2012;26: 1095–1101.
144 Black DS, Cole SW, Irwin MR, et al: Yogic meditation reverses NF-kappaB and IRF-related transcriptome dynamics in leukocytes of family dementia caregivers in a randomized controlled trial. Psychoneuroendocrinology 2012, Epub ahead of print.

Charles L. Raison, MD
Department of Psychiatry, College of Medicine
John and Doris Norton School of Family and Consumer Sciences, University of Arizona
1501 N. Campbell Ave, Tucson, AZ 85724–5137 (USA)
E-Mail craison@email.arizona.edu

Halaris A, Leonard BE (eds): Inflammation in Psychiatry.
Mod Trends Pharmacopsychiatry. Basel, Karger, 2013, vol 28, pp 49–60 (DOI: 10.1159/000343967)

Virus Infection as a Cause of Inflammation in Psychiatric Disorders

Karl Bechter

Bezirkskrankenhaus, Abteilung für Psychotherapie und Psychosomatik, Ulm University, Günzburg, Germany

Abstract

Many neurotropic viruses exist and may cause classical inflammation but also low-level neuroinflammation. However, viruses may be dormant within the CNS and become active later. The role of neurotropic virus infections in the causation of psychiatric disorders may be underestimated, because the diagnostic approach to the CNS is difficult and to dormant infections in general, but especially within the CNS. Evidence is increasing that infections increase the risk of psychiatric disorders, not only prenatal infections but also infections during the lifetime. The question how low level neuroinflammation may be involved in severe psychiatric disorders like affective and schizophrenic spectrum disorders is intriguing but remains to be studied. Experimental data clearly show that low-level neuroinflammation can be induced by viruses, but the definitions of inflammation and low level neuroinflammation appear to be blurred and apparently the previous classical definition of inflammation has to be widened. Virus infection itself or virus-related products or virus-induced autoimmunity may play a role in disease pathogenesis. More sensitive diagnostic approaches from neuroimaging and CSF investigations may hold the key to a better understanding and definition of CNS viral infections as an etiopathogenetic subgroup of severe psychiatric disorders.

Viruses are ubiquitous infectious agents which for the major part contain only genetic information for replication in the host cell. Virus infections may induce classical inflammation, but many viruses may induce little or no inflammation and seem to be of overall low pathogenicity for the infected host. The latter aspect is an evolutionary advantage, because adaptation to the host may increase the overall likelihood of being able to replicate for a long period, whereas highly aggressive agents tend to eradicate not only the host but in parallel reduce their own chances of replication. These evolutionary aspects are a precondition for complex interactions between the infecting agent and the host and therefore a priori many virus infections and their consequences tend to be complicated, and may often become dormant though may reactivate at some time point or lead to slight dysfunction over time [1]. The etiopathogenesis of psychiatric disorders to a considerable extent remains unclear or even unknown, although a large subgroup of psychiatric disorders, especially the so-called endogenous or idiopathic psychoses (like schizophrenia, bipolar disorder, severe depression), seems to be based on biological grounds, and a biological core in the pathophysiology and pathogenesis seems to prevail. A possible role of low-level neuroinflammation (LLNI) in the pathogenesis of these disorders, induced by viruses and other infectious agents, was outlined in the mild encephalitis hypothesis, which was proposed to differentiate from classical encephalitis in that only very short-lived small classical lesions in the CNS may arise or even no classical inflammation may be observed; instead, there was exclusively LLNI, which remained to be more clearly defined [2]. This hypothesis was exemplified in Borna disease virus (BDV) infection, for which there is still lack of proof in humans. BDV represents a frequent cause of meningoencephalitis in horses and sheep in Middle Europe and can infect nearly all species (overview about various aspects in [3]). Evidence has accumulated that BDV infection is also a rare cause of neurological and psychiatric disorders [4–7], although a hype of publications claiming to have identified BDV sequences had to be corrected as having been due to laboratory contamination [8]. This recent experience with the difficulties being confronted when trying to identify virus association in etiopathogenetically unclear psychiatric or neurological disorders, illustrates the methodological difficulties involved, making understandable, at least in part, the difficulties of finding the paths into research on the viral causes of psychiatric disorders, and to identify LLNI. The idea that viruses may underlie etiologically unknown psychiatric disorders is not new. At the beginning of the 20th century and over decades, a virus hypothesis (bacteria and viruses were only beginning to be differentiated) was a mainstream idea, but proof remained open and thus the ideas disappeared more and more. Only recently can a reappraisal of such a hypothesis be observed, because more and more studies are demonstrating a risk increase from infections for psychiatric disorders. A most striking result, which can be considered proof of the concept, was a large epidemiological study carried out over 30 years in Denmark, demonstrating an additive risk increase for various psychoses due to severe infections and autoimmune disorders [9]. It is of notice here that leading immunologists now think that infections generally may contribute to autoimmune disorders, though proof for this is still lacking (cf. the long debate about virus infections in multiple sclerosis, now supported by recent histopathological findings [10]).

Here the current situation of epidemiological and pathophysiological findings in the psychiatric research field will be outlined beginning with a look back into the long history on the subject.

A Look Back into the History

The idea that viruses may cause psychiatric disorders has been voiced for nearly a century, but the proof of a causal role for viruses was difficult or only successful in rare single cases [11]. Nevertheless, a large number of virus infections are known to rarely cause psychiatric disease. The most frequently identified infection is herpes simplex virus encephalitis which can induce a range of psychiatric syndromes, usually accompanied by severe organic psychosyndromes and neurological symptoms [12]. Today, it is often reported that the elucidation (and the diagnosis) of organic causes of psychiatric disorders does not appear to be very complicated or challenging. However, this is a misinterpretation [13].

When looking into the details of the history of psychiatric disorders which are now clearly associated with infections, it appears that also in history the path to an etiological diagnosis was rather difficult. For example, the elucidation of the etiology of general paresis was characterized by long controversial discussions in the scientific field, and the majority of researchers, not least the histopathologists, rejected even the hypothesis that syphilis may be an important risk factor [cf. 14]. One of the earliest ideas that syphilis may represent a risk factor for general paresis was proposed by Esmarch and Jessen [15], initiated by single case observations. The first statistical data (a term not yet used at the time) were provided by Erich Mendel in the 1880s. His conclusion that syphilis may be an underestimated risk factor in addition to other risk factors, which were considered psychological and genetic, was heavily criticized so he took a step back. Only after seven decades from the first hypothesis to the detection of spirochetes by Noguchi and Moore in 1911, and after the introduction of lumbar puncture by Quincke and the Wassermann reaction (representing an antibody test), the reluctance against a causality hypothesis between syphilis and general paresis or tabes dorsalis was declining. Another important insight was that a similar histopathological picture, inflammatory vs. degenerative type of general paresis, was nearly identically seen with experimental toxoplasma infection. Thus, the idea emerged that a general type of pathogenesis was associated with CNS infections was used to explain the histopathological picture but not the specific infection itself. In fact, in the present knowledge about syphilis, a number of questions remain open with regard to the degenerative and inflammatory aspects of infectious diseases. In retrospect, the clinical categorization of general paresis by Mendel with regard to psychopathology was perfect but rather arbitrary and negligible concerning the understanding of its underlying pathogenesis. The important aspect not solved up to now is the overall low pathogenicity of syphilis infection, in that only some of the infected subjects may develop CNS disease and that often only after long latent period. The limited knowledge about these aspects was surely one reason to undertake unethical studies like the Tuskagee study and a similar study in Guatemala (by US researchers), for which President Obama recently apologized: to observe untreated syphilis over the long-term and to possibly learn why and how the persistent infection may at some time point or another become pathogenic for the infected host. Though unethical, these studies demonstrate the unsolved scientific questions involved in persistent infections or pseudoscientific thinking. A rather similar problem prevails with late Lyme disease, and the ethical problems to neglect or overestimate were repeatedly debated hotly [cf. 16]. Apparently, we do not have the choice to make research easy, but have to deal with complicated questions involving complex research approaches.

These prevailing problems to understand the pathogenesis of CNS infections are further highlighted in three well-studied infections including viral infection (rabies, cerebral malaria, and African trypanosomiasis), which demonstrates in detail, as brains and a lot of studies are available, that the pathogenesis of these CNS infections over time is extremely complex and variable, that similarities but also differences exist, and that insights into one disease can only in part be applied to another type of CNS infection as each agent has different consequences because of the details of its pathogenesis [17].

Such complicated backgrounds and differences in pathogenesis have to be kept in mind when dealing with the question of CNS inflammation caused by viruses possibly underlying psychiatric disorders. A view that one size fits all can be assumed to be very unlikely. Furthermore, in virus infections the interaction with the host is generally rather close, as viruses mainly represent genetic information, and it is no surprise that many neurotropic viruses exist, most of which with overall low pathogenicity, posing a major difficulty for research because only in a minority of the infected subjects does a certain disease develop [18]. Nevertheless, new viruses are being detected every year and may be of possible relevance for human disease. Furthermore, one should be aware that a number of infectious agents may not have been identified yet. A recent large study in California, suggested that even during the acute phase of meningoencephalitis there is only a less than 50% chance of identifying the respective etiology of the disease [19].

Borna Disease Virus Infection

Over many years, our work has been dealing with the possible role of Borna disease virus (BDV) in neurological and psychiatric disorders of humans [7]. BDV is the strongest known neurotropic virus. The research proved to be rather complicated and the interpretation difficult as the likely low overall pathogenicity together with the general unspecificity of symptoms in all known organic causes of a psychiatric disorder generally poses still unsolved problems for research. From this perspective and experience the issue of virus infection as a cause of inflammation is reconsidered and reviewed here.

From our research on the possible role of BDV infection for human neurological and psychiatric disorders we accumulated evidence in thousands of patients tested with the indirect immunofluorescence assay for BDV antibodies and found a prevalence of about 5% in our psychiatric sample, whereas only about 3% was found in surgical controls [2, 7]. The specificity of the human BDV antibodies was confirmed, recognizing linear epitopes of BDV proteins [20]. Attempts were made to isolate virus from CSF and postmortem brains which to date remained unsuccessful [cf. 4 and unpubl.]. Nevertheless, we found cases, though very few, in which BDV antibodies were produced within the intrathecal spaces (increased antibody index = AI), demonstrating that the specific agent was BDV and was likely to have induced CNS infection [21, 22]. This method to quantitatively validate antibodies by comparing blood to CSF levels is now well established and considered valid to indicate any specific type of encephalitis [23, 24]. Nevertheless, proof of human BD is yet to come. Overall, we suggest that a maximum of 3% of psychiatric cases might be causally related to BDV infection but very few cases fulfilled the probable criteria of BDV-associated mild encephalitis only [cf. 21, and unpubl.].

An interesting aspect is that it has now been shown that BDV sequences are integrated into the human genome [25, 26] and these findings raise new questions on the possible pathomechanisms involved [27]. A similar scenario is with another endogenous retrovirus, HERV-W, which has been found in both, in multiple sclerosis and in schizophrenia, the mechanisms involved seemed to relate to inflammation [28–30]. One should keep in mind that viruses in general may be able to cause CNS dysfunction, for example neuronal dysfunction without any sign of inflammation, which was termed the molecular anatomy of virus disease [31]. The question how to differentiate LLNI from noninflammatory dysfunction due to virus infection is, however, open.

Definitions of Inflammation and Neuroinflammation

The definition of inflammation has widened during the last years from results of extensive experimental inflammation research. Now, inflammation is understood as 'a complex set of interactions among soluble factors and cells that can arise in any tissue in response to traumatic, infectious, postischemic, toxic or autoimmune injury' [32]. This broad understanding of inflammation is apparently justified, but has only been established preliminarily in the clinic not forgetting the always limited possibilities in the clinical diagnostic approach for practical and ethical reasons as compared to the experimental approach. Nevertheless, a considerable number of new markers of inflammation were also established in the clinic in the last years, e.g. approaches like proteomics are rather promising to improve diagnostic sensitivity. But several types and aspects of nonclassical inflammation were only recently recognized and only partly classified, e.g. a long-known type relevant for example in leprosy was renamed 'parainflammation', being characterized mainly by fibrosis [33]. With regard to neuroinflammation, the situation is even more difficult and preliminary. Not only is the CNS not easily accessible for examination in the patient but it is also difficult in experiments. Nevertheless, it seems a plausible logic that LLNI may be quite frequent as recent insights have made it clear that immune privilege is not maintained by an absence of immunology instead by more complex immune activity within the CNS than in other tissues, considered now as CNS endogenous immunity [34]. The term LLNI is now often found in research [35] as well as clinical papers but its description is still unclear. Definition of LLNI remains preliminary yet similar to the view previously outlined with the mild encephalitis hypothesis for psychiatric purposes [cf. 2]. Also, a difference between acute and chronic neuroinflammation is seen in experiments and in the clinic [36], both of which may have important consequences for CNS function. But again, accepted criteria to differentiate chronic from acute neuroinflammation are not available. In clinical neurology, a convention seems that acute meningitis is considered chronic after 4 weeks, whereas some other neuroinflammatory disorders are considered chronic after 6 months. Even when classifying criteria were consensus, one difficulty still remains: not only time but also the type of neuroinflammation appears to change from the acute to chronic type and had to be differentiated by histological appearance including its relationship to neurodegeneration [36]. Also, the term subacute encephalitis may be found in the literature, a category widely lacking any further description though intuitively presenting between acute and chronic encephalitis.

The clinical approach to the diagnosis of neuroinflammation is mainly based on neuroimaging and cerebrospinal fluid (CSF) investigation. Neuroimaging, especially MRI, is sensitive for many purposes when focal lesions or meningeal involvement characterizes neuroinflammation, whereas diffuse or LLNI or local inflammation of the gray matter is difficult to detect [37]. For psychiatric disorders, the hypothesized inflammatory contribution to pathogenesis may be impossible to be detected by neuroimaging: comparing the findings at the beginning of meningoencephalitis in neurological disorders, a stage which can be considered an LLNI situation, it is apparent that imaging for these situations is rather insensitive [38]. The most sensitive method and therefore often referred to as the gold standard to diagnose neuroinflammation is CSF investigation. However, CSF analysis requires sophisticated experience and training in the interpretation of the findings as well as well-validated laboratory methods. Analysis of the CSF and blood has to be performed in parallel, preferably in the morning, for proper interpretation [24]. Acute encephalitis with accompanying meningitis can be sensitively detected by CSF investigation, whereas chronic encephalitis sometimes may be difficult or insensitive, as can be inferred from the experience with neurological inflammatory disorders [39]. Nevertheless, applying modern CSF diagnostic methods, we recently found in about 40% of psychiatric patients with affective or schizophrenic disorders some CSF pathology which was shown to be either definite LLNI or suggestive LLNI [21]. In addition, with newly developed methods of refined CSF cell analysis by differentiating CSF cells according to their surface markers, we found CSF cell activation in one third of the cases despite normal overall CSF counts, the activation patterns resembling those of chronic or acute meningoencephalitis [40]. With proteomic analysis, the range and sensitivity to detect CSF pathologies may be larger [41], thus appears to be promising for psychiatric disorders [42, 43], but these methods need to be validated in larger clinical approaches. The question whether such findings in psychiatric patient groups can be considered LLNI responses, and how far the findings match the findings in beginning classical neuroinflammatory disorders with neurological symptoms, as well as other questions remain to be addressed.

The limited sensitivity of methods to detect neuroinflammation is, however, only one aspect in the problem. For the rather well investigated example of HIV encephalitis (HIVe) the problem becomes more general. The clinical definition of HIVe only matches well with the histopathological definition when a wide definition of histopathological HIVe is applied: HIVe is diagnosed only when three focal lesions are detected in the postmortem brain, although apparently even two classical inflammatory foci detected within the postmortem brain would theoretically be considered as HIVe. The experience was that histopathological strictly defined HIVe does not match well with clinical HIVe symptoms [44]. The compromise found which works well is from a theoretical point however flawed. From a theoretical point, one should consider that even one single inflammatory focus should be termed HIVe, and when taking into consideration that the slices to investigate the brain left a lot of space for undetected lesions which can be inferred from simple considerations, the flaw is big. The problem behind the need for such a flawed compromise apparently is the relatively weak correlation between symptoms and LLNI and brain dysfunction: psychiatric symptoms may not become apparent even when several inflammatory foci prevail within the brain. A similar compromise was made with the original mild encephalitis hypothesis [2]: low-grade brain inflammation, difficult to detect by available diagnostic methods, was considered core and sufficient to induce brain dysfunction resulting in an unspecific psychiatric syndrome, the type of which depended on pre-existing factors able together to induce affective and schizophrenic and other psychiatric spectrum disorders. Furthermore, LLNI in ME was considered to be variable over time, though usually rather appearing as a subacute or chronic syndrome, giving room for change of symptoms over time.

Neuroinflammation and Psychiatric Symptoms: A Difficult Relationship

The nonspecificity of psychopathology is a general difficulty for the clinical approach: the diagnosis of organic or (specifically) inflammation-related psychiatric disorder depends nearly exclusively on the appropriate methodology. But even neuroimaging often fails to detect inflammatory lesions in certain brain regions specific for this psychopathology. This is described in cases of limbic encephalitis [45]. And as chronic or subacute LLNI may in themselves induce rather unspecific symptoms one can conclude that the type of psychopathological appearance may more be shaped by pre-existing independent factors and not the LLNI process itself. This might explain that an association of neuroinflammation in the individual case with regard to its clinical outcome is extremely hampered, and may only be possible with observations over long time periods.

We could show this in one remarkable case of therapy-resistant depression which remitted completely with antibiotic treatment [46]: during the chronic phase of 2 years of depression, the CSF pathology was detected only at an advanced stage, and the etiology could not be verified with specific antibodies within the CSF because there are no available assays that have been validated for CSF. Antibiotic treatment was based on clinical evaluation, was nevertheless successful and the overall evaluation of all findings over time (more than 6 years) supported the classification as an infection-induced autoimmune LLNI causing therapy-resistant depression.

This type of general diagnostic difficulty is not confined to inflammatory organic causes of psychiatric disorders: we found intriguing cases in which personality disorder was associated with arachnoid cysts, disorders being improved after neurosurgery of the arachnoid cyst [47]. Similar problems, however, may be more difficult even with CNS-relevant virus infections, as a general disadvantage of present methods to assess virus-specific LLNI in vivo is due to their limited sensitivity. So, improvement of the quality of methods to assess specific LLNI in vivo is requested.

Most extensive experience in neurological inflammatory disorders and CSF investigation is available with acute meningoencephalitis with clear established rules [23]. Nevertheless, even the modern most sensitive methods for detecting agents by PCR are sensitive for only a few days after the onset of meningoencephalitis, with some exceptions [24]. Furthermore, the sensitivity of the methods depends on the type of infection. Thus a negative PCR may often not exclude a specific etiology. The current status of the methodology was demonstrated in a large recent study, the California Encephalitis Project, showing that in cases of definite or probable lymphocytic meningoencephalitis, slightly less than 50% of the cases could be classified by their etiology [19]; it remains open whether this is related to unknown agents or to the insensitivity of the methods or both together. These findings from a focused study in general match the broad clinical experience in neurology: to detect neuroinflammation requires various methods applied in parallel in the individual case, and the respective sensitivity of the methods has to be considered in the light of the single case history and the phase of the disease. Over time, the symptomatology may change apparently due to changes of dysfunction caused by the underlying neuroinflammatory process. This has been clearly demonstrated in cases with a relatively rapid or subacute course, like the recently detected cases of limbic encephalitis, including varying psychiatric syndromes and schizophrenia-like psychoses [48]. The scientific progress in these cases of antibody-mediated encephalitis was the recent development of laboratory methods to show these antibodies, and as these cases usually demonstrate these antibodies in blood, whereas CSF may be normal, the approach is rather well applicable but also special [cf. 45].

Where Does CNS Inflammation Begin?

At this point, some recent general new insights into the immune surveillance of CNS in health and disease should be recognized: the role of the CSF and especially of CSF cells for immune surveillance of the CNS was previously underestimated. CSF cells seem on the one hand to representing an interface between systemic immunity and CNS-specific immunity [49], but on the other hand appear to be more directly involved in CNS functioning in health and disease than previously assumed [50, 51]. The view that the role of meninges in CNS immunity was underestimated is also increasingly becoming evident, and matches on the one hand long-standing clinical findings, e.g. meningeal enhancement known as a sensitive sign of meningoencephalitis more sensitive than parenchymal lesions [37], and on the other hand recent experimental findings, which in in vivo imaging demonstrate that inflammation begins at the meningeal vasculature and only later involves the parenchyma [52, 53].

It should also be noted here that the previous assumption of a general breakdown of the blood-brain barrier (BBB) in meningoencephalitis is not further tenable and that CSF abnormalities found in first-line meningoencephalitis demonstrate a disturbance of the blood-CSF barrier (BCB [cf. 23]) but not necessarily the BBB. Nevertheless, there might be a parallelism of BCB dysfunction and BBB dysfunction; the latter, however, has to be demonstrated in neuroimaging and cannot, contrary to previous assumptions, be assessed with CSF examination.

The BBB itself was further subdivided according to functional aspects into a BBB for solutes and a BBB for cells [54]. The BCB was anatomically defined as involving the choroid plexus plus the circumventricular organ [55]. Both barriers, the BBB and the BCB, are important for cell trafficking from blood into the brain and the CSF [56], but vivid cell trafficking was especially shown at the choroid plexus [57, 58].

Previously, especially CD4 and CD8 T cells had been studied, but the innate immune system and its respective cells were still unknown. It now appears that in many infections, including a number of virus infections, natural killer cells play an important role in combating virus infection [59], and, therefore, the plausibility and importance of nonspecific CSF findings as important diagnostic markers is underlined.

Special Aspects in the Control of Neuroinflammation

Beyond these aspects, some recent experimental findings directly show the long-known importance of neuroprotection in context with defined neuroinflammation. Moderate ethanol preconditioning of rat brain cultures engenders neuroprotection against dementia-inducing neuroinflammatory proteins [60]. Beta-amyloid peptide-induced inflammatory process could be inhibited by double-stranded RNA-dependent protein kinase, which is important in virus infections [61]. The adaptive immune response inside the CNS was shown to be a complex process of interactions between immune and neural cells with pro- and anti-inflammatory reactions present in parallel, though in general mimicking what occurs with inflammation in other organs and tissues [62]. Neuroinflammation may, however, represent a rather localized inflammation not even involving all other parts of the brain, but may nevertheless induce dysfunction at the cognitive level [63]. Also, an increasing number of players appear to be involved in neuroinflammatory processes, which can only selectively be touched here, e.g. micro-RNAs may be involved [64]; platelets as systemic players, though long known, are now recognized to be also involved in pathologies of the CNS like Alzheimer's disease and multiple sclerosis [65].

Inflammation induced by infection may potentiate tau pathological features and thus enhance a local toxin within the CNS, and established psychopharmacological substances, like lithium, may reduce inflammatory activity [66]. In chronic autoimmune encephalomyelitis, regulatory T cell responses may be paralyzed and the cellular arm of neuroprotection is thus hampered [67].

In an intriguing model of neuropsychiatric lupus erythematodes, it was attempted to differentiate cases with CNS involvement during acute systemic lupus erythematodes from cases of non-CNS involvement: a behavioral disorder as well as a cognitive deficit could be mediated by the antibody neurotoxicity in the amygdalae which was, however, transient, lasting only minutes after administration of a BBB breaching drug (epinephrine), but which alone did not cause an inflammatory response in the brain. Neuronal damage could nevertheless occur in the absence of an inflammatory response, depending on both the presence of neurotoxic antibody in the serum and opening of the BBB, the location depending on the agent used to open the BBB [68].

Thus, neuroinflammation and neuroprotection, but also the systemic aspects of inflammation, may be involved in parallel [69, 70].

Pathomechanisms in Neuroinflammation, Possibly Virus-Induced

Even in well-studied diseases, the role of virus for neuroinflammation is under debate. For example, herpes simplex virus type 1 appears to be involved in Alzheimer's disease but whether this can be termed 'causative' remains open yet [71]. In a rather well investigated most frequent neuroinflammatory disorder, multiple sclerosis, the immune attack is coordinated across the BBB but within the CNS requires a local pro-inflammatory milieu which is contributed to by macrophages and microglia to mediate the effector function of T and B cells which ultimately leads to damage of the myelin sheath, oligodendrocytes, and neurons [72]. Now, Epstein-Barr virus infection seems to play a more important role in the pathogenesis than previously shown as it was detected to be highly prevalent in multiple sclerosis plaques [10].

That a similar scenario may be relevant in human virus disease is plausible as can be seen in rare cases with LLNI when assessed repeatedly over time, e.g. in human herpesvirus-6-associated encephalitis in cases treated with cyclophosphamide and total body radiation before allogeneic stem cell transplantation [73]: the virus responded well to antiviral treatment when detected early and treatment was begun early. As HHV6, a neurotropic virus, is latent in most adults, these findings are of outstanding interest and teaching because of the possibility to carefully follow and repeatedly investigate the case by imaging, CSF and serum examinations.

When trying to apply similar experiences to the treatment of psychoses which may match LLNI as proposed in the mild encephalitis hypothesis and which is now being given more and more support, it appears from the experiences with limbic encephalitis which perfectly matches the scenario, that the key is more specific and sensitive laboratory and imaging methods. Antibody-mediated encephalitis may indeed represent a treatable cause of schizophrenia [48], although this has to be shown in more detail regarding its efficacy and preferred treatment. Interestingly, in the majority of schizophrenia cases presenting with CNS autoantibodies, no tumor was found in contrast to classical encephalitis syndromes with autoantibodies. Thus, the trigger for antibody-mediated encephalitis remained open.

Milder courses of encephalitis were also studied in late lyme neuroborreliosis including late Lyme disease [16, 74]. Lyme disease rarely presents with evidence of focal CNS damage or clearly abnormal MRI, one may see exclusively nonspecific small white matter abnormalities in a few patients or nothing at all. In addition, the difficulties of clinical investigation and of appropriate treatment are demonstrated in an interesting single case with Lyme disease of the brainstem [75]: the regions involved in this case were very well visible as being abnormal, the CSF demonstrating a strong increase of inflammatory markers and both evidence of systemic and intrathecal antibody production against *Borrelia burgdorferi*. However, the patient did not remit under antibiotic treatment, and only improved with a stronger antibiotic regimen over 6 weeks, definitely showing that the previously suggested diagnosis of brain stem neoplasm was false and it was indeed a Lyme disease inflammatory lesion of the brain stem (and in general, in bacterial and viral infections) it should be recognized that both pathogen-specific and nonspecific autoimmune aspects may prevail in parallel and contribute to the pathology and symptoms induced [76].

Conclusions

In the last 20 years a lot of further data in clinical and epidemiological investigations and in laboratory and experimental approaches have been accumulated, clearly suggesting a considerable potential of virus infections to induce psychiatric disorders. The nonspecificity of psychiatric syndromes has long been known and was true for neuroinflammatory causes as outlined in the mild encephalitis hypothesis [2]. A major hypothesis was that prenatal or perinatal infections may hold the key to understand schizophrenia as well as some other psychiatric syndromes [cf. 77]. Many infectious agents have now been detected and many of those seem to be relevant at any age during the lifetime, but may vary with age and not least by gene-environmental factors and changes within the immune system [13]. Thus, the possibility exists that LLNI induced by virus infections during the lifetime and/or by other infectious agents or triggers of autoimmunity may be considerably more important than previously assumed. This view is supported by very recent findings like antibody-mediated encephalitis, now mainly discussed with regard to schizophrenia syndromes. From larger studies on limbic encephalitis, it was clear that limbic encephalitis can induce a variety of psychiatric syndromes which may change over time from a prodromal pattern during the early phases of limbic encephalitis to more severe psychosyndromes. These findings from recent studies nearly perfectly support the outline given some years ago with the mild encephalitis hypothesis and as exemplified for the BDV infection assumed to demonstrate one rather rare example [cf. 78]. One may expect that many infectious agents and especially neurotropic viruses should be investigated and research into these viruses should be enhanced considerably.

References

1 Dworkin J, Shah IM: Exit from dormancy in microbial organisms. Nat Rev Microbiol 2010;8:890–896.
2 Bechter K: Mild encephalitis underlying psychiatric disorder: a reconsideration and hypothesis exemplified on Borna disease. Neurol Psychiatry Brain Res 2001;9:55–70.
3 Carbone KM (ed): Borna Disease Virus and Its Role in Neurobehavioral Disease. Washington, ASM Press, 2002.
4 Herzog S, et al: Molecular characterization of Borna disease virus from naturally infected animals and possible links to human disorders. Arch Virol Suppl 1997;13:183–190.
5 Richt JA, et al: Failure to detect Borna disease virus infection in peripheral blood leukocytes from humans with psychiatric disorders. J Neurovirol 1997;3: 174–178.
6 Richt JA, et al: Borna disease virus infection in animals and humans. Emerg Infect Dis 1997;3:343–352.
7 Bechter K: Borna disease virus. Mögliche Ursache neurologischer und psychiatrischer Störungen des Menschens; in Hippius H, Janzarik W, Müller C (eds): Monographien aus dem Gesamtgebiete Psychiatrie. Darmstadt, Steinkopff, 1998.
8 Durrwald R, et al: Meta-analysis of putative human bornavirus sequences fails to provide evidence implicating Borna disease virus in mental illness. Rev Med Virol 2007;17:181–203.
9 Benros M, et al: Autoimmune diseases and severe infections as risk factors for schizophrenia. A 30-year population-based register study. Am J Psychiatry 2011;168:1303–1310.
10 Meier UC, et al: Translational Mini-Review Series on B cell subsets in disease. B cells in multiple sclerosis: drivers of disease pathogenesis and Trojan horse for Epstein-Barr virus entry to the central nervous system? Clin Exp Immunol 2012;167:1–6.
11 O'Reilly RL: Viruses and schizophrenia. Aust NZ J Psychiatry 1994;28:222–228.
12 Lishman WA (ed): Organic Psychiatry. The Psychological Consequences of Cerebral Disorder, ed 3. Oxford, Blackwell Science, 1998.
13 Bechter K: Diagnosis of infectious or inflammatory psychosyndromes. Open J Neurol 2012;6(suppl. 1-M6):113–118.
14 Bechter K, Hodgkiss A: Research strategies in 'slow' infections in psychiatry. Hist Psychiatry 1995;6(24 Pt 4):503–511.
15 Esmarch F, Jessen W: Syphilis und Geistesstörung. Allg Z Psychiatr 1857; 14:20–36.
16 Fallon BA, et al: Regional cerebral blood flow and metabolic rate in persistent Lyme encephalopathy. Arch Gen Psychiatry 2009;66:554–563.
17 Bentivoglio M, Mariotti R, Bertini G: Neuroinflammation and brain infections: historical context and current perspectives. Brain Res Rev 2011;66:152–173.
18 Johnson RT (ed): Viral Infections of the Nervous System. New York, Raven, 1982.
19 Glaser CA, et al: Beyond viruses: clinical profiles and etiologies associated with encephalitis. Clin Infect Dis 2006;43: 1565–1577.
20 Billich C, et al: High-avidity human serum antibodies recognizing linear epitopes of Borna disease virus proteins. Biol Psychiatry 2002;51:979–987.
21 Bechter K, et al: Cerebrospinal fluid analysis in affective and schizophrenic spectrum disorders: identification of subgroups with immune responses and blood-CSF barrier dysfunction. J Psychiatr Res 2010;44:321–330.
22 Bechter K, et al: Investigations of cerebrospinal fluid in Borna disease virus seropositive psychiatric patients. Eur Psychiatry 1995;10:250–258.
23 Reiber H, Peter JB: Cerebrospinal fluid analysis: disease-related data patterns and evaluation programs. J Neurol Sci 2001;184:101–122.
24 Wildemann B, Oschmann P, Reiber H (eds): Laboratory Diagnosis in Neurology. Stuttgart, Thieme, 2010.
25 Belyi VA, Levine AJ, Skalka AM: Unexpected inheritance: multiple integrations of ancient bornavirus and ebolavirus/marburgvirus sequences in vertebrate genomes. PLoS Pathog 2010;6:e1001030.
26 Horie M, et al: Endogenous non-retroviral RNA virus elements in mammalian genomes. Nature 2010;463:84–87.
27 Feschotte C: Virology: Bornavirus enters the genome. Nature 2010;463:39–40.
28 Perron H, Lang A: The human endogenous retrovirus link between genes and environment in multiple sclerosis and in multifactorial diseases associating neuroinflammation. Clin Rev Allergy Immunol 2010;39:51–61.
29 Perron H, et al: Endogenous retrovirus type W GAG and envelope protein antigenemia in serum of schizophrenic patients. Biol Psychiatry 2008; 64:1019–1023.

30 Perron H, et al: Human endogenous retrovirus type W envelope expression in blood and brain cells provides new insights into multiple sclerosis disease. Mult Scler 2012, doi: 10.1177/1352458512441381.
31 Oldstone MB: Molecular anatomy of viral disease. The George Cotzias memorial lecture 1986. Neurology 1987;37: 453–460.
32 Nathan C: Points of control in inflammation. Nature 2002;420:846–852.
33 Medzhitov R: Origin and physiological roles of inflammation. Nature 2008;454: 428–435.
34 McGeer PL, McGeer EG: The inflammatory response system of brain: implications for therapy of Alzheimer and other neurodegenerative diseases. Brain Res Brain Res Rev 1995;21:195–218.
35 O'Callaghan JP, Sriram K, Miller DB: Defining 'neuroinflammation'. Ann NY Acad Sci 2008;1139:318–330.
36 Amor S, et al: Inflammation in neurodegenerative diseases. Immunology 2010; 129:154–169.
37 Osborn AG, Salzman KL, Barkovich AJ (eds): Diagnostic Imaging: Brain. Philadelphia, Lippincott/Williams & Wilkins, 2010.
38 Bechter K: Neuroimaging in the diagnosis of encephalitis as a cause of psychoses. Neurol Croat 2005;54(suppl 2): 59–63.
39 Reiber H, Otto M, Bechter K: Crash Course: Cerebrospinal Fluid Diagnostics for Psychiatrists and Neurologists. Cerebrospinal fluid analysis for diagnosis of non-inflammatory, dementive and psychiatric diseases. Acta Neuropsychiatr 2009;21(suppl 2):58–61.
40 Maxeiner HG, et al: Flow cytometric analysis of T cell subsets in paired samples of cerebrospinal fluid and peripheral blood from patients with neurological and psychiatric disorders. Brain Behav Immun 2009;23:134–142.
41 Schwarz E, et al: Identification of a biological signature for schizophrenia in serum. Mol Psychiatry 2012;17:494–502.
42 Kranaster L, et al: Cerebrospinal fluid diagnostics in first-episode schizophrenia. Eur Arch Psychiatry Clin Neurosci 2011;261:529–530.
43 Leweke FM, et al: Antibodies to infectious agents in individuals with recent onset schizophrenia. Eur Arch Psychiatry Clin Neurosci 2004;254:4–8.
44 Cherner M, et al: Neuropathologic confirmation of definitional criteria for human immunodeficiency virus-associated neurocognitive disorders. J Neurovirol 2007;13:23–28.
45 Vincent A, et al: Autoantibodies associated with diseases of the CNS: new developments and future challenges. Lancet Neurol 2011;10:759–772.
46 Bechter K, et al: Therapieresistente Depression mit Fatigue. Fall einer vermutlichen streptokokkenassoziierten Autoimmunkrankheit. Nervenarzt 2007;78:338, 340–341.
47 Bechter K, et al: Personality disorders improved after arachnoid cyst neurosurgery, then rediagnosed as 'minor' organic personality disorders. Psychiatry Res 2010;184:196–200.
48 Lennox BR, Coles AJ, Vincent A: Antibody-mediated encephalitis: a treatable cause of schizophrenia. Br J Psychiatry 2012;200:92–94.
49 Schwartz M, Shechter R: Protective autoimmunity functions by intracranial immunosurveillance to support the mind: the missing link between health and disease. Mol Psychiatry 2010;15: 342–354.
50 Derecki NC, et al: Regulation of learning and memory by meningeal immunity: a key role for IL-4. J Exp Med 2010;207: 1067–1080.
51 Schwartz M, Kipnis J: A conceptual revolution in the relationships between the brain and immunity. Brain Behav Immun 2011;25:817–819.
52 Ransohoff RM: Immunology: In the beginning. Nature 2009;462:41–42.
53 Bartholomaus I, et al: Effector T cell interactions with meningeal vascular structures in nascent autoimmune CNS lesions. Nature 2009;462:94–98.
54 Bechmann I, Galea I, Perry VH: What is the blood-brain barrier (not)? Trends Immunol 2007;28:5–11.
55 Wolburg, H, Paulus W: Choroid plexus: biology and pathology. Acta Neuropathol 2010;119:75–88.
56 Engelhardt B, Sorokin L: The blood-brain and the blood-cerebrospinal fluid barriers: function and dysfunction. Semin Immunopathol 2009;31:497–511.
57 Ransohoff RM, Kivisakk P, Kidd G: Three or more routes for leukocyte migration into the central nervous system. Nat Rev Immunol 2003;3:569–581.
58 Goverman J: Autoimmune T cell responses in the central nervous system. Nat Rev Immunol 2009;9:393–407.
59 Newman KC, Riley EM: Whatever turns you on: accessory-cell-dependent activation of NK cells by pathogens. Nat Rev Immunol 2007;7:279–291.
60 Collins MA, et al: Moderate ethanol preconditioning of rat brain cultures engenders neuroprotection against dementia-inducing neuroinflammatory proteins: possible signaling mechanisms. Mol Neurobiol 2010;41:420–425.
61 Couturier J, et al: Prevention of the beta-amyloid peptide-induced inflammatory process by inhibition of double-stranded RNA-dependent protein kinase in primary murine mixed co-cultures. J Neuroinflammation 2011;8:72.
62 Pedemonte E, et al: Mechanisms of the adaptive immune response inside the central nervous system during inflammatory and autoimmune diseases. Pharmacol Ther 2006;111:555–566.
63 Hein AM, et al: Sustained hippocampal IL-1beta overexpression impairs contextual and spatial memory in transgenic mice. Brain Behav Immun 2010; 24:243–253.
64 Lukiw WJ, et al: Upregulation of micro RNA-146a (miRNA-146a), a marker for inflammatory neurodegeneration, in sporadic Creutzfeldt-Jakob disease (sCJD) and Gerstmann-Straussler-Scheinker (GSS) syndrome. J Toxicol Environ Health A 2011;74:1460–1468.
65 Nurden AT: Platelets, inflammation and tissue regeneration. Thromb Haemost 2011;105(suppl 1):S13–S33.
66 Sy M, et al: Inflammation induced by infection potentiates tau pathological features in transgenic mice. Am J Pathol 2011;178:2811–2822.
67 Matsumoto Y, et al: Paralysis of CD4(+) CD25(+) regulatory T cell response in chronic autoimmune encephalomyelitis. J Neuroimmunol 2007;187:44–54.
68 Diamond B, et al: Immunity and acquired alterations in cognition and emotion: lessons from SLE. Adv Immunol 2006;89:289–320.
69 Bokemeyer M, et al: Evidence for neuroinflammation and neuroprotection in HCV infection-associated encephalopathy. Gut 2011;60:370–377.

70 Senzolo M, et al: Neuropsychological alterations in hepatitis C infection: the role of inflammation. World J Gastroenterol 2011;17:3369–3374.
71 Itzhaki RF, Wozniak MA: Herpes simplex virus type 1 in Alzheimer's disease: the enemy within. J Alzheimers Dis 2008;13:393–405.
72 Hemmer B, et al: Multiple sclerosis – a coordinated immune attack across the blood brain barrier. Curr Neurovasc Res 2004;1:141–150.
73 Gorniak RJ, et al: MR imaging of human herpesvirus-6-associated encephalitis in 4 patients with anterograde amnesia after allogeneic hematopoietic stem-cell transplantation. AJNR Am J Neuroradiol 2006;27:887–891.
74 Pachner AR, Steiner I: Lyme neuroborreliosis: infection, immunity, and inflammation. Lancet Neurol 2007;6: 544–552.
75 Kalina P, et al: Lyme disease of the brainstem. Neuroradiology 2005;47:903–907.
76 Kuenzle S, et al: Pathogen specificity and autoimmunity are distinct features of antigen-driven immune responses in neuroborreliosis. Infect Immun 2007;75: 3842–3847.
77 Brown AS: The environment and susceptibility to schizophrenia. Prog Neurobiol 2011;93:23–58.
78 Bechter K: Updating the mild encephalitis hypothesis of schizophrenia. Prog Neuropsychopharmacol Biol Psychiatry, in press; doi: 10.1016/j.pnpbp.2012. 1006.1019.

Prof. Dr. Karl Bechter
Bezirkskrankenhaus, Abteilung für Psychotherapie und Psychosomatik
Ludwig-Heilmeyer-Strasse 2
DE–89312 Günzburg (Germany)
E-Mail karl.bechter@bkh-guenzburg.de

Halaris A, Leonard BE (eds): Inflammation in Psychiatry.
Mod Trends Pharmacopsychiatry. Basel, Karger, 2013, vol 28, pp 61–74 (DOI: 10.1159/000343968)

Inflammation, Neurotoxins and Psychiatric Disorders

Aye-Mu Myint

Psychiatric Hospital, Ludwig-Maximilians University, Munich, Germany

Abstract

Although immune reactions are necessary to defend against danger signals, the mediator molecules such as cytokines can be detrimental to the organism if the exposure is longer than necessary or in certain abnormal concentrations. The neuroprotection and neurotoxicity induced by the interaction between certain cytokines and the metabolites from tryptophan catabolism, the neuroactive kynurenines, which is partly influenced by corticosteroid action plays an important role in several neurotransmissions such as serotonergic, dopaminergic and glutamatergic transmissions and receptor functions such as N-methyl-D-aspartate receptor or α_7-nicotinic-acetylcholine receptor. While the molecules in normal concentrations are essential to the normal glial-neuronal interaction, any changes that induce imbalance in the network between those molecules could disturb the interaction. The role of this network in major psychiatric disorders such as depression, bipolar disorder and schizophrenia and future therapeutic roles are discussed in this manuscript.

Although immune reactions are essential body defense mechanism, the mediators secreted or produced by the immune cells along these reactions are not cent per cent positive or harmless to the surrounding cells and tissues or the organism. Moreover, these molecules induce not only direct effects but also through interaction with other endocrine, metabolic and biochemical systems. This network orchestrates to keep the body systems in balance. Any changes that disturb the balance can be detrimental to the organism. Especially when those reactions are not resolved in due course and the immune mediator molecules are increased in the cellular environment, the negative outcome could be the result. Neuroprotection and neurotoxicity through the immune-endocrine-metabolic-neurochemical network in the central nervous system (CNS) plays a major role in pathophysiology major psychiatric disorders.

Immune Parameters and Neurotoxicity in Major Psychiatric Disorders

In case of immune activation in neurotoxicity, several molecules play roles. The molecules which are frequently reported in terms of major psychiatric disorders are the cytokines such as IL-1, IL-6 and tumor necrosis factor-α (TNFα) and all these have both positive and negative impacts on CNS.

All types of cells in the brain including astrocytes, microglia, oligodendrocytes, neurons and endothelial cells produce IL-1 although microglia are the primary source [1]. The addition of nano-molar (nM) range concentrations of IL-1 to astrocytes culture could induce release of secondary inflammatory cytokines such as IL-6, IL-8 and TNFα, adhesion molecules [2–4], chemokines, prostaglandins E_2 and F_2 (PGE_2 and PGF_2), neurotoxic factors such as nitric oxide (NO) [5–7], and especially, by IL-1β, neuroprotective factors such as nerve growth factor (NGF) [8]. Since astrocytes express IL-1 receptors, IL-1 could activate the signaling pathways, nuclear factor kappa-β (NFκβ), mitogen-activated protein kinases (MAPKs), c-Jun N-terminal kinase (JNK) and extracellular-signal-regulated kinase (ERK1/2) [9, 10]. IL-1 increased the intracellular calcium ion in astrocytes [11] and it could also induce morphology associated with astrogliosis [12]. To the neurons, IL-1 can be both neurotoxic and neuroprotective. IL-1 can induce neurotoxicity either via the induction of neurotoxins secretion from astrocytes or by direct toxicity. In the in-vitro pure neuronal culture system, very low picomolar (pM) level of IL-1 could induce exacerbation of oxygen glucose deprivation (OGD) [13], α-amino-3-hydroxy-5-methyl-4-isoxazolepropionic acid (AMPA)- [14] or n-methyl-D-aspartate (NMDA)-induced [15] cell deaths, whereas higher nM level of IL-1 can protect all those cell deaths and it is NGF-dependent [13, 14, 16]. Based on those results, one could hypothesize that the balance between the effect of neurotoxins from the surrounding cells or direct toxicity through AMPA and NMDA receptors and the effect of neurotrophic factors.

Similarly, IL-6 can also induce both neurotrophic effects and neurodegeneration. In unstimulated culture, IL-6 is secreted in low amount although its production is upregulated by many inflammatory stimuli [17]. Unlike IL-1 and TNFα, IL-6 does not enhance production of adhesion molecule [18] but excites the adrenocorticotrophic hormone (ACTH) secretion to enhance cortisol action to suppress secretion of IL-1 and TNFα to suppress inflammation [19]. However, in case of neuronal injury, IL-6 enhances neuronal degeneration by increasing expression of IL-6 and IL-6 receptor-α (IL-6Rα) [20]. It is summarized in a review that in CNS, IL-6 is both inflammatory and anti-inflammatory [21]. IL-6 was proposed as developmental neurotrophic factor [21] and neurodegenerative factor which is involved in Alzheimer's disease (AD) pathology [22–24] although its neurotoxic effect is not as obvious as TNFα or IL-1 because of its inability to directly induce inducible nitric oxide synthase (iNOS) in CNS [25].

It was reported in a review that most of resident cells in CNS, such as astrocytes, microglia and neurons produced TNFα when there is physiological or pathological stimuli [26] whereas in normal intact CNS, neurons are the only cell type that produce low level of TNFα [27]. In in vitro culture, TNFα is proliferative to astrocytes [28] and it enhances differentiation of immature neuroblastoma [29]. However, higher concentrations of TNFα are toxic to the neuronal culture and oligodendrocytes [26]. While TNFα is a potent inducer of production of other cytokines such as IL-1 and IL-6, adhesion molecules, glutamate, NO and arachidonic acid from astrocytes and microglia [26], and stimulate caspace-3 activation in hippocampus neurons and induce apoptosis [30], it also induces production of neurotrophic factors such as NGF and brain-derived neurotrophic factor (BDNF) [31] and neuroprotection via neuron-specific upregulation of cAMP response element-binding protein (CREB) [32]. Some studies reported TNFα to be involved in the pathological process of most of the CNS infections and neurodegenerative diseases such as multiple sclerosis (MS) and AD [26].

Considering all above findings together, it can be concluded that based on the concentration and condition, the immune molecules can be neurotrophic or neurotoxic and the effects depend on the interaction of different molecules.

Depression

Evidence implicating a role for the pro-inflammatory cytokines, in the aetiology of depression has been provided by studies on the changes in IL-1, IL-6 and TNFα in depressed patients [33–35] and also by the effects of IFNα on non-depressed individuals being treated for hepatitis C [36] or malignant melanoma [37]. These changes in IFNα treatment appear to be a consequence of the neurotransmitter and endocrine changes induced by the cytokines rather than the pathological condition of the diseases [38]. In the initial studies linking depression with an abnormality of the immune system, neutrophil function [39], impaired mitogen stimulated lymphocyte proliferation [40] and reduced NK cell activity [41] in untreated depressed patients showed that those changes largely returned to normal once the patient recovered from the depression. Later research into the immune changes occurring in depression has concentrated on cytokines, soluble cytokine receptors and plasma acute-phase proteins. The positive acute-phase proteins (e.g. α_1-acid glycoprotein, C-reactive protein) have been shown to increase while the negative acute phase proteins (e.g. retinol-binding protein) decreased in depression, changes that are known to be a consequence of the action of IL-6 on liver function [42]. Further evidence of immune activation in depressed patients is provided by the studies showing that the plasma concentration of IL-1, IL-6, IFNγ, soluble IL-6 and IL-2 receptors, and the IL-1 receptor antagonist, are raised and these changes are correlated with a rise in plasma positive acute phase proteins [43]. Effective antidepressant treatments largely attenuated such immune changes [44, 45]. In addition to the increases in pro-inflammatory cytokines, there is also evidence of an increased number of T helper, T memory, activated T cells and B cells that act as a source of the plasma cytokines [46, 47]. From these changes, it would appear that in depression there is an imbalance between the pro-inflammatory and the anti-inflammatory arms of the immune system. Those findings paved the way to the hypothesis that the T helper type 1 (Th1) pro-inflammatory cytokines (such as IFNγ) becoming predominant over the Th2 anti-inflammatory cytokines (e.g. IL-4). A recent study has shown that the Th3 cytokine, transforming growth factor-β_1 (TGFβ_1), which regulates the balance between Th1 and Th2, is increased in depressed patients following effective antidepressant treatment [34].

These inflammatory changes occur not only in the periphery but also in the brain parenchyma. The elevated microglial density in major depression and schizophrenia was reported by Steiner et al. [48]. It is well known that microglia produce inflammatory mediators such as PGE_2, cyclo-oxygenase-1 and cyclo-oxygenase-2 (COX-1 and COX-2), pro-inflammatory cytokines and inducible nitric oxide synthase (iNOS) [49–51]. The PGE_2 content of the saliva, serum and CSF of depressed patients are reported to be increased [52, 53] and antidepressants have been shown to inhibit the secretion of pro-inflammatory cytokines and to reduce the synthesis of PGE_2 [54]. Those findings indicate the possible involvement of COX-1 and COX-2 in the pathophysiology of depression. Clinical studies have suggested that the COX-2 inhibitor celecoxib has positive effects on cognitive function in depressed patients [55, 56].

Taking all the above findings and reports together, the chronic imbalance between pro- and anti-inflammatory response in the blood and CNS of depressed patients would result in chronic exposure to neurotoxic molecules and that might be one of the key pathophysiological mechanisms involved in some changes such as brain volume loss and conditions such as association between depression and dementia.

Bipolar Disorder

Unlike depression and schizophrenia, less studies have been carried out in inflammation in bipolar disorder (BD). However, similar to depression,

BD is also associated with increased plasma levels of pro-inflammatory cytokines, such as IL-6, TNFα and IL-1β [57, 58]. Both signaling molecules of IL-1, NFκβ, and IL-1 receptor antagonist (IL-1Ra) protein and mRNA levels are increased in postmortem frontal cortex of bipolar patients [59]. An increased acute-phase protein level such as for CRP [60] is also associated with BD. The cytokine levels are found to be altered during symptomatic and asymptomatic phases [61]. Reversal of elevated IL-6 level with remission is also reported [62]. Regarding the anti-inflammatory cytokine, IL-10 is reported to be increased only in the early stage of the disorder [63]. Regarding the balances between Th1, Th2 and Th3 cytokines in bipolar mania patients, the ratio between Th1 and Th3 cytokines (IFNγ/TGFβ_1) and the ratio between Th2 and Th3 cytokines (IL-4/TGFβ1) are significantly higher than controls and with remission Th3 cytokine is increased and the ratios are reverted to normal [64].

Regarding the neurotrophic factors, the reduction in BDNF is well documented. Several studies have reported decreases in BDNF levels which are associated with severity of the symptoms in the acute phases of both medicated and unmedicated bipolar mania and depression [65–67]. The change in BDNF is stage dependent [57].

The role of add on anti-inflammatory medication with celecoxib in bipolar depressive or mixed episodes does not have as clear an advantage as in the case of major depression although celecoxib add on treatment could reduce the clinical symptom scores better at the time point of 2 weeks after the baseline compared to placebo add on. Celecoxib might induce a rapid onset of antidepressant action [68].

Although there is some evidence of inflammatory changes in bipolar disorder, the role played by neurotoxic changes induced by imbalanced pro- and anti-inflammatory mechanism is not conclusive.

Schizophrenia

In schizophrenia, most of the cytokine changes are not very consistent. The most consistent findings are with IL-6. Most of the studies reported increased blood levels of IL-6 [69–71] although there are some contradictory findings [72, 73]. The soluble IL-6 receptor levels in both serum and CSF were associated with severity of symptoms in paranoid schizophrenia [74]. The findings in IL-2 are also controversial [72, 75, 76]. Regarding IL-1, the serum levels and mRNA expression in peripheral blood mononuclear cells are increased in drug-naïve first-episode patients. The increased mRNA levels of IL-1 are associated with increased NFκβ expression [77]. Regarding the balance between Th1, Th2 and Th3, the ratio between Th1/Th2 is increased in the plasma of drug-naïve and drug-free patients and was reversed by 6 weeks of antipsychotic treatment [78]. In this study, Th3 cytokine was increased at baseline and reversed after 6 weeks of treatment. Although there is no consistent finding, some reports are available regarding the advantage of celecoxib add-on medication on symptom scores reduction [79–81].

In the brain, in the prefrontal cortex area of schizophrenia patients, the mRNA levels of IL-1Ra are decreased although IL-1β showed no change [82]. The convincing finding in the brain of schizophrenia patients regarding immune activation in the brain is glial activation [83]. Glial activation in the brain with associated inflammatory markers changes would induce neurotoxicity and frequent consistent findings on brain volume loss in schizophrenia patients [84–87] might be the possible consequent.

Immune-Neuroendocrine-Tryptophan Metabolism Interaction and Neurotoxicity in Major Psychiatric Disorders

Recent in vivo evidence suggests that stress-induced elevation of glucocorticoids also induces immune activation within the CNS through microglia activation and proliferation and loss in the number and volume of astrocytes [88]. Animal studies show that stress induces an enhanced expression of proinflammatory factors such as IL-1β [89], macrophage migration inhibitory factor (MIF) [90–92] and COX-2 [93] in the brain. Elevation of these proinflammatory factors is accompanied by dendritic atrophy and neuronal death within the hippocampus [94, 95]. These detrimental effects of glucocorticoids in the CNS are mediated by a rise in extracellular glutamate [96, 97] and subsequent overstimulation of the NMDA receptor (NMDA-R) which results in excitotoxic neuronal damage [98]. It was demonstrated that restraint-induced psychological stress stimulates the proliferation of microglia, which was prevented by blockade either of corticosterone synthesis, the glucocorticoid receptor or the NMDA receptor [99]. These data show that stress-induced microglia proliferation is mediated by corticosterone-induced and NMDA-R-mediated action within the CNS. Major psychiatric disorders are somehow associated with increased cortisol secretion. The neurotoxicity through the network of inflammation, corticosteroid and NMDA-R activation in psychiatric disorders is recently explained by the enhanced tryptophan catabolism [100].

Tryptophan is an essential amino acid which is obtained from a dietary source. Serotonin is synthesized from about 1% of the available tryptophan in the body. About 99% of tryptophan is metabolized in the liver by tryptophan 2,3-dioxygenase (TDO) [101]. The TDO activity is mainly controlled by the tryptophan level itself and therefore, its activity is generally stable. The central availability of tryptophan mainly depends on the competition by the large amino acids at the transport across BBB and partially depends on the cerebral demand [102]. After tryptophan is catabolized into kynurenine (KYN) by TDO, it is further catabolized into 3-hydroxy-kynurenine (3HK) by kynurenine-3-mono-oxygenase (KMO) enzyme, and then to 3-hydroxyanthranilic acid (HAA) through the action of kynureninase. After that the catabolism continues either into the complete oxidation pathway and forms adenosine triphosphate (ATP) which occurs mainly in the liver, or into quinolinic acid (QUIN) which is finally degraded into nicotinamide adenine dinucleotide (NAD). From the complete oxidation pathway, picolinic acid (PIC) is also formed in a small quantity. In physiological conditions, the catabolism goes mainly to ATP formation and only a minor portion goes to NAD formation [103]. In the normal state, to achieve a normal NAD requirement, QUIN synthesis occurs only transiently in the liver and it does not accumulate in the hepatocytes [104]. KYN can also be catabolized by the kynurenine aminotransferases (KATs) into kynurenic acid (KYNA). Tryptophan metabolism in the liver is more or less stable and age and gender has an influence on this metabolism [103].

Since KYN itself could be transported across the blood-brain barrier, on top of the kynurenine formed in the brain by tryptophan breakdown, extra KYN is available from the periphery for further kynurenine metabolism in the brain. Sixty percent of brain KYN was contributed from the periphery [105]. In the brain, tryptophan catabolism occurs mainly in the astrocytes and microglia [106–108]. Although some neurons also possess indoleamine 2,3-dioxygenase (IDO) and/or TDO2 [109], neurons are not the main sites of the kynurenine pathway in the brain. While the astrocytes are shown to produce mainly KYNA because of the lack of a KMO enzyme, microglia and macrophages mainly produce QUIN [110–112]. The astrocytes metabolize QUIN produced by the neighboring microglia [111]. In the brain, in a physiological condition without immune challenge, as in the liver, this kynurenine pathway may serve mainly for glycogen storage and synthesis of the small amount of NAD required for the central nervous system [103].

In case of inflammation which activates the enzyme IDO in the extrahepatic tissues, such as lungs, placenta, kidneys, spleen, blood and the

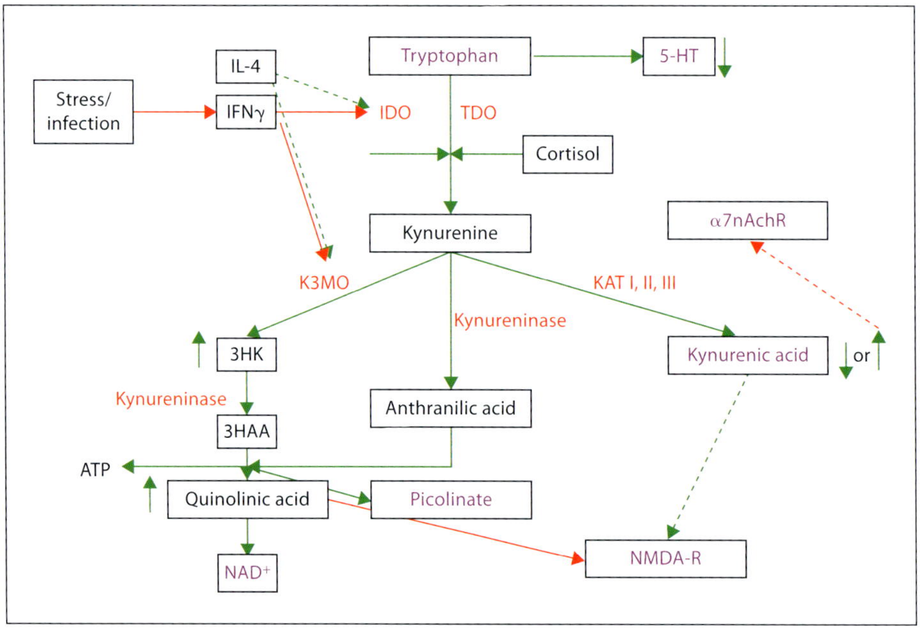

Fig. 1. Immune-tryptophan metabolism-neuroendocrine interaction in inflammation.

brain [113, 114], the extrahepatic tryptophan metabolism shifts the tryptophan metabolism away from the liver [115]. In this case, tryptophan breakdown through the KYN pathway occurs mainly in the blood and lymphoid tissues [116]. The IDO activity is enhanced by pro-inflammatory cytokines such as interferon-γ (IFNγ) [117, 118], and is inhibited by the anti-inflammatory cytokine IL-4 [119]. In case of stress or related conditions where cortisol secretion is enhanced, TDO activity is also further enhanced by glucocorticoids [120, 121]. In this case, the KYN formation becomes much higher than under physiological conditions. Since the liver cell uptake of KYN is not efficient for extrahepatic KYN, further KYN catabolism mainly occurs extrahepatically. The activity of KMO is also enhanced by pro-inflammatory cytokines [114]. Therefore, in case of inflammation, formation of 3HK becomes enhanced much faster than KYNA formation and the balance between formation of 3HK and KYNA shifted to the arm of the metabolism with 3HK and QUIN (fig. 1). In the presence of inflammation, activated monocytes are also found to be the robust producers of QUIN [122].

In the brain, the enhanced tryptophan breakdown induces low availability of tryptophan for serotonin synthesis. Moreover, in case of IDO activation, serotonin is degraded also by IDO into formyl-5-hydroxykynuramine (f5OHKYM) [123], and therefore less serotonin is available for optimal serotonergic neurotransmission. With enhanced tryptophan breakdown, an extra amount of peripheral KYN becomes available for further catabolism in the brain especially in the astrocytes and microglia. The KYN metabolites contribute directly to the neuroprotective-neurodegenerative changes in the brain through direct effects on several neurotransmissions. The 3HK causes neuronal apoptosis [124], while QUIN, an NMDA-R agonist [125], causes excitotoxic neurodegenerative changes [126] and astrocytes apoptosis [127]. However, KYNA is the NMDA-R antagonist [128] and is protective against the excitotoxicity of QUIN [129].

In pro-inflammatory state, although the balance between 3HK and KYNA might generally shift to the 3HK arm and because of the general increase of KYN, the formation of KYNA may also be higher than in the normal state. A well-balanced increase may counteract the negative effects of QUIN on NMDA-R and the receptor function homeostasis will be maintained. However, since KYNA is also antagonist of all ionotropic excitatory amino acid receptors activities [128], its abnormal accumulation beyond physiological level could induce glutamatergic hypofunctioning and might disturb cognitive function [130]. Moreover, KYNA is an antagonist of the α7-nicotinic acid acetylcholine receptor (α7nAchR) [131] and influences the expression of non-α7nAchR [131] which in turn disturbs the α7- and non-α7nAchR-mediated release of noradrenaline (NA), dopamine (DA) and acetylcholine (Ach) [100]. In addition, antagonism of α7nAchR could impair microglial phagocytosis since activation of this receptor was demonstrated to enhance microglia amyloid-β phygocytosis [132].

The above findings indicated the importance of a well-balanced network of kynurenines and any disturbance could induce neurotoxic changes.

Depression

In case of cytokine therapy-induced depression, such as IFNα therapy-induced depression, an increase in IL-6 and a decrease in KYNA shows a significant association with the development of depressive symptoms since IFNα could induce a 15-kDa protein which would further increase IFNγ production and activate IDO and KMO [133]. Another study also showed an increased KYN pathway with an increase in both KYNA and QUIN in the CSF of IFNα-treated patients [134]. The pro-inflammatory state in major depression without IFNα treatment would also activate IDO and KMO enzyme activities since other pro-inflammatory cytokines such as IL-1 could also enhance IDO and KMO activities [135]. This might shift the KYN metabolism to the 3HK and QUIN arm. It was proposed that the increase in those toxic metabolites imbalanced to formation of KYNA might prime the astrocyte-microglial-neuronal network to be vulnerable to environmental factors such as stress. It was also proposed that the imbalanced KYN pathway-induced impaired glial-neuronal network might contribute to the recurrent and chronic nature of major depression [136].

In patients with major depression who are drug naïve or medication free for at least 4 months, an imbalance between those neuroprotective and neurotoxic pathways with a lower protective metabolite level has been demonstrated [137]. The ratio between KYNA and KYN (KYNA/KYN) which indicated how much KYN would be degraded into KYNA was significantly lower in depressed patients than healthy controls. Moreover, a 6-week medication with selective serotonin reuptake inhibitors (SSRIs) could not revert the metabolic imbalance in the KYN pathway back to normal. We have hypothesized that such an uncorrected imbalance with higher neurotoxic metabolites to KYNA ratios, in the long term, might induce neurodegenerative changes and that in turn might induce chronicity and treatment resistance. A possible important role of KYNA in depression was supported by a recent genetic study which demonstrates that a single nucleotide polymorphism (SNP) from the KAT-III gene is associated with major depression and bipolar depression [138].

In major depression, there are evidences of neurodegenerative changes and loss of astrocytes [139]. In a recent study, it was observed that in the anterior cingulate cortex in postmortem brain of depressed patients who committed suicide, the number of QUIN-positive microglia are increased compared to control brains [140]. It is hypothesized that such changes might occur in the brain of depressed patients during the course of the disease and these further induce vulnerability of the neuron-glial network to biological or environmental challenges. In addition, QUIN-induced astrocytes apoptosis or astrocytes loss would further disturb glutamate metabolism and glutamatergic neurotransmission since astrocytes play a key role in glutamate homeostasis.

Not only in adult depression but also in adolescent depression, the kynurenines seem to play a role. A recent study reported that in magnetic resonance (MR) spectroscopy in melancholic depressed adolescents, the choline levels which indicated the turnover of cells showed a positive correlation with the serum KYN and HAA/KYN ratio [141]. Moreover, it was reported in this study that serum KYN and HAA/KYN were positively correlated with depression scores. It could be concluded that the shift in KYN pathway is more to the arm of 3HK, HAA and QUIN is also involved in adolescent melancholic depression.

This evidence indicates that changes in KYN metabolism could induce depressive behavior or depression via toxic effects of KYN downstream metabolites.

Bipolar Disorder

Only very few studies have reported on the changes in tryptophan metabolites in patients with bipolar disorder. In the plasma, tryptophan levels are reduced in bipolar mania patients and initial plasma tryptophan and tryptophan index which indicates the the index of tryptophan to competing large amino acids is associated with a response to treatment [142] which indirectly demonstrates the role of tryptophan depletion. Another study on TDO2 and KYN in the anterior cingulate cortex showed increased TDO2 expression and KYN levels in the brains of bipolar patients [143] and the nicotinimide levels in this brain area are associated with suicide [144]. Our recent postmortem study on QUIN immunopositive cells showed a slight increase in QUIN-positive microglia in the subgenual-anterior cingulate cortex in bipolar depressed patients who committed suicide [140]. In our recent genetic study, an SNP from KAT-III also showed a significant association with bipolar depression [138]. Based on the above findings in bipolar disorder, the neurotoxicity due to immune activation and kynurenine-related mechanisms would be mainly associated to bipolar depression.

Schizophrenia

In cases of schizophrenia, a study on postmortem brain tissue in different cortical regions revealed increased KYNA levels in schizophrenic samples compared to a control sample, particularly in the prefrontal cortex [145]. In another investigation in the amygdala, an insignificant increase of KYNA in medicated schizophrenics was observed [143]. Those postmortem studies raised a question as to whether the increase in KYNA might be associated with antipsychotic medication. Since the increased levels of KYNA were also observed in the CSF of drug-naive first-episode schizophrenic patients [146], this could not be have been caused by antipsychotic treatment. It was hypothesized that accumulation of KYNA in the brain may lead to schizophrenic symptoms [147]. Since the prefrontal cortex abnormality is involved in the pathophysiology of schizophrenia [148], the increase in KYNA might be involved in the pathophysiological mechanism. The effect of KYNA on both NMDA-R and α7nAchR might be involved in the development of positive symptoms and cognitive impairment in schizophrenia.

However, not only the positive symptoms and cognitive impairment but also the negative symptoms and loss of brain volume [149] are components of the psychopathology of schizophrenia. Unfortunately, in schizophrenia, most of the studies concentrated only on KYNA and its role in the development of positive symptoms and cognitive impairment. Yao's group has reported on 3HK [150, 151] although no clear change was observed. Nevertheless, this group has demonstrated the association between OHK and total symptoms score at the time of recruitment (1), and response of positive symptoms to 4-week neuroleptic treatment in first-episode neuroleptic naïve schizophrenia patients (2). The associations between negative symptoms, brain volume changes and potentially neurotoxic metabolites in the pathophysiology have been ignored in schizophrenia research. Our recent finding in medication-naive schizophrenia patients indicated increased plasma 3HK and decreased plasma KYNA compared to healthy controls [152] and it was reversed by 6 weeks of antipsychotic treatment. This indicates that the induced neurotoxicity is the accumulation of 3HK due to the increased formation from enhanced KMO activity resulting in an inflammatory state.

Future Perspectives and Conclusion

The findings discussed above are clear evidence that the interaction between immune activation, cortisol and the tryptophan pathway is involved in the pathophysiology of major psychiatric disorders. The suppression of inflammation or manipulation of tryptophan metabolism is of interest for future therapeutic development. However, it is important to consider the possible occurrence of imbalance between different molecules which could lead to untoward effects when a particular enzyme or molecule is manipulated. Therefore, this type of manipulation should be carried out only when there are clear indications such as evidence of change in the metabolites or ratios between metabolites as biomarkers. Moreover, close monitoring of those changes during therapy would also be necessary. Application of immune parameters and related metabolites as biomarkers through a personalized medicine approach would be a realistic approach for the future of psychiatry.

Acknowledgement

The work of the author is partially supported by EU 7th Framework Program Collaborative Project 'MOODINFLAME' (Grant No. 22963).

References

1 Pinteaux E, Trotter P, Simi A: Cell-specific and concentration-dependent actions of interleukin-1 in acute brain inflammation. Cytokine 2009;45:1–7.
2 Aloisi F, Borsellino G, Care A, Testa U, Gallo P, Russo G, Peschle C, Levi G: Cytokine regulation of astrocyte function: in-vitro studies using cells from the human brain. Int J Dev Neurosci 1995; 13:265–274.
3 Vitkovic L, Chatham JJ, da Cunha A: Distinct expressions of three cytokines by IL-1-stimulated astrocytes in vitro and in AIDS brain. Brain Behav Immun 1995;9:378–388.
4 Kyrkanides S, Olschowka JA, Williams JP, Hansen JT, O'Banion MK: TNF alpha and IL-1beta mediate intercellular adhesion molecule-1 induction via microglia-astrocyte interaction in CNS radiation injury. J Neuroimmunol 1999; 95:95–106.
5 Peterson PK, Hu S, Salak-Johnson J, Molitor TW, Chao CC: Differential production of and migratory response to beta chemokines by human microglia and astrocytes. J Infect Dis 1997;175: 478–481.
6 O'Banion MK, Miller JC, Chang JW, Kaplan MD, Coleman PD: Interleukin-1 beta induces prostaglandin G/H synthase-2 (cyclooxygenase-2) in primary murine astrocyte cultures. J Neurochem 1996;66:2532–2540.
7 Molina-Holgado E, Ortiz S, Molina-Holgado F, Guaza C: Induction of COX-2 and PGE(2) biosynthesis by IL-1beta is mediated by PKC and mitogen-activated protein kinases in murine astrocytes. Br J Pharmacol 2000;131:152–159.
8 Juric DM, Carman-Krzan M: Interleukin-1 beta, but not IL-1 alpha, mediates nerve growth factor secretion from rat astrocytes via type I IL-1 receptor. Int J Dev Neurosci 2001;19:675–683.
9 Parker LC, Luheshi GN, Rothwell NJ, Pinteaux E: IL-1 beta signalling in glial cells in wildtype and IL-1RI deficient mice. Br J Pharmacol 2002;136:312–320.
10 Andre R, Pinteaux E, Kimber I, Rothwell NJ: Differential actions of IL-1 alpha and IL-1 beta in glial cells share common IL-1 signalling pathways. Neuroreport 2005;16:153–157.
11 Beskina O, Miller A, Mazzocco-Spezzia A, Pulina MV, Golovina VA: Mechanisms of interleukin-1beta-induced Ca^{2+} signals in mouse cortical astrocytes: roles of store- and receptor-operated Ca^{2+} entry. Am J Physiol Cell Physiol 2007;293:C1103–C1111.
12 Pita I, Jelaso AM, Ide CF: IL-1beta increases intracellular calcium through an IL-1 type 1 receptor mediated mechanism in C6 astrocytic cells. Int J Dev Neurosci 1999;17:813–820.

13 Pringle AK, Niyadurupola N, Johns P, Anthony DC, Iannotti F: Interleukin-1beta exacerbates hypoxia-induced neuronal damage, but attenuates toxicity produced by simulated ischaemia and excitotoxicity in rat organotypic hippocampal slice cultures. Neurosci Lett 2001;305:29–32.
14 Bernardino L, Xapelli S, Silva AP, Jakobsen B, Poulsen FR, Oliveira CR, Vezzani A, Malva JO, Zimmer J: Modulator effects of interleukin-1beta and tumor necrosis factor-alpha on AMPA-induced excitotoxicity in mouse organotypic hippocampal slice cultures. J Neurosci 2005; 25:6734–6744.
15 Viviani B, Bartesaghi S, Gardoni F, Vezzani A, Behrens MM, Bartfai T, Binaglia M, Corsini E, Di Luca M, Galli CL, Marinovich M: Interleukin-1beta enhances NMDA receptor-mediated intracellular calcium increase through activation of the Src family of kinases. J Neurosci 2003;23:8692–8700.
16 Pizzi M, Goffi F, Boroni F, Benarese M, Perkins SE, Liou HC, Spano P: Opposing roles for NF-kappa B/Rel factors p65 and c-Rel in the modulation of neuron survival elicited by glutamate and interleukin-1beta. J Biol Chem 2002; 277:20717–20723.
17 Mizuno T, Sawada M, Suzumura A, Marunouchi T: Expression of cytokines during glial differentiation. Brain Res 1994;656:141–146.
18 Shrikant P, Weber E, Jilling T, Benveniste EN: Intercellular adhesion molecule-1 gene expression by glial cells. Differential mechanisms of inhibition by IL-10 and IL-6. J Immunol 1995;155: 1489–1501.
19 Akira S, Taga T, Kishimoto T: Interleukin-6 in biology and medicine. Adv Immunol 1993;54:1–78.
20 Hirota H, Kiyama H, Kishimoto T, Taga T: Accelerated Nerve Regeneration in Mice by upregulated expression of interleukin (IL) 6 and IL-6 receptor after trauma. J Exp Med 1996;183:2627–2634.
21 Gadient RA, Otten UH: Interleukin-6 (IL-6)–a molecule with both beneficial and destructive potentials. Prog Neurobiol 1997;52:379–390.
22 Blum-Degen D, Muller T, Kuhn W, Gerlach M, Przuntek H, Riederer P: Interleukin-1 beta and interleukin-6 are elevated in the cerebrospinal fluid of Alzheimer's and de novo Parkinson's disease patients. Neurosci Lett 1995;202:17–20.
23 Hull M, Fiebich BL, Lieb K, Strauss S, Berger SS, Volk B, Bauer J: Interleukin-6-associated inflammatory processes in Alzheimer's disease: new therapeutic options. Neurobiol Aging 1996;17: 795–800.
24 Hull M, Strauss S, Berger M, Volk B, Bauer J: The participation of interleukin-6, a stress-inducible cytokine, in the pathogenesis of Alzheimer's disease. Behav Brain Res 1996;78:37–41.
25 Munoz-Fernandez MA, Cano E, O'Donnell CA, Doyle J, Liew FY, Fresno M: Tumor necrosis factor-alpha (TNF-alpha), interferon-gamma, and interleukin-6 but not TNF-beta induce differentiation of neuroblastoma cells: the role of nitric oxide. J Neurochem 1994; 62:1330–1336.
26 Munoz-Fernandez MA, Fresno M: The role of tumour necrosis factor, interleukin 6, interferon-gamma and inducible nitric oxide synthase in the development and pathology of the nervous system. Prog Neurobiol 1998;56:307–340.
27 Breder CD, Tsujimoto M, Terano Y, Scott DW, Saper CB: Distribution and characterization of tumor necrosis factor-alpha-like immunoreactivity in the murine central nervous system. J Comp Neurol 1993;337:543–567.
28 Barna BP, Estes ML, Jacobs BS, Hudson S, Ransohoff RM: Human astrocytes proliferate in response to tumor necrosis factor alpha. J Neuroimmunol 1990;30: 239–243.
29 Obregon E, Punzon MC, Gonzalez-Nicolas J, Fernandez-Cruz E, Fresno M, Munoz-Fernandez MA: Induction of adhesion/differentiation of human neuroblastoma cells by tumour necrosis factor-alpha requires the expression of an inducible nitric oxide synthase. Eur J Neurosci 1997;9:1184–1193.
30 Zhao X, Bausano B, Pike BR, Newcomb-Fernandez JK, Wang KK, Shohami E, Ringger NC, DeFord SM, Anderson DK, Hayes RL: TNF-alpha stimulates caspase-3 activation and apoptotic cell death in primary septo-hippocampal cultures. J Neurosci Res 2001;64:121–131.
31 Cheng K, Sawamura Y, Sakuma S, Tada M, Sudo M, Aida T, Abe H: Antiproliferative effect of tumor necrosis factor-alpha on human glioblastoma cells linked with cell cycle arrest in G1 phase. Neurol Med Chir (Tokyo) 1994;34:274–278.
32 Saha RN, Ghosh A, Palencia CA, Fung YK, Dudek SM, Pahan K: TNF-alpha preconditioning protects neurons via neuron-specific up-regulation of CREB-binding protein. J Immunol 2009;183: 2068–2078.
33 Maes M: Cytokines in major depression. Biol Psychiatry 1994;36:498–499.
34 Myint AM, Leonard BE, Steinbusch HW, Kim YK: Th1, Th2, and Th3 cytokine alterations in major depression. J Affect Disord 2005;88:167–173.
35 Rothermundt M, Arolt V, Peters M, Gutbrodt H, Fenker J, Kersting A, Kirchner H: Inflammatory markers in major depression and melancholia. J Affect Disord 2001;63:93–102.
36 Hosoda S, Takimura H, Shibayama M, Kanamura H, Ikeda K, Kumada H: Psychiatric symptoms related to interferon therapy for chronic hepatitis C: clinical features and prognosis. Psychiatry Clin Neurosci 2000;54:565–572.
37 Schafer M, Messer T, Wegner U, Schmid-Wendtner MH, Volkenandt M: Psychiatric side effects during adjuvant therapy with interferon-alphain patients with malignant melanoma. Clinical evaluation as well as diagnostic and therapeutic possibilities (in German). Hautarzt 1999;50:654–658.
38 Valentine AD, Meyers CA, Kling MA, Richelson E, Hauser P: Mood and cognitive side effects of interferon-alpha therapy. Semin Oncol 1998;25(1 Suppl 1): 39–47.
39 O'Neill B, Leonard BE: Abnormal zymosan-induced neutrophil chemiluminescence as a marker of depression. J Affect Disord 1990;19:265–272.
40 Kronfol Z, House JD: Depression, hypothalamic-pituitary-adrenocortical activity, and lymphocyte function. Psychopharmacol Bull 1985;21:476–478.
41 Irwin M, Smith TL, Gillin JC: Low natural killer cytotoxicity in major depression. Life Sci 1987;41:2127–2133.
42 Maes M, Scharpe S, Van Grootel L, Uyttenbroeck W, Cooreman W, Cosyns P, Suy E: Higher alpha 1-antitrypsin, haptoglobin, ceruloplasmin and lower retinol binding protein plasma levels during depression: further evidence for the existence of an inflammatory response during that illness. J Affect Disord 1992;24:183–192.

43 Sluzewska A, Rybakowski J, Bosmans E, Sobieska M, Berghmans R, Maes M, Wiktorowicz K: Indicators of immune activation in major depression. Psychiatry Res 1996;64:161–167.
44 Sluzewska A, Rybakowski JK, Laciak M, Mackiewicz A, Sobieska M, Wiktorowicz K: Interleukin-6 serum levels in depressed patients before and after treatment with fluoxetine. Ann NY Acad Sci 1995;762:474–476.
45 Maes M, Song C, Lin AH, Bonaccorso S, Kenis G, De Jongh R, Bosmans E, Scharpe S: Negative immunoregulatory effects of antidepressants: inhibition of interferon-gamma and stimulation of interleukin-10 secretion. Neuropsychopharmacology 1999;20:370–379.
46 Birmaher B, Rabin BS, Garcia MR, Jain U, Whiteside TL, Williamson DE, al-Shabbout M, Nelson BC, Dahl RE, Ryan ND: Cellular immunity in depressed, conduct disorder, and normal adolescents: role of adverse life events. J Am Acad Child Adolesc Psychiatry 1994;33: 671–678.
47 Maes M, Lambrechts J, Suy E, Vandervorst C, Bosmans E: Absolute number and percentage of circulating natural killer, non-MHC-restricted T cytotoxic, and phagocytic cells in unipolar depression. Neuropsychobiology 1994; 29:157–163.
48 Steiner J, Bielau H, Brisch R, Danos P, Ullrich O, Mawrin C, Bernstein HG, Bogerts B: Immunological aspects in the neurobiology of suicide: elevated microglial density in schizophrenia and depression is associated with suicide. J Psychiatr Res 2008;42:151–157.
49 Chen S, Averett NT, Manelli A, Ladu MJ, May W, Ard MD: Isoform-specific effects of apolipoprotein E on secretion of inflammatory mediators in adult rat microglia. J Alzheimers Dis 2005;7:25–35.
50 Campuzano O, Castillo-Ruiz MM, Acarin L, Castellano B, Gonzalez B: Distinct pattern of microglial response, cyclooxygenase-2, and inducible nitric oxide synthase expression in the aged rat brain after excitotoxic damage. J Neurosci Res 2008;86:3170–3183.
51 Depboylu C, Weihe E, Eiden LE: COX1 and COX2 expression in non-neuronal cellular compartments of the rhesus macaque brain during lentiviral infection. Neurobiol Dis 2011;42:108–115.
52 Ohishi K, Ueno R, Nishino S, Sakai T, Hayaishi O: Increased level of salivary prostaglandins in patients with major depression. Biol Psychiatry 1988;23: 326–334.
53 Linnoila M, Whorton AR, Rubinow DR, Cowdry RW, Ninan PT, Waters RN: CSF prostaglandin levels in depressed and schizophrenic patients. Arch Gen Psychiatry 1983;40:405–406.
54 Mtabaji JP, Manku MS, Horrobin DF: Actions of the tricyclic antidepressant clomipramine on responses to pressor agents. Interactions with prostaglandin E_2. Prostaglandins 1977;14:125–132.
55 Muller N, Schwarz MJ, Dehning S, Douhe A, Cerovecki A, Goldstein-Muller B, Spellmann I, Hetzel G, Maino K, Kleindienst N, Moller HJ, Arolt V, Riedel M: The cyclooxygenase-2 inhibitor celecoxib has therapeutic effects in major depression: results of a double-blind, randomized, placebo controlled, add-on pilot study to reboxetine. Mol Psychiatry 2006;11:680–684.
56 Akhondzadeh S, Jafari S, Raisi F, Nasehi AA, Ghoreishi A, Salehi B, Mohebbi-Rasa S, Raznahan M, Kamalipour A: Clinical trial of adjunctive celecoxib treatment in patients with major depression: a double blind and placebo controlled trial. Depress Anxiety 2009;26: 607–611.
57 Berk M, Kapczinski F, Andreazza AC, Dean OM, Giorlando F, Maes M, Yucel M, Gama CS, Dodd S, Dean B, Magalhaes PV, Amminger P, McGorry P, Malhi GS: Pathways underlying neuroprogression in bipolar disorder: focus on inflammation, oxidative stress and neurotrophic factors. Neurosci Biobehav Rev 2011;35:804–817.
58 Drexhage RC, Knijff EM, Padmos RC, Heul-Nieuwenhuijzen L, Beumer W, Versnel MA, Drexhage HA: The mononuclear phagocyte system and its cytokine inflammatory networks in schizophrenia and bipolar disorder. Expert Rev Neurother 2010;10:59–76.
59 Rao JS, Harry GJ, Rapoport SI, Kim HW: Increased excitotoxicity and neuroinflammatory markers in postmortem frontal cortex from bipolar disorder patients. Mol Psychiatry 2010;15: 384–392.
60 Dickerson F, Stallings C, Origoni A, Boronow J, Yolken R: Elevated serum levels of C-reactive protein are associated with mania symptoms in outpatients with bipolar disorder. Prog Neuropsychopharmacol Biol Psychiatry 2007;31:952–955.
61 Goldstein BI, Kemp DE, Soczynska JK, McIntyre RS: Inflammation and the phenomenology, pathophysiology, comorbidity, and treatment of bipolar disorder: a systematic review of the literature. J Clin Psychiatry 2009;70:1078–1090.
62 Kim YK, Jung HG, Myint AM, Kim H, Park SH: Imbalance between pro-inflammatory and anti-inflammatory cytokines in bipolar disorder. J Affect Disord 2007;104:91–95.
63 Kauer-Sant'Anna M, Kapczinski F, Andreazza AC, Bond DJ, Lam RW, Young LT, Yatham LN: Brain-derived neurotrophic factor and inflammatory markers in patients with early- vs. late-stage bipolar disorder. Int J Neuropsychopharmacol 2009;12:447–458.
64 Kim YK, Myint AM, Lee BH, Han CS, Lee SW, Leonard BE, Steinbusch HW: T-helper types 1, 2, and 3 cytokine interactions in symptomatic manic patients. Psychiatry Res 2004;129:267–272.
65 Cunha AB, Frey BN, Andreazza AC, Goi JD, Rosa AR, Goncalves CA, Santin A, Kapczinski F: Serum brain-derived neurotrophic factor is decreased in bipolar disorder during depressive and manic episodes. Neurosci Lett 2006; 398:215–219.
66 Machado-Vieira R, Dietrich MO, Leke R, Cereser VH, Zanatto V, Kapczinski F, Souza DO, Portela LV, Gentil V: Decreased plasma brain derived neurotrophic factor levels in unmedicated bipolar patients during manic episode. Biol Psychiatry 2007;61:142–144.
67 Fernandes BS, Gama CS, Kauer-Sant'Anna M, Lobato MI, Belmonte-de-Abreu P, Kapczinski F: Serum brain-derived neurotrophic factor in bipolar and unipolar depression: a potential adjunctive tool for differential diagnosis. J Psychiatr Res 2009;43:1200–1204.
68 Nery FG, Monkul ES, Hatch JP, Fonseca M, Zunta-Soares GB, Frey BN, Bowden CL, Soares JC: Celecoxib as an adjunct in the treatment of depressive or mixed episodes of bipolar disorder: a double-blind, randomized, placebo controlled study. Hum Psychopharmacol 2008;23: 87–94.

69 Maes M, Bosmans E, Calabrese J, Smith R, Meltzer HY: Interleukin-2 and interleukin-6 in schizophrenia and mania: effects of neuroleptics and mood stabilizers. J Psychiatr Res 1995;29:141–152.
70 Lin A, Kenis G, Bignotti S, Tura GJ, De Jong R, Bosmans E, Pioli R, Altamura C, Scharpe S, Maes M: The inflammatory response system in treatment-resistant schizophrenia: increased serum interleukin-6. Schizophr Res 1998;32:9–15.
71 Garcia-Miss Mdel R, Perez-Mutul J, Lopez-Canul B, Solis-Rodriguez F, Puga-Machado L, Oxte-Cabrera A, Gurubel-Maldonado J, Arankowsky-Sandoval G: Folate, homocysteine, interleukin-6, and tumor necrosis factor alfa levels, but not the methylenetetrahydrofolate reductase C677T polymorphism, are risk factors for schizophrenia. J Psychiatr Res 2010; 44:441–446.
72 Kim YK, Kim L, Lee MS: Relationships between interleukins, neurotransmitters and psychopathology in drug-free male schizophrenics. Schizophr Res 2000;44: 165–175.
73 Singh B, Bera NK, Nayak CR, Chaudhuri TK: Decreased serum levels of interleukin-2 and interleukin-6 in Indian Bengalee schizophrenic patients. Cytokine 2009;47:1–5.
74 Muller N, Dobmeier P, Empl M, Riedel M, Schwarz M, Ackenheil M: Soluble IL-6 receptors in the serum and cerebrospinal fluid of paranoid schizophrenic patients. Eur Psychiatry 1997; 12:294–299.
75 Licinio J, Seibyl JP, Altemus M, Charney DS, Krystal JH: Elevated CSF levels of interleukin-2 in neuroleptic-free schizophrenic patients. Am J Psychiatry 1993; 150:1408–1410.
76 Kim YK, Lee MS, Suh KY: Decreased interleukin-2 production in Korean schizophrenic patients. Biol Psychiatry 1998;43:701–704.
77 Song XQ, Lv LX, Li WQ, Hao YH, Zhao JP: The interaction of nuclear factor-kappa B and cytokines is associated with schizophrenia. Biol Psychiatry 2009;65: 481–488.
78 Kim YK, Myint AM, Lee BH, Han CS, Lee HJ, Kim DJ, Leonard BE: Th1, Th2 and Th3 cytokine alteration in schizophrenia. Prog Neuropsychopharmacol Biol Psychiatry 2004;28:1129–1134.
79 Muller N, Ulmschneider M, Scheppach C, Schwarz MJ, Ackenheil M, Moller HJ, Gruber R, Riedel M: COX-2 inhibition as a treatment approach in schizophrenia: immunological considerations and clinical effects of celecoxib add-on therapy. Eur Arch Psychiatry Clin Neurosci 2004; 254:14–22.
80 Bresee CJ, Delrahim K, Maddux RE, Dolnak D, Ahmadpour O, Rapaport MH: The effects of celecoxib augmentation on cytokine levels in schizophrenia. Int J Neuropsychopharmacol 2006;9:343–348.
81 Akhondzadeh S, Tabatabaee M, Amini H, Ahmadi Abhari SA, Abbasi SH, Behnam B: Celecoxib as adjunctive therapy in schizophrenia: a double-blind, randomized and placebo-controlled trial. Schizophr Res 2007;90:179–185.
82 Toyooka K, Watanabe Y, Iritani S, Shimizu E, Iyo M, Nakamura R, Asama K, Makifuchi T, Kakita A, Takahashi H, Someya T, Nawa H: A decrease in interleukin-1 receptor antagonist expression in the prefrontal cortex of schizophrenic patients. Neurosci Res 2003;46:299–307.
83 Doorduin J, de Vries EF, Willemsen AT, de Groot JC, Dierckx RA, Klein HC: Neuroinflammation in schizophrenia-related psychosis: a PET study. J Nucl Med 2009; 50:1801–1807.
84 Harvey I, Ron MA, Du Boulay G, Wicks D, Lewis SW, Murray RM: Reduction of cortical volume in schizophrenia on magnetic resonance imaging. Psychol Med 1993;23:591–604.
85 Garey LJ, Ong WY, Patel TS, Kanani M, Davis A, Mortimer AM, Barnes TR, Hirsch SR: Reduced dendritic spine density on cerebral cortical pyramidal neurons in schizophrenia. J Neurol Neurosurg Psychiatry 1998;65:446–453.
86 Asami T, Bouix S, Whitford TJ, Shenton ME, Salisbury DF, McCarley RW: Longitudinal loss of gray matter volume in patients with first-episode schizophrenia: DARTEL automated analysis and ROI validation. Neuroimage 2012;59: 986–996.
87 Rais M, Cahn W, Schnack HG, Hulshoff Pol HE, Kahn RS, van Haren NE: Brain volume reductions in medication-naive patients with schizophrenia in relation to intelligence quotient. Psychol Med 2012;42:1847–1856.
88 Czeh B, Simon M, Schmelting B, Hiemke C, Fuchs E: Astroglial plasticity in the hippocampus is affected by chronic psychosocial stress and concomitant fluoxetine treatment. Neuropsychopharmacology 2006;31:1616–1626.
89 Pugh CR, Nguyen KT, Gonyea JL, Fleshner M, Wakins LR, Maier SF, Rudy JW: Role of interleukin-1 beta in impairment of contextual fear conditioning caused by social isolation. Behav Brain Res 1999;106:109–118.
90 Bacher M, Meinhardt A, Lan HY, Dhabhar FS, Mu W, Metz CN, Chesney JA, Gemsa D, Donnelly T, Atkins RC, Bucala R: MIF expression in the rat brain: implications for neuronal function. Mol Med 1998;4:217–230.
91 Niino M, Ogata A, Kikuchi S, Tashiro K, Nishihira J: Macrophage migration inhibitory factor in the cerebrospinal fluid of patients with conventional and optic-spinal forms of multiple sclerosis and neuro-Behcet's disease. J Neurol Sci 2000;179(S 1–2):127–131.
92 Suzuki T, Ogata A, Tashiro K, Nagashima K, Tamura M, Yasui K, Nishihira J: Japanese encephalitis virus up-regulates expression of macrophage migration inhibitory factor (MIF) mRNA in the mouse brain. Biochim Biophys Acta 2000;1517:100–106.
93 Madrigal JL, Garcia-Bueno B, Moro MA, Lizasoain I, Lorenzo P, Leza JC: Relationship between cyclooxygenase-2 and nitric oxide synthase-2 in rat cortex after stress. Eur J Neurosci 2003;18: 1701–1705.
94 Sapolsky RM: A mechanism for glucocorticoid toxicity in the hippocampus: increased neuronal vulnerability to metabolic insults. J Neurosci 1985;5: 1228–1232.
95 Woolley CS, Gould E, McEwen BS: Exposure to excess glucocorticoids alters dendritic morphology of adult hippocampal pyramidal neurons. Brain Res 1990;531: 225–231.
96 Aucott JN: Glucocorticoids and infection. Endocrinol Metab Clin North Am 1994;23:655–670.
97 Stein-Behrens BA, Lin WJ, Sapolsky RM: Physiological elevations of glucocorticoids potentiate glutamate accumulation in the hippocampus. J Neurochem 1994; 63:596–602.

98 Takahashi T, Kimoto T, Tanabe N, Hattori TA, Yasumatsu N, Kawato S: Corticosterone acutely prolonged N-methyl-d-aspartate receptor-mediated Ca^{2+} elevation in cultured rat hippocampal neurons. J Neurochem 2002;83:1441–1451.
99 Nair A, Bonneau RH: Stress-induced elevation of glucocorticoids increases microglia proliferation through NMDA receptor activation. J Neuroimmunol 2006;171:72–85.
100 Myint AM: Kynurenines: from the perspective of major psychiatric disorders. FEBS J 2012;279:1375–1385.
101 Watanabe Y, Fujiwara M, Yoshida R, Hayaishi O: Stereospecificity of hepatic L-tryptophan 2,3-dioxygenase. Biochem J 1980;189:393–405.
102 Fernstrom JD: Effects on the diet on brain neurotransmitters. Metabolism 1977;26:207–223.
103 Leklem JE: Quantitative aspects of tryptophan metabolism in humans and other species: a review. Am J Clin Nutr 1971; 24:659–672.
104 Bender DA: Effects of a dietary excess of leucine and of the addition of leucine and 2-oxo-isocaproate on the metabolism of tryptophan and niacin in isolated rat liver cells. Br J Nutr 1989;61: 629–640.
105 Gal EM, Sherman AD: L-kynurenine: its synthesis and possible regulatory function in brain. Neurochem Res 1980;5: 223–239.
106 Grant RS, Naif H, Espinosa M, Kapoor V: IDO induction in IFN-gamma activated astroglia: a role in improving cell viability during oxidative stress. Redox Rep 2000;5:101–104.
107 Grant RS, Kapoor V: Murine glial cells regenerate NAD, after peroxide-induced depletion, using either nicotinic acid, nicotinamide, or quinolinic acid as substrates. J Neurochem 1998; 70:1759–1763.
108 Heyes MP, Achim CL, Wiley CA, Major EO, Saito K, Markey SP: Human microglia convert l-tryptophan into the neurotoxin quinolinic acid. Biochem J 1996; 320:595–597.
109 Miller CL, Llenos IC, Dulay JR, Barillo MM, Yolken RH, Weis S: Expression of the kynurenine pathway enzyme tryptophan 2,3-dioxygenase is increased in the frontal cortex of individuals with schizophrenia. Neurobiol Dis 2004;15:618–629.
110 Guillemin GJ, Smythe G, Takikawa O, Brew BJ: Expression of indoleamine 2,3-dioxygenase and production of quinolinic acid by human microglia, astrocytes, and neurons. Glia 2005;49: 15–23.
111 Guillemin GJ, Kerr SJ, Smythe GA, Smith DG, Kapoor V, Armati PJ, Croitoru J, Brew BJ: Kynurenine pathway metabolism in human astrocytes: a paradox for neuronal protection. J Neurochem 2001; 78:842–853.
112 Guillemin GJ, Smith DG, Kerr SJ, Smythe GA, Kapoor V, Armati PJ, Brew BJ: Characterisation of kynurenine pathway metabolism in human astrocytes and implications in neuropathogenesis. Redox Rep 2000;5:108–111.
113 Heyes MP, Saito K, Major EO, Milstien S, Markey SP, Vickers JH: A mechanism of quinolinic acid formation by brain in inflammatory neurological disease. Attenuation of synthesis from L-tryptophan by 6-chlorotryptophan and 4-chloro-3-hydroxyanthranilate. Brain 1993;116:1425–1450.
114 Mellor AL, Munn DH: Tryptophan catabolism and T-cell tolerance: immunosuppression by starvation? Immunol Today 1999;20:469–473.
115 Moffett JR, Blinder KL, Venkateshan CN, Namboodiri MA: Differential effects of kynurenine and tryptophan treatment on quinolinate immunoreactivity in rat lymphoid and non-lymphoid organs. Cell Tissue Res 1998;293:525–534.
116 Moffett JR, Namboodiri MA: Tryptophan and the immune response. Immunol Cell Biol 2003;81:247–265.
117 Carlin JM, Borden EC, Sondel PM, Byrne GI: Biologic-response-modifier-induced indoleamine 2,3-dioxygenase activity in human peripheral blood mononuclear cell cultures. J Immunol 1987;139:2414–2418.
118 Yasui H, Takai K, Yoshida R, Hayaishi O: Interferon enhances tryptophan metabolism by inducing pulmonary indoleamine 2,3-dioxygenase: its possible occurrence in cancer patients. Proc Natl Acad Sci USA 1986;83:6622–6626.
119 Musso T, Gusella GL, Brooks A, Longo DL, Varesio L: Interleukin-4 inhibits indoleamine 2,3-dioxygenase expression in human monocytes. Blood 1994;83: 1408–1411.
120 Knox WE: Two mechanisms which increase in vivo the liver tryptophan peroxidase activity: specific enzyme adaptation and stimulation of the pituitary adrenal system. Br J Exp Pathol 1951;32:462–469.
121 Salter M, Pogson CI: The role of tryptophan 2,3-dioxygenase in the hormonal control of tryptophan metabolism in isolated rat liver cells. Effects of glucocorticoids and experimental diabetes. Biochem J 1985;229:499–504.
122 Chiarugi A, Calvani M, Meli E, Traggiai E, Moroni F: Synthesis and release of neurotoxic kynurenine metabolites by human monocyte-derived macrophages. J Neuroimmunol 2001;120: 190–198.
123 Pertz H, Back W: Synthesis and resolution of chiral ring-opened serotonin analogs of the 5-hydroxykynuramine type. Pharm Acta Helv 1988;63:128–131.
124 Okuda S, Nishiyama N, Saito H, Katsuki H: 3-Hydroxykynurenine, an endogenous oxidative stress generator, causes neuronal cell death with apoptotic features and region selectivity. J Neurochem 1998;70: 299–307.
125 Bender DA, McCreanor GM: Kynurenine hydroxylase: a potential rate-limiting enzyme in tryptophan metabolism. Biochem Soc Trans 1985;13:441–443.
126 Schwarcz R, Whetsell WO Jr, Mangano RM: Quinolinic acid: an endogenous metabolite that produces axon-sparing lesions in rat brain. Science 1983;219: 316–318.
127 Guillemin GJ, Wang L, Brew BJ: Quinolinic acid selectively induces apoptosis of human astrocytes: potential role in AIDS dementia complex. J Neuroinflammation 2005;2:16.
128 Perkins MN, Stone TW: An iontophoretic investigation of the actions of convulsant kynurenines and their interaction with the endogenous excitant quinolinic acid. Brain Res 1982;247: 184–187.
129 Kim JP, Choi DW: Quinolinate neurotoxicity in cortical cell culture. Neuroscience 1987;23:423–432.
130 Olney JW, Labruyere J, Wang G, Wozniak DF, Price MT, Sesma MA: NMDA antagonist neurotoxicity: mechanism and prevention. Science 1991;254: 1515–1518.

131 Hilmas C, Pereira EF, Alkondon M, Rassoulpour A, Schwarcz R, Albuquerue EX: The brain metabolite kynurenic acid inhibits alpha7 nicotinic receptor activity and increases non-alpha7 nicotinic receptor expression: physiopathological implications. J Neurosci 2001;21: 7463–7473.
132 Takata K, Kitamura Y, Saeki M, Terada M, Kagitani S, Kitamura R, Fujikawa Y, Maelicke A, Tomimoto H, Taniguchi T, Shimohama S: Galantamine-induced amyloid-{beta} clearance mediated via stimulation of microglial nicotinic acetylcholine receptors. J Biol Chem 2010; 285:40180–40191.
133 Wichers MC, Koek GH, Robaeys G, Verkerk R, Scharpe S, Maes M: IDO and interferon-alpha-induced depressive symptoms: a shift in hypothesis from tryptophan depletion to neurotoxicity. Mol Psychiatry 2005;10:538–544.
134 Raison CL, Dantzer R, Kelley KW, Lawson MA, Woolwine BJ, Vogt G, Spivey JR, Saito K, Miller AH: CSF concentrations of brain tryptophan and kynurenines during immune stimulation with IFN-alpha: relationship to CNS immune responses and depression. Mol Psychiatry 2010;15:393–403.
135 Zunszain PA, Anacker C, Cattaneo A, Choudhury S, Musaelyan K, Myint AM, Thuret S, Price J, Pariante CM: Interleukin-1beta: a new regulator of the kynurenine pathway affecting human hippocampal neurogenesis. Neuropsychopharmacology 2012;37: 939–949.
136 Myint AM, Kim YK: Cytokine-serotonin interaction through IDO: a neurodegeneration hypothesis of depression. Med Hypotheses 2003;61:519–525.
137 Myint AM, Kim YK, Verkerk R, Scharpe S, Steinbusch H, Leonard B: Kynurenine pathway in major depression: evidence of impaired neuroprotection. J Affect Disord 2007;98:143–151.
138 Claes S, Myint AM, Domschke K, Del-Favero J, Entrich K, Engelborghs S, De Deyn P, Mueller N, Baune B, Rothermundt M: The kynurenine pathway in major depression: haplotype analysis of three related functional candidate genes. Psychiatry Res 2011;188:355–360.
139 Rajkowska G, Miguel-Hidalgo JJ, Wei J, Dilley G, Pittman SD, Meltzer HY, Overholser JC, Roth BL, Stockmeier CA: Morphometric evidence for neuronal and glial prefrontal cell pathology in major depression. Biol Psychiatry 1999;45: 1085–1098.
140 Steiner J, Walter M, Gos T, Guillemin GJ, Bernstein HG, Sarnyai Z, Mawrin C, Brisch R, Bielau H, Meyer zu Schwabedissen L, Bogerts B, Myint AM: Severe depression is associated with increased microglial quinolinic acid in subregions of the anterior cingulate gyrus: evidence for an immune-modulated glutamatergic neurotransmission? J Neuroinflammation 2011;8:94.
141 Gabbay V, Liebes L, Katz Y, Liu S, Mendoza S, Babb JS, Klein RG, Gonen O: The kynurenine pathway in adolescent depression: preliminary findings from a proton MR spectroscopy study. Prog Neuropsychopharmacol Biol Psychiatry 2010;34:37–44.
142 Myint AM, Kim YK, Verkerk R, Park SH, Scharpe S, Steinbusch HW, Leonard BE: Tryptophan breakdown pathway in bipolar mania. J Affect Disord 2007;102:65–72.
143 Miller CL, Llenos IC, Dulay JR, Weis S: Upregulation of the initiating step of the kynurenine pathway in postmortem anterior cingulate cortex from individuals with schizophrenia and bipolar disorder. Brain Res 2006;1073–1074:25–37.
144 Miller CL, Llenos IC, Cwik M, Walkup J, Weis S: Alterations in kynurenine precursor and product levels in schizophrenia and bipolar disorder. Neurochem Int 2008;52:1297–1303.
145 Schwarcz R, Rassoulpour A, Wu HQ, Medoff D, Tamminga CA, Roberts RC: Increased cortical kynurenate content in schizophrenia. Biol Psychiatry 2001;50: 521–530.
146 Erhardt S, Blennow K, Nordin C, Skogh E, Lindstrom LH, Engberg G: Kynurenic acid levels are elevated in the cerebrospinal fluid of patients with schizophrenia. Neurosci Lett 2001;313:96–98.
147 Erhardt S, Schwieler L, Engberg G: Kynurenic acid and schizophrenia. Adv Exp Med Biol 2003;527:155–165.
148 Andreasen NC, Rezai K, Alliger R, Swayze VW, 2nd, Flaum M, Kirchner P, Cohen G, O'Leary DS: Hypofrontality in neuroleptic-naive patients and in patients with chronic schizophrenia. Assessment with xenon 133 single-photon emission computed tomography and the Tower of London. Arch Gen Psychiatry 1992;49:943–958.
149 Takahashi T, Wood SJ, Soulsby B, McGorry PD, Tanino R, Suzuki M, Velakoulis D, Pantelis C: Follow-up MRI study of the insular cortex in first-episode psychosis and chronic schizophrenia. Schizophr Res 2009;108:49–56.
150 Yao JK, Dougherty GG Jr, Reddy RD, Keshavan MS, Montrose DM, Matson WR, Rozen S, Krishnan RR, McEvoy J, Kaddurah-Daouk R: Altered interactions of tryptophan metabolites in first-episode neuroleptic-naive patients with schizophrenia. Mol Psychiatry 2010; 15:938–953.
151 Condray R, Dougherty GG, Keshavan MS, Reddy RD, Haas GL, Montrose DM, Matson WR, McEvoy J, Kaddurah-Daouk R, Yao JK: 3-Hydroxykynurenine and clinical symptoms in first-episode neuroleptic-naive patients with schizophrenia. Int J Neuropsychopharmacol 2011;14:756–767.
152 Myint AM, Schwarz M, Verkerk R, Scharpe S, Steinbusch HWM, Leonard BE, Kim YK: Imbalance of kynurenine metabolites in drug naive schizophrenia. Brain Behav Immun 2011;25:6.

Dr. Aye-Mu Myint
Psychiatric Hospital, Ludwig-Maximilians University
Nussbaumstrasse 7
DE–80336 Munich (Germany)
E-Mail ayemu.myint@med.uni-muenchen.de

Halaris A, Leonard BE (eds): Inflammation in Psychiatry.
Mod Trends Pharmacopsychiatry. Basel, Karger, 2013, vol 28, pp 75–89 (DOI: 10.1159/000343970)

Essential Fatty Acids as Potential Anti-Inflammatory Agents in the Treatment of Affective Disorders

Cai Song[a,b]

[a]Chinese Academy Engineering Institute for the Development of Endangered Medicinal Resources in Southwest China, Guangxi Botanic Garden of Medicinal Plants, Nanning, Guangxi, PR China;
[b]Department of Psychology, Dalhousie University, Halifax, N.S., Canada

Abstract

In the last two decades, many studies have demonstrated that increased inflammatory responses occur in patients with affective disorders, which could contribute to the dysfunction of neuroendocrine and neurotransmission in these diseases. The evidence consequently suggests that drugs or natural products with anti-inflammatory properties should be developed to treat the illness. Among several candidates that modulate the neuro-immune network, omega-(n)-3 essential unsaturated fatty acids have emerged as effective treatments for affective disorders. Because these fatty acids can influence membrane structure, reduce inflammation and regulate protein and gene expressions, their effects seem promising. In this chapter, the newest findings that support the inflammatory theory of affective disorders have been first selected and reviewed. Secondly, the role of n–3 fatty acids in the immune system and brain has been introduced and then the possible mechanisms by which n–3 fatty acids attenuated symptoms of depression are discussed by the insight of the results from the animal studies. Finally, the clinical effects of n–3 fatty acids on major, bipolar and postpartum depression are summarized. Furthermore, the chapter points out the limitations and future research in this area.

Inflammation and Affective Disorders

Immune aging, auto-immune diseases, environment pollution, mental stress and un-balanced diets could all increase inflammatory responses. Inflammatory markers include immune cell activities, pro-inflammatory cytokines, acute-phase proteins, chemokines and adhesion molecules. It is well known that pro-inflammatory cytokines, antibodies and activated macrophages and T lymphocytes can pass through the blood-brain barrier. White cell-produced proinflammatory cytokines can cause the dysfunction of hypothalamic-pituitary-adrenal axis [1] and serotonergic or noradrenalinergic neurotransmission [2], changes which are similar to those observed in depressed patients. There is convincing evidence that inflammatory processes in the brain are involved in the development and progression of affective disorders. Brain inflammatory processes may not only result from peripheral infection of bacterial or viral agents but may also be activated by the production of endogenous neurotoxic agents or variations in the process of lipid and protein oxidation [3]. However, the emerging literature is largely based on demonstrating the mechanisms underlying the association between pro-inflammatory cytokines and affective disorders [4]. Since the main focus of this review is on omega-3 fatty acids in the treatment of affective disorders, the following section selectively reviews the newest findings that further support the inflammation theory of affective diseases.

Increase in the Production of Proinflammatory Cytokines

Pro-inflammatory cytokines, such as IL-1, IL-6, IL-12, interferons (IFNs) and TFNα are released by activated immune cells, which can influence the brain function relevant to human behavior. Both clinical and experimental studies have demonstrated that administration of proinflammatory cytokines can induce the symptoms of affective disorders. In the past, increases in proinflammatory cytokines have been extensively studied in affective disorders. However, it is still unknown whether different types of affective disorders are associated with certain immune changes or with specific proinflammatory cytokines. So far, a few studies have explored this direction. For example, Carpenter et al. [5] explored the association between plasma IL-6 production and childhood maltreatment in a group of 69 healthy adult subjects without depression or post-traumatic stress disorder. They found that the adults who were maltreated in childhood showed a higher degree of inflammatory response to acute stress challenge. The study of Janelidze et al. [6] demonstrated elevated levels of plasma IL-6 and TNF-α concentrations in the group of suicide attempters, while lowered plasma IL-2 concentrations in 17 nonsuicidal depressed patients were found when compared to 16 healthy controls. Thus, the authors suggested that cytokine levels in the blood might identify suicide attempters in depressed patients. The results also indicate that different immune changes may occur in different types of depressed patients. However, the mechanism involved in cytokine releases is unknown. To better understand the profile of pro- and anti-inflammatory cytokines, Lehto et al. [7] conducted a human clinical study to explore the depression-related alterations in the serum levels of anti-inflammatory markers, e.g. IL-1 receptor antagonist (RA) and IL-10 and IL-6, in 416 participants with and without depressive symptoms. The results demonstrated that the depressive symptomatology was related to higher levels of IL-1RA and IL-6 but not IL-10. Because the authors did not measure IL-1 concentrations, it is hard to know whether the increase in IL-1RA is the compensatory mechanism during a depression-related inflammatory state. Recent clinical findings from patients with IFNα treatment of hepatitis C further explore the relationship between inflammation and affective disorders. In 74 patients who received treatment, 20% showed moderate-to-severe symptoms of depression, and about 30% had minor depressive symptoms as scored by the Hamilton Depression Rating Scale and Zung Self-Rating Depression Scale [8]. Then, in 29 patients Friebe et al. [9] demonstrated that IFNα-induced depressive symptoms were dose dependent. Raison et al. [10] and Kennis et al. [11] further revealed the inflammation mechanism in serotonin metabolism and neurotrophin expression, respectively. In 24 patients with hepatitis C, monoamine metabolites, corticotropin-releasing hormone and proinflammatory parameters in the cerebrospinal fluid samples were analyzed after chronic IFNα treatment. IFNα-treated patients exhibited significant elevations in CSF IL-6 and MCP-1, both of which were highly correlated with CSF IFNα concentrations. Of the immunologic and neurotransmitter variables, CSF concentrations of the serotonin metabolite 5-hydroxyindoleacetic acid (5-HIAA) were decreased. CSF concentrations of IL-6, but not IFNα or MCP-1, were negatively correlated with CSF 5-HIAA. Furthermore, Kenis et al. [11] reported that depressive symptoms were associated with a decline in the serum level of blood-derived neurotropic factor (BDNF) during the course of treatment, and were significantly and inversely associated with total MADRS score. These important data suggest that a peripherally administered cytokine can activate CNS inflammatory responses in humans by interacting with serotonin metabolism and BDNF, which are the markers for depression.

To explore the role of inflammation in different types of affective disorders, Suvisaari et al. [12] conducted a population-based study to compare the inflammatory changes in patients with depression, schizophrenia and controls. They found that schizophrenia had significantly higher sIL-2Rα, IL-1RA and C-reactive protein (CRP) levels, while patients with affective psychosis showed higher TNFα compared to their matched controls. sIL-2Rα, a marker of T cell activation, was associated with depressive symptoms, schizophrenia, and affective psychosis.

To further understand cytokine functions, the changes in cytokine receptors have been studied. Hope et al. [13] determined whether sTNF-R1, IL-1RA, IL-6, high-sensitivity C-reactive protein (hsCRP), osteoprotegerin (OPG) and von Willebrand factor (vWf) were associated with affective states and mood symptoms in 112 bipolar patients, 153 schizophrenic patients and 239 healthy controls. In bipolar patients, all the inflammatory markers were lowest in the depressed state, with significant group differences after control for confounders with respect to TNF-R1, IL-1Ra, OPG and IL-6. sTNF-R1 was positively correlated with the item elevated mood, whereas sad mood was negatively correlated with OPG, IL-1RA and IL-6. Compared to controls, the neutral mood group had significantly higher levels of OPG and IL-6, and the elevated mood group had higher levels of TNF-R1 and vWf. There were no significant associations between affective states or symptoms in SCZ. These results at least indicate that different proinflammatory cytokines and receptors are associated with the different symptoms in the affective disorder.

Changes in Cell-Mediated Immune Activation

Cell-mediated immunity also plays an important role in affective disorders. Maes [14], in several reviews, provides abundant evidence to support the hypothesis of depression. These changes include increased serum levels of the soluble IL-2 receptor (sIL-2R) and the sCD8 molecule, increased numbers and percentages of T cells bearing T cell activation markers such as CD2+CD25+, CD3+CD25+ and HLA-DR+, increased stimulated production of IFNγ, and higher neopterin and sTNFR-1 or sTNFR-2 levels. Glucocorticoid resistance in immune cells also occurred.

In 38 bipolar patients and 22 matched controls, a monocyte pro-inflammatory status and Th1- and Th2-related immune function were determined. In the patients, serum sCD25 levels and percentages of anti-inflammatory CD4(+)CD25(high) FoxP3+ regulatory T cells were higher. The percentages of Th1, Th2 and Th17 cells were normal. Of the pro-inflammatory monocyte cytokines, CCL2 and PTX3 were raised in the serum. However, the monocyte pro-inflammatory state and the raised percentages of CD4(+)CD25(high) FoxP3(+) regulatory T cells occurred independently from each other [15]. These data show an enhancement of pro-inflammatory monocyte and anti-inflammatory T cell forces in bipolar disorder (BD) patients.

At the cellular and molecular levels, inflammation triggers receptors expressed on myeloid cells 1 (TREM-1). Its adaptor molecule DAP12 and its transcription factor (TF) PU.1 are important key genes in inflammation and are expressed in activated monocytes and microglia. Weigelt et al. [16] studied the gene expression of TREM-1, DAP12, PU.1, ATF3 and EGR3 in the monocytes of 73 patients with severe psychiatric disorders (27 recent-onset schizophrenia, 22 bipolar and 24 major depressed patients) and of 79 healthy controls. TREM-1 gene expression was increased in the monocytes of schizophrenia and bipolar patients and tended to be increased in the monocytes of depressed patients. PU.1 expression levels were increased in the monocytes of depressed patients, but not in those of schizophrenia and bipolar patients. TREM-1 expression levels correlated in particular to ATF3 and EGR3 expression levels, DAP12 expression levels correlated in particular to PU.1 and the data demonstrated that monocytes are in a pro-inflammatory state in severe psychiatric conditions.

In another study with 23 bipolar mania and 23 matched controls, Tsai et al. [17] showed that the lymphocyte proliferation to PHA and the plasma sIL-2R levels of bipolar patients were much higher in acute mania than in consequent remission, but no change was found in sIL-6R concentrations. These elevations were not due to differences in medication status. Only in acute mania were the plasma sIL-2R levels of patients significantly higher than in control subjects. A positive correlation between the changes of manic severity and plasma sIL-2R levels was observed. Remitted bipolar patients and normal control subjects did not differ in any of these measures.

In seasonal depressed patients, macrophage activity was significantly increased during the winter season, which was associated with increase concentrations of proinflammatory cytokines IL-1β and Th1 cytokine IFNγ, while mitogen-stimulated lymphocyte proliferation and Th2 cytokine IL-10 were decreased in the patients when compared to matched controls. After light therapy, the improvement in cellular and humoral functions was correlated with mood recovery [Song et al., in prep.]. The mechanisms of these immune disorders were related to overproduction of melatonin during seasonal light change [18].

Other Inflammation Markers

The most exciting and interesting finding in the past several years is the inflammation-indoleamine 2,3-dioxygenase(IDO)-tryptophan/serotonin pathway in depression [19–21]. Inflammatory responses, stress-induced glucocorticoid secretion and overintake precursors of inflammatory substances such as omega-6 fatty acids can induce IDO production. IDO can break down tryptophan into its catabolites (TRYCATs), such as kynurenine and quinolinic acid, which are depressogenic and anxiogenic. TRYCATs can activate oxidative pathways, cause mitochondrial dysfunctions, and have neuroexcitatory and neurotoxic effects that may also lead to neurodegeneration. The TRYCAT pathway is also activated following the induction of tryptophan 2,3-dioxygenase (TDO) by glucocorticoids, which are elevated in depression. There is evidence that activation of IDO reduces plasma tryptophan and increases TRYCAT synthesis in depressive illness and that TDO activation may play a role as well. The development of depressive symptoms during IFNα-based immunotherapy is strongly associated with IDO activation, increased production of detrimental TRYCATs and lowered levels of tryptophan. Women show greater IDO activation and TRYCAT production following immune challenge than men. In the early puerperium, IDO activation and TRYCAT production are associated with the development of affective symptoms. Clinical depression is accompanied by lowered levels of neuroprotective TRYCATs or increased levels or neurotoxic TRYCATs and lowered plasma tryptophan, which is associated with indices of immune activation and glucocorticoid hypersecretion. Lowered tryptophan and increased TRYCATs can induce T cell unresponsiveness and therefore may exert a negative feedback on the primary inflammatory response in depression. It is concluded that activation of the TRYCAT pathway by IDO and TDO may be associated with the development of depressive symptoms through tryptophan depletion and the detrimental effects of TRYCATs.

In addition, even though many studies have demonstrated that the relationship between depressive symptoms and peripheral blood TRP, KYN and KA, especially after treatment with IFNα, whether peripheral blood changes in these IDO catabolites are manifest in the brain and whether they are related to brain cytokine responses and/or behavior is unknown. Thus, Raison et al. [22] measured TRP, KYN, QUIN and KA in CSF and blood along with CSF concentrations of relevant cytokines, chemokines and soluble cytokine receptors in 27 patients with hepatitis C after approximately 12 weeks of either treatment with IFNα (n = 16) or no treatment (n = 11). Depressive symptoms were assessed using the Montgomery-Asberg Depression Rating Scale. IFNα significantly increased peripheral blood KYN, which was accompanied by a marked increase in CSF KYN. Increased CSF KYN was in turn associated with significant increases in CSF QUIN and KA. Despite significant decreases in peripheral blood TRP, IFNα had no effect on CSF TRP concentrations. Increases in CSF KYN and QUIN were correlated with increased CSF IFNα, sTNF-R2 and monocyte chemoattractant protein-1 as well as increased depressive symptoms. This study strongly supports that peripheral inflammation activated IDO with central cytokine responses, resulting in increased brain KYN and QUIN, which correlated with depressive symptoms. The study of Gabbay et al. [23] determined the possible role of the kynurenine pathway in 20 adolescents with major depression, 30 adolescents with non-depression, and 22 healthy adolescents, respectively. The results indicated that KYN:TRP ratios were increased and TRP levels were decreased in adolescents with major depression compared to non-depression adolescents and healthy controls [23].

Moreover, Kohl et al. [24] pointed out that low postpartal mood has a relationship with continuously low serum tryptophan after delivery due to an increased degradation to kynurenine. Laugeray et al. [25] studied the role of both the peripheral and cerebral kynurenine pathways in the pathophysiology of depressive disorders in a murine model created by unpredictable chronic mild stress. The kynurenine pathway has a close relationship with the levels of glucocorticoids (GC)/proinflammatory cytokines and altered neurotransmission, which are similar to those observed in depressed patients. Also, quinolinic acid and kynurenic acid can regulate the glutamate concentration in the brain. Unpredictable chronic mild stress (UCMS) also induced a depressive-like syndrome and increased TRP catabolism along the KP in the periphery, and 5-HT and KYN were strongly negatively correlated in all the brain structures of control mice and UCMS mice except for the hippocampus. They also revealed that KYN was preferentially metabolized along the QUIN pathway at the subcortical level (amygdala/striatum), while the QUIN pathway was decreased at the cortical level (cingulate cortex). Thus, they proposed that such pathway alterations could take part in the cortical/subcortical glutamatergic alterations in depressed patients [25].

Besides many studies on proinflammatory cytokines and cellular immune changes in affective disorders, there are not enough studies to explore the role of the antigen-antibody mechanism. Recently, Severance et al. [26] reported that antibodies to bovine milk caseins were involved in bipolar disorder. In 75 subjects with bipolar disorder and 65 controls they found that individuals with bipolar disorder had significantly elevated anti-casein IgG compared to controls. Casein IgG measures were correlated with the severity of mania. Lithium and/or antipsychotic medication did not significantly affect the anti-casein IgG level. In summary, the above evidence has further demonstrated that inflammation at the cellular and humoral levels correlated with the development of affective disorders.

Omega-3 Fatty Acids, Inflammation and Brain Function

As mentioned above, an elevated level of proinflammatory markers is consistently found in depressive subjects, which indicates that inflammation may play a key role in the etiology of affective disorders. Thus, increasing concern has been drawn to the development of anti-inflammatory agents in the treatment of affective disorders. Compelling evidence indicated that long-chain polyunsaturated fatty acids (PUFAs) have anti-inflammatory effects [27].

PUFAs are a group of lipids, which can be only synthesized from dietary precursors. There are two types of omega (n) fatty acids, (n)–3 α-linolenic (ALA), docosahexaenoic acid (DHA) and eicosapentaenoic acid (EPA), and n–6 linoleic fatty acids (LA), γ-linolenic acid (GLA) and arachidonic acid (AA). Free PUFAs from food are able to cross the blood-brain barrier and have actions on specific binding sites in the brain. These fatty acids are important lipid components of cell membranes. n–3 and n–6 fatty acids in the brain regulate the membrane fluidity, neurotransmitter transmission, membrane-bound enzymes and cellular signal transduction [28–30]. Thus, the incorporation of n–3 and n–6 fatty acids in neuronal membranes can influence neurotransmission, the interaction between genes and proteins, and signal transduction [31].

The difference between n–3 fatty acids and n–6 fatty acids is the location of their double bonds in the fatty acid carbon chain. Because n–6 fatty acid metabolite eicosanoids are the precursors of inflammatory mediators, overintake of n–6 fatty acids may trigger inflammation and autoimmune diseases. Also, an increasing intake of n–6 fatty acid can interfere with the metabolism and synthesis of DHA and EPA, both of which have anti-inflammatory effects. Therefore, n–3 and n–6 fatty acids cooperate and compete with each other to maintain the function of the cell membrane [31]. In the western diet, n–3 fatty acids are relatively underrepresented, while high saturated fats and n–6 fatty acids are enriched. Therefore, the Western dietary patterns have a reverse effect on human health, which is associated with inflammation and some inflammation-induced diseases.

In the past decade, many studies have demonstrated the mechanism by which n–3 fatty acids inhibit inflammation, improve brain function and treat psychiatric diseases. Epidemiological evidence has established that ingestion of n–3 PUFAs, abundant in fish oils, has profound effects on many human disorders and diseases, including affective disorders. The change in PUFA concentrations in the membranes is significantly related to neuronal, glial and immune cell functions [3]. Thus, the deficit of n–3 PUFAs has been associated with neurological and affective diseases. For example, societies that consume a small amount of omega-3 PUFAs appear to have a higher prevalence of major depressive disorder. In addition, depressive patients have showed a lower level of omega-3 PUFAs, and the antidepressant effect of PUFAs has been reported in a number of clinical trials [32].

Possible Mechanisms Revealed by Experimental Studies

The classic hypotheses of depression are noradrenaline or serotonin deficiency, hypercortisolism or dysfunction of the HPA axis. New hypotheses are inflammation, oxidants and neurotrophic deficiency. Experiments from animal models have demonstrated that n–3 fatty acids treat affective disorder from all these aspects. The effect of n–3 fatty acids on the relationship between inflammation and serotonergic function was studied in rats. Animals were fed diets with or without the n–3 fatty acid precursor ALA during perinatal development, and a subset of rats fed the ALA– diet were switched to the ALA+ diet postweaning. In adulthood, plasma IL-6, TNFα, and CRP levels were measured. Additionally, indices of liver n–6 fatty acid biosynthesis, erythrocyte fatty acid composition, and regional brain monoamine turnover were determined. Indices of liver delta-6 desaturase activity were upregulated in n–3-deficient rats, and were associated with greater erythrocyte membrane arachidonic acid (AA, 20:4 n–6) composition. Plasma IL-6, TNFα, and CRP concentrations were significantly greater in n–3-deficient rats relative to controls. The

5-HIAA:5-HT ratio was significantly increased in the frontal cortex, hypothalamus, and ventral striatum of n–3-deficient rats relative to controls. Changes in membrane n–3 and n–6 fatty acid compositions, elevations in plasma IL-6 and TNFα, and increased central 5-HT turnover were all prevented by normalization of n–3 fatty acid status. Erythrocyte DHA was inversely correlated, and AA and the AA/DHA and AA/EPA were positively correlated, with plasma IL-6, TNFα and CRP levels. Plasma IL-6 levels were positively correlated with 5-HIAA:5-HT ratios in all brain regions. These data provide evidence for a functional link between n–3 fatty acid deficiency, elevated peripheral inflammatory signaling, and increased central 5-HT turnover [33]. Furthermore, a recent study was performed to evaluate the involvement of 5-HT receptors, particularly of 5-HT(1A), and hippocampal BDNF expressions in a force-swim model of rats with and without feeding of fish oil. Fish oil treatment showed a robust antidepressant effect. Increased BDNF expression in the cortex and hippocampus and reduced animal immobility during force swimming. Also, in 90-day-old rats, 5-HT content in the hippocampus and cortex was increased, whereas 5-HIAA formation was diminished in the fish oil group. Fish oil-induced changes were reversed by the 5-HT(1A) receptor antagonist WAY 100135, which indicates that the 5-HT1A receptor is involved in n–3 fatty acid central effects [34]. Park et al. [35] also demonstrated that when the 0.5% and 1% DHA+EPA groups were compared with that in the 0% n–3 group, DHA+EPA combination significantly reversed force-swim-induced immobility, increased serum 5-HT concentrations and hippocampal c-AMP response element binding protein (CREB) expression. Conversely, IL-6 expression was significantly reduced.

With regard to the neurotrophic deficiency in depression, that prolonged dietary manipulation induces biochemical changes in the brain was studied in mice. After feeding a safflower oil (SAF) diet (n–3-restricted, n–6-adequate) or a perilla oil (PER) diet (containing adequate amounts of n–3 and n–6 fatty acids) for 8 weeks from weaning, the DHA contents and p38 MAPK activities in the cerebral cortex, striatum and hippocampus were significantly lower in the SAF group [36]. The neurotrophin BDNF contents and protein kinase C (PKC) activities in the cerebral cortex and in the striatum, but not in the hippocampus, were significantly lower in the SAF group. These data indicate that the biochemical changes induced by the dietary restriction of ALA have a time lag in the striatum and cortex, suggesting that the signal is transmitted through decreased p38 MAPK activity and BDNF content and ultimately decreased PKC activity [36].

As an n–3 fatty acid, EPA has similar functions as DHA in the brain, while it is of also important in regulating immune and inflammation functions [37]. EPA can reduce the AA level on cell membrane and prostaglandin E_2 (PGE_2) synthesis. Interestingly, animals fed a high AA diet or treated with PGE_2 presented sickness behaviors of anorexia, low activity, and changes in sleep pattern and attention which are similar to the somatic symptoms of depression in humans. EPA concentrations are very low in the brain, but it seems more effective in the treatment of psychiatric disease. It has been considered to be highly utilized or acts as a message in the cells and brain [38]. EPA can also modulate the neurochemical pathways, the activity of the HPA axis and the promotion of neurogenesis, which influence behavior and cognitive functions such as regulating mood, stress responses and memory [39–41]. In a rat model of inflammation induced by central subchronic administration of IL-1β, a 7-week EPA treatment significantly attenuated the IL-1-induced activation of the HPA axis, limbic noradrenaline reduction, increased PGE_2 and proinflammatory cytokines, and reduced animal anxiety and stress-like behavior [42–46]. As well EPA effects on depression-like changes were studied in a model of depression; olfactory bulbectomized (OB) rats who exhibited a stress-like response in the open-field, impaired memory in the Morris water maze, elevated CRF expression and secretion of corticosterone. In the OB rat, mRNA expression of the nerve growth factor (NGF) was significantly lower in the hippocampus, while phospholipase A_2, an enzyme breaking n–6 fatty acids to the inflammatory mediator, was higher in the hypothalamus; this change was associated with increased IL-1β and PGE_2 in the serum and brain. EPA treatments normalized these behavioral impairments and reduced CRF expression and corticosterone secretion. EPA also reduced the serum concentrations of IL-1β and PGE_2, but reversed NGF reduction. Similar to the effects of EPA, the anti-inflammatory drug celecoxib significantly reduced blood PGE_2, IL-1β, and corticosterone concentrations and increased NGF expression in OB rats. Furthermore, anti-NGF treatment blocked the EPA effects on behavior. These results suggest that an interaction exists between inflammation and NGF in the depression model. EPA may improve depression via its anti-inflammation properties and the upregulation of NGF [47].

At the cellular and molecular levels, Song's team demonstrated that EPA protected neurons against neurotoxicity via anti-inflammation, upregulated anti-oxidant and anti-apoptosis gene expressions, and upregulated neurotrophins and TrKB receptors [48–51]. These experimental data strongly support that n–3 fatty acids could treat affective diseases in many pathological aspects. Furthermore, Lu et al. [52] identify the molecular mechanisms through which DHA modulates oxidative reactions and inflammatory cytokine production in microglial and neuronal cells. The study showed that DHA reduced expressions of TNFα, IL-6, nitric oxide synthase, and cyclo-oxygenase-2, induced by IFNγ, and induced upregulation of heme oxygenase-1 (HO-1) in BV-2 microglia. The inhibitory effect of DHA on nitric oxide production was abolished by the HO-1 inhibitor zinc protoporphyrin IX. In addition, DHA caused AKT and ERK activation in a time-dependent manner, and the DHA-induced HO-1 upregulation could be attenuated by PI-3 kinase/AKT and MEK/ERK inhibitors. DHA also increased IKKα/β phosphorylation, IκBα phosphorylation, and IκBα degradation, whereas both nuclear factor-κB and IκB protease inhibitors could inhibit DHA-induced HO-1 expression. The other major n–3 PUFA, EPA, showed similar effects as DHA on inflammation and HO-1 in repeated key experiments. This study strongly supports the inflammation hypothesis of depression and clinical studies supporting the antidepressant effects of n–3 PUFAs.

If deficient n–3 fatty acids are involved in inflammation and depression, the improvement of depressive symptoms should be associated with the normalization of n–3 fatty acids after antidepressant treatments. An experiment in rats injected with a low dose of LPS for 6 days showed that n–6 AA, PGE_2 and AA metabolites were largely increased in rat brains. Six weeks after lithium treatment, these increments were prevented and 17-hydroxy-DHA formation was increased. This study demonstrated that lithium could reduce LPS-induced sickness behavior, which was associated with the increase in brain n–3 fatty acids, indicating a new and potentially important therapeutic action of lithium [53]. A summary of the mechanisms by which PUFAs improve affective diseases is shown in figure 1.

Omega-3 Fatty Acids as Effective Treatment for Affective Disorders

The dietary intake of n–3 fatty acids in fish oil, such as EPA and DHA, has been associated with a range of mental health conditions [54]. Increasing evidence has indicated that n–3 fatty acids could decrease some symptoms of affective disorders, including major depressive disorder (MDD), BD and postnatal depression. Moreover, some biochemical studies showed that the decreasing levels of n–3 fatty acids were found in the erythrocyte membranes of depressed patients [33,

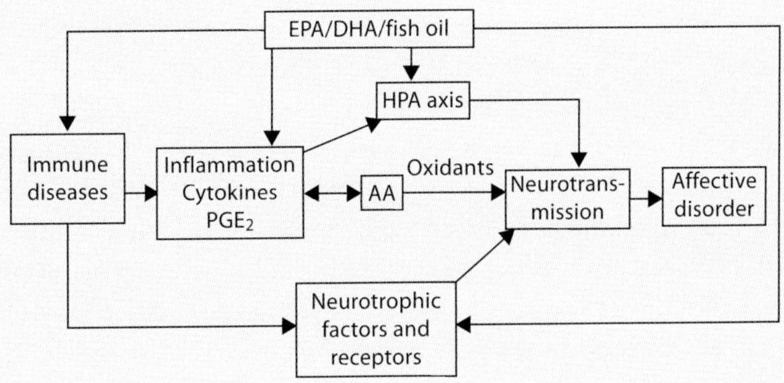

Fig. 1. The possible pathway of n–3 fatty acids in the treatment of affective disorders.

55–57]. Most epidemiologic and tissue compositional studies provide evidence to support that n–3 fatty acid supplementation has a protective effect on mood disorders, especially EPA and DHA [58]. The potential therapeutic role of n–3 dietary supplementation has also been investigated extensively in clinical studies of major depressive disorder, bipolar disorder and postpartum depression. However, few clinical trials have explored the correlative changes between inflammation and depressive symptoms during n–3 fatty acid treatments.

Major Depressive Disorder

It has been reported that populations with a high intake of fish that is a good resource for n–3 fatty acids have a lower incidences of MDD [59]. In the study of Frasure-Smith et al. [60], depressed patients were found to have considerable lower concentrations of total n–3 fatty acids and DHA, and higher ratios of AA to DHA, AA to EPA, and n–3 to n–6 than controls. A study by Colangelo et al. [61] reported that a lower intake of n–3 fatty acids has been associated with chronic depressive symptoms in females. Chiu et al. [55] revealed that total erythrocyte n–3 fatty acid concentrations were associated with a previous depression in old people. Thus, lower concentrations of n–3 fatty acids or ALA in erythrocyte membranes could be a good predictor for previous recurrent depression in old people.

The intake of DHA and EPA can reduce depressive symptoms and the risk of progressing from mild cognitive impairment (MCI) to dementia [62]. Thus, the antidepressant effect of n–3 fatty acids has been attracting interest. The short-term efficacy of n–3 fatty acids in decreasing the symptoms of MDD has been investigated to date. A double-blind, randomized, controlled, 8-week, parallel-group trial was carried out in 8 Canadian academic and psychiatric clinics. 423 adult outpatients with MCI (Mini-International Neuropsychiatric Interview, version 5.0.0, criteria) lasting at least 4 weeks, including 40.3% taking antidepressants at baseline, were randomly assigned to 8 weeks of 1,050 mg/day of EPA and 150 mg/day of DHA or matched sunflower oil placebo (2% fish oil). The primary outcome was the self-report Inventory of Depressive Symptomatology (IDS-SR 30), and the secondary outcome was the clinician-rated Montgomery-Åsberg Depression Rating Scale (MADRS). For patients without comorbid anxiety disorders (CAD) (n = 204), n–3 supplementation was superior to placebo in improving depressive symptoms, while the improvement in CAD was near significance [63]. Rondanelli et al. [64] conducted a 2-month, randomized, double-blind, placebo-controlled trial in a group of 46 depressed women aged 66–95 years. With supplementation of 2.5 g/day of n–3 fatty acids including 1.67 g of EPA and 0.83 g of

DHA, the depressive symptoms of patients are significantly alleviated.

Several publications in the past indicated that EPA seems more effective in the treatment of MDD compared to DHA [65, 66]. Sublette et al. [65] reviewed the 15 trials including 915 subjects, and made the conclusion that supplements of PUFAs containing more than 60% EPA in total EPA + DHA at a dose range of 200–2,200 mg/day effectively decreased the symptoms of depression. Furthermore, depressive symptoms were found to decrease significantly in 21 studies when the subjects were supplied with a mixture of EPA and DHA including more than 50% EPA in 13 studies and pure ethyl-EPA in 8 studies [66].

However, in elderly patients with depression, DHA was not only the same as EPA in reducing depressive symptoms but also more effective in improving cognitive function and life quality when a 6-month, double-blind, randomised controlled trial was conducted. A total of 50 people aged >65 years with MDD were allocated to receive a supplement rich in EPA (1.67 g EPA + 0.16 g DHA/day; n = 17), DHA (1.55 g DHA + 0.40 g EPA/day; n = 18) or the n–6 PUFA linoleic acid (LA; 2.2 g/day; n = 15) [67]. Treatment allocation was by minimization based on age, sex and depressive symptoms (Geriatric Depression Scale; GDS). Compared with the LA group, GDS scores improved in the EPA and DHA groups and verbal fluency (Initial Letter Fluency) in the DHA group, but not in EPA group. Improved GDS scores were correlated with increased DHA plus EPA. Improved self-reported physical health was associated with increased DHA.

Bipolar Disorder

Reduced concentrations of DHA and AA have been reported in postmortem brains of patients with BD. Additionally, an increased prevalence of BD has been related to low dietary intake of fish, and dietary supplements containing fish products or DHA have been reported to ameliorate BD symptoms. These observations suggest that brain lipid metabolism, particularly involving DHA, is disturbed in BD. The fatty acid composition in postmortem orbitofrontal cortex (OFC) of 35 patients with BD and 35 age-matched normal controls have been studied by gas chromatography. After correction for multiple comparisons, DHA, AA and stearic acid compositions were significantly lower and cis-vaccenic acid composition significantly higher in the OFC of BD patients relative to normal controls. Based on metabolite:precursor ratios, significant elevations in AA, stearic acid and palmitic acid conversion/metabolism were observed in the OFC of BD, and were inversely correlated with DHA compositions. Deficits in OFC DHA and AA composition, and elevations in AA metabolism were numerically (but not significantly) greater in drug-free BD patients relative to patients treated with mood-stabilizers or antipsychotic medications. These results add to a growing body of evidence implicating n–3 fatty acid deficiency as well as the OFC in the patho-etiology of BD [68].

However, Balanzá-Martínez et al. [69] reported that in the postmortem OFC from 8 BD patients and 6 matched controls, there was no statistically significant difference in mean concentrations of DHA or AA within stable lipids. Fractional esterified AA and DHA concentrations also did not differ significantly between groups. Some fatty acid concentrations were found in low-abundant cholesterylesters. These results do not support the hypothesis of disturbed brain lipid concentrations, including concentrations of AA and DHA, in BD. On the other hand, n–3 fatty acids have also been used to treat BD. Osher et al. [70] conducted a clinical study in 12 BD patients diagnosed by DSM-IV with supplement of 1.5–2 g/day of the n–3 fatty acid EPA for up to 6 months. The results indicated that the intake of EPA can significantly reduce symptoms of BD and no serious side effects were found in this study. Moreover, some studies also try to use n–3 fatty acids as adjunctive treatment for BD. Keck et al. [71] conducted a 4-month, randomized, placebo-controlled adjunctive trial. EPA 6 g/day was used in the treatment of BD and rapid cycling BD. However, there is no overall evidence of efficacy for adjunctive treatment at this dose in patients with either condition. The dose of EPA used could be too higher in this trial.

More interesting, a clinical trial explored serum n–3 and n–6 PUFA levels in BD and determined possible associations with suicidal risk, including suicidal history and relevant personality factors that have been associated with suicidality. 27 bipolar subjects diagnosed by the NEO-PI and the five personality factors structured interviews were used to verify diagnosis and assess suicidal history [72]. The ratio of ALA:EPA was negatively correlated, while the ratio of EPA:DHA was positively associated with agreeableness and extraversion. Finally, serum concentrations of AA were significantly lower in subjects with a history of suicide attempt compared to non-attempters. The data suggest that specific lipid profiles, which are controlled by an interaction between diet and genetics, were correlated with suicidal history and personality factors related to suicidal risk.

In addition, studies on the effects of n–3 fatty acids on juvenile BD was carried out in 18 children and adolescents who received supplements containing 360 mg/day EPA and 1,560 mg/day DHA for 6 weeks in an open-label study. Mania, depression and global function were assessed before and after supplementation. Red blood cell EPA and DHA concentrations were significantly higher following supplementation. Clinician ratings of mania and depression were significantly lower and global functioning significantly higher after supplementation. Parent ratings of internalizing and externalizing behaviors were also significantly lower following the treatment [73]. Whereas, similar efficacy of EPA treating MDD was reported in BD patients, while DHA seemed ineffective as well. Using the PubMed database, a meta-analysis of double-blind, placebo and randomized controlled trials studied the effects of treatments containing DHA, EPA, or both in BD patients. The analysis from 28 studies met the above inclusion criteria, revealed a significant effect of higher levels of baseline depression and lower supplement DHA:EPA ratio on therapeutic efficacy. However, symptoms of depression were not significantly reduced in 3 studies using pure DHA or in 4 studies using supplements containing greater than 50% DHA. In contrast, symptoms of depression were significantly reduced in 13 studies using supplements containing greater than 50% EPA and in 8 studies using pure ethyl-EPA [72].

Postpartum Depression

MDD during pregnancy and postpartum are associated with significant maternal and neonatal morbidity. Although antidepressants are readily used in pregnancy, studies have raised concern regarding neurobehavioral outcomes in exposed infants. n–3 fatty acid supplementation, most frequently from fish oil, has emerged as a possible treatment or prevention strategy for MDD in nonpregnant individuals, and may have beneficial effects in pregnant women. Published observational studies in the psychiatric literature suggest that maternal DHA deficiency may lead to the development of MDD in pregnancy and postpartum, while intervention trials also suggest a clinical benefit for supplementation with EPA in MDD. A systematic review was conducted to assess the possible association between PUFA supplementation and intake in the perinatal period and the risk of maternal perinatal depression. In 10 articles, three longitudinal cohort studies, five randomized controlled trials and two pilot trials that met the selection criteria were reviewed. Six found no association, two found mixed results, and two found a positive association between n–3 PUFAs and reduced incidence of maternal perinatal depression. Some of the larger studies and those that found a positive effect were more likely to be using higher doses, close to 2 g of DHA+EPA, and began the supplementation earlier in pregnancy.

n–3 Fatty Acids, Inflammation and Affective Disorder

As mentioned above, inflammation plays an important role in depression and n–3 fatty acids may improve depression via anti-inflammatory effects and HPA regulation. To demonstrate the mechanism by which n–3 fatty acids benefit to affective disorders through anti-inflammatory effects, Dinan et al. [74] studied 20 DSM-IV major depressives who had failed to antidepressant SSRI treatment for 6 weeks, 14 subjects who responded to the SSRI and 24 healthy controls. Blood AA, a precursor of proinflammatory mediators, levels were elevated in both the responders and the nonresponders. IL-6 was elevated in a similar manner. The EPA:AA ratio in the 3 groups was controls 0.08:0.01, responders 0.08:0.01, nonresponders 0.04:0.01. These differences are significant. AA and IL-6 were highly correlated in both responders and nonresponders but not in healthy volunteers. The findings support that major depression is associated with a pro-inflammatory phenotype and increased n–6 fatty acid AA. However, Kiecolt-Glaser et al. [76] measured plasma cortisol and serum IL-1β and IL-6 in patients with MDD with 1,000 mg EPA alone or in combination with 20 mg fluoxetine on HPA axis activity and inflammatory cytokine production, and then compared the changes in these variables with those of treating with 20 mg fluoxetine alone. 42 patients were included in the analysis. The results showed that plasma cortisol decreased significantly after 8 weeks of intervention without a significant difference among the groups. There was no interaction between group and response to treatment over time in the cortisol response. Serum concentrations of IL-1β and IL-6 did not change significantly after intervention. In conclusion, EPA alone or in combination with fluoxetine, as well as fluoxetine alone decreased serum cortisol after 8 weeks of treatment in patients with MDD without any significant effect of response to treatment. Serum IL-1β and IL-6 did not change significantly after intervention. These findings suggest that EPA may exert its therapeutic effects through the reduction of cortisol.

In 43 older adults (mean age = 66.67 years) PUFAs and TNFα, IL-6, and sIL-6r were measured. Depressive symptoms were assessed by the Center for Epidemiological Studies Depression Scale. Higher n–6:n–3 ratios were associated with progressively elevated TNFα and IL-6 levels as depressive symptoms increased. Higher levels of sIL-6r were associated with higher n–6:n–3 ratios. Six individuals who met the criteria for major depressive disorder had higher n–6:n–3 ratios and TNFα, IL-6, and sIL-6r levels than those who did not meet the criteria. However, whether n–3 fatty acid treatment could reverse the proinflammatory cytokine levels and depressive symptoms is unknown [75].

To determine whether n–3 fatty acids decrease proinflammatory cytokine production and depressive and anxiety symptoms in healthy young adults, a placebo-controlled, double-blind 12-week trial compared n–3 supplementation with placebo. Serial blood samples from 68 medical students were taken during lower-stress periods and on days before an exam. The blood was incubated with and without LPS stimulation. The students received either n–3 (2,085 mg EPA and 348 mg DHA/day) or placebo capsules. Compared to controls, those students who received n–3 showed a 14% decrease in LPS stimulated IL-6 production and a 20% reduction in anxiety symptoms, without significant change in depressive symptoms. The plasma n–6:n–3 ratio in the treatment group showed that decreasing n–6:n–3 ratios led to lower anxiety and reductions in stimulated IL-6 and TNFα production, as well as marginal differences in serum TNFα. These data suggest that n–3 supplementation can reduce inflammation and anxiety, as well as the reduction in anxiety symptoms is associated with n–3 supplementation [77].

Conclusion, Limitation and Future Research

In summary, the evidence reviewed by this chapter suggests that (1) inflammation is causally related to depression, (2) n–3 inhibits, while n–6 enhances inflammation, and (3) n–3 fatty acids, especially EPA, is an effective treatment for depression. However, the clinical evidence is not sufficient to answer whether the benefits of n–3 to depression are from their anti-inflammatory effects. Future studies could focus on (1) whether n–3 fatty acid-decreased inflammatory change is associated with the improvement of depressive symptoms, or (2) whether effective antidepressants normalize PUFA concentrations and metabolism in the brain, especially the limbic system.

The heterogeneity of above results can be explained by dissimilar study designs, including differences in study duration, time period of measurements and number of participants, and in the varied dosages and types of supplemental PUFAs. These should be further studied.

Acknowledgement

The author thanks Ms Qiong Jiang for her assistant in searching and adding references.

References

1 Song C, Wang H: Cyokines mediated inflammation and decreased neurogenesis in animal models of depression. Prog Neuropsychopharmacol Biol Psychiatry 2011;9;35:760–768.
2 Song C: The effect of thymectomy and IL-1 on memory: implications for the relationship between immunity and depression. Brain Behav Immun 2002;16:557–568.
3 Song C, Zhao S: Omega-3 fatty acid eicosapentaenoic acid. A new treatment for psychiatric and neurodegenerative diseases: a review of clinical investigations. Expert Opin Investig Drugs 2007;16:1627–1638.
4 Madeline L, Joanna KS, Sidney HK: Inflammatory biomarkers in depression: an opportunityfor novel therapeutic interventions. Curr Psychiatry Rep 2011;13:316–320.
5 Carpenter LL, Gawuga CE, Tyrka AR, et al: Association between plasma IL-6 response to acute stress and early-life adversity in healthy adults. Neuropsychopharmacology 2010;35:2617–2623.
6 Janelidze S, Mattei D, Westrin D, et al: Cytokine levels in the blood may distinguish suicide attempters from depressed patients. Brain Behav Immun 2011;25: 335–339.
7 Lehto SM, Niskanen L, Miettola J, et al: Serum anti-inflammatory markers in general population subjects with elevated depressive symptoms. Neurosci Lett 2010;484:201–205.
8 Pavlović Z, Delić D, Marić NP, et al: Depressive symptoms in patients with hepatitis C treated with pegylated interferon alpha therapy: a 24-week prospective study. Psychiatr Danub 2011;23: 370–377.
9 Friebe A, Horn M, Schmidt F, et al: Dose-dependent development of depressive symptoms during adjuvant interferon-{alpha} treatment of patients with malignant melanoma. Psychosomatics 2010;51:466–473.
10 Raison CL, Borisov AS, Majer M, et al: Activation of central nervous system inflammatory pathways by interferon-alpha: relationship to monoamines and depression. Biol Psychiatry 2009;65: 296–303.
11 Kenis G, Prickaerts J, van Os J, et al: Depressive symptoms following interferon-α therapy: mediated by immune-induced reductions in brain-derived neurotrophic factor? Int J Neuropsychopharmacol 2011;14:247–253.
12 Suvisaari J, Loo BM, Saarni SE, et al: Inflammation in psychotic disorders: a population-based study. Psychiatry Res 2011;189:305–311.
13 Hope S, Dieset I, Agartz I, et al: Affective symptoms are associated with markers of inflammation and immune activation in bipolar disorders but not in schizophrenia. J Psychiat Res 2011;45:1608–1616.
14 Maes M: Depression is an inflammatory disease, but cell-mediated immune activation is the key component of depression. Prog Neuro-Psychopharmacol Biol Psychiatr 2011;35:664–675.
15 Drexhage RC, Hoogenboezem TH, Versnel MA, et al: The activation of monocyte and T cell networks in patients with bipolar disorder. Brain Behav Immun 2011;25:1206–1213.
16 Weigelt K, Carvalho LA, Drexhage RC, et al: TREM-1 and DAP12 expression in monocytes of patients with severe psychiatric disorders. EGR3, ATF3 and PU.1 as important transcription factors. Brain Behav Immun 2011;25:1162–1169.
17 Tsai SY, Chen KP, Yang YY, et al: Activation of indices of cell-mediated immunity in bipolar mania. Biol Psychiatry 1999;45:989–994.
18 Lam RW, Song C, Yatham LN: Does neuroimmune dysfunction mediate seasonal mood changes in winter depression? Med Hypotheses 2004;63:567–573.
19 Song C, Lin A, Bonaccorso S, et al: The inflammatory response system and the tryptophan availability to the brain of patients with major depression and sleep disorders. J Affect Disord 1998; 49:211–219.

20 Leonard BE, Myint A: The psychoneuroimmunology of depression. Hum Psychopharmacol 2009;24:165–175.
21 Maes M, Leonard BE, Myint AM, et al: The new '5-HT' hypothesis of depression: cell-mediated immune activation induces indoleamine 2,3-dioxygenase, which leads to lower plasma tryptophan and an increased synthesis of detrimental tryptophan catabolites (TRYCATs), both of which contribute to the onset of depression. Prog Neuropsychopharmacol Biol Psychiatry 2011;35:702–721.
22 Raison CL, Dantzer R, Kelley KW, et al: CSF concentrations of brain tryptophan and kynurenines during immune stimulation with IFN-alpha: relationship to CNS immune responses and depression. Mol Psychiatry 2010;15:393–403.
23 Gabbay V, Klein RG, Katz Y, et al: The possible role of the kynurenine pathway in adolescent depression with melancholic features. J Child Psychol Psychiatry 2010;51:935–943.
24 Kohl C, Walch T, Huber R, et al: Measurement of tryptophan, kynurenine and neopterin in women with and without postpartum blues. J Affect Disord 2005;86:135–142.
25 Laugeray A, Launay JM, Callebert J, et al: Peripheral and cerebral metabolic abnormalities of the tryptophan-kynurenine pathway in a murine model of major depression. Behav Brain Res 2010;210:84–91.
26 Severance EG, Dupont D, Dickerson FB, et al: Immune activation by casein dietary antigens in bipolar disorder. Bipol Dis 2010;12:834–842.
27 Ibrahim A, Mbodji K, Hassan A, et al: Anti-inflammatory and anti-angiogenic effect of long chain n-3 polyunsaturated fatty acids in intestinal microvascular endothelium. Clin Nutr 2011;30:678–687.
28 Ramakrishnan U: Fatty acid status and maternal mental health. Child Nutr 2011;7(suppl 2):99–111.
29 Gupta S, Knight AG, Gupta S, et al: Saturated long-chain fatty acids activate inflammatory signaling in astrocytes. J Neurochem 2012;120:1060–1071.
30 Weylandt KH, Chiu CY, Gomolka B, et al: Omega-3 fatty acids and their lipid mediators: Towards an understanding of resolvin and protectin formation omega-3 fatty acids and their resolvin/protectin mediators. Prostaglandins Other Lipid Mediat 2012;97:73–82.
31 Pascoe MC, Crewther SG, Carey LM, Crewther DP: What you eat is what you are. A role for polyunsaturated fatty acids in neuroinflammation induced depression? Clin Nutr 2011;30:407–415.
32 Freeman MP: Omega-3 fatty acids in major depressive disorder. J Clin Psychiatry 2009;70(suppl 5):7–11.
33 McNamara RK, Jandacek R, Rider T, et al: Selective deficits in erythrocyte docosahexaenoic acid composition in adult patients with bipolar disorder and major depressive disorder. J Affect Disord 2010;126:303–311.
34 Vines A, Delattre AM, Lima MM, et al: The role of 5-HT_1A receptors in fish oil-mediated increased BDNF expression in the rat hippocampus and cortex: a possible antidepressant mechanism. Neuropharmacology 2012;62:184–191.
35 Park Y, Moon HJ, Kim SH: n-3 polyunsaturated fatty acid consumption produces neurobiological effects associated with prevention of depression in rats after the forced swimming test. J Nutr Biochem 2012;23:924–928.
36 Miyazawa D, Yasui Y, Yamada K, et al: Biochemical responses to dietary α-linolenic acid restriction proceed differently among brain regions in mice. Biomed Res 2011;4:237–245.
37 Su KP: Biological mechanism of antidepressant effect of omega–3 fatty acids. How does fish oil act as a 'mind-body interface'? Neurosignals 2009;17: 144–152.
38 Chen CT, Liu Z, Ouellet M, Calon F, Bazinet RP: Rapid beta-oxidation of eicosapentaenoic acid in mouse brain: an in situ study. Protaglandins Leukot Essent Fatty Acids 2009;80:157–163.
39 Song C, Li XW, Leonard BE, Horrobin DF: Effects of dietary n-3 or n-6 fatty acids on interleukin-1beta-induced anxiety, stress and inflammatory responses in rats. J Lipid Res 2003;44:1984–1991.
40 Song C, Leonard BE, Horrobin DF: Ethyl-eicosapentaenoic acid but not soybean oil reverses central interleukin-1-induced anxiety, stress and inflammatory response in rats. Stress 2004;7:43–54.
41 Song C, Horrobin DF: Omega-3 fatty acid ethyl-eicosapentaenoic acid but not soybean oil attenuates memory impairment induced by central IL-1beta administration. J Lipid Res 2004;45:1112–1121.
42 Song C, Phillips AG, Leonard BE: Interleukin 1 beta enhances conditioned fear memory but impairs spatial learning in rats: possible involvement of glucocorticoids and monoamine function in the amygdala and hippocampus. Eur J Neurosci 2003;18:1739–1743.
43 Song C, Li XW, Leonard BE, Horrobin DF: Effects of dietary n–3 or n–6 fatty acids on interleukin-1beta-induced anxiety, stress and inflammatory responses in rats. J Lipid Res 2003;44:1984–1991.
44 Song C, Phillips AG, Leonard BE, Horrobin DF: Omega-3 fatty acid and RU 486 reverse IL-1-induced memory impairment in rats: a glucocorticoid involved mechanism. Mol Psychiatry 2004;9:630–638.
45 Song C, Leonard BE: Effects of acute and sub-chronic administration of Interleukin-1β on the behaviour, immune and neurotransmitter function in the rat. Pharmacopsychiatry 2006;39:88–99.
46 Song C, Li XW, Kang ZJ, Kadotomi Y: Omega-3 fatty acid ethyl-eicosapentaenoate attenuates IL-1-induced changes in dopamine and metabolites in the shell of the nucleus accumbens: involved with PLA2 activity and corticosterone secretion. Neuropsychophamacology 2007;32:736–744.
47 Song C, Zhang XY, Manku M: Increased phospholipase A2 activity and inflammatory response but decreased nerve growth factor expression in a depression model olfactory bulbectomized rats: effects of chronic ethyl-eicosapentaenoate treatment. J Neurosci 2009;29:14–22.
48 Kou W, Luchtman D, Song C: Eicosapentaenoic acid (EPA) increases cell viability and expression of neurotrophin receptors in retinoic acid and brain-derived neurotrophic factor differentiated SH-SY5Y cells. Eur J Nutr 2008;47:104–113.
49 Meng Q, Luchtman DW, El Bahh B, Zidichouski JA, Yang J, Song C: Ethyl-eicosapentaenoate modulates changes in neurochemistry and brain lipids induced by parkinsonian neurotoxin 1-methyl-4-phenylpyridinium in mouse brain slices. Eur J Pharmacol 2010;15; 649:127–134.
50 Taepavarapruk P, Song C: Reductions of acetylcholine release and nerve growth factor expression are correlated with memory impairment induced by interleukin-1β administrations: effects of omega-3 fatty acid EPA treatment. J Neurochem 2010;112:1054–1064.

51 Luchtman DW, Meng Q, Song C: Ethyl-eicosapentaenoate (E-EPA) attenuates motor impairments and inflammation in the MPTP-probenecid mouse model of Parkinson's disease. Behav Brain Res 2012;226:386–396.
52 Lu DY, Tsao YY, Leung YM, Su KP: Docosahexaenoic acid suppresses neuroinflammatory responses and induces heme oxygenase-1 expression in BV-2 microglia: implications of antidepressant effects for ω–3 fatty acids. Neuropsychopharmacology 2010;35:2238–2248.
53 Basselin M, Kim HW, Chen M, Ma K, et al: Lithium modifies brain arachidonic and docosahexaenoic metabolism in rat lipopolysaccharide model of neuroinflammation. J Lipid Res 2010; 51:1049–1056.
54 Peet M, Stokes C: Omega-3 fatty acids in the treatment of psychiatric disorders. Drugs 2005;65:1051–1059.
55 Chiu CC, Frangou S, Chang CJ, et al: Associations between n-3 PUFA concentrations and cognitive function after recovery from late-life depression. Am J Clin Nutr 2012;95:420–427.
56 Frangou S, Lewis M, McCrone P: Efficacy of ethyl-eicosapentaenoic acid in bipolar depression: randomised double-blind placebo-controlled study. Br J Psychiatry 2006;188:46–50.
57 Peet M, Murphy B, Shay J, et al: Depletion of omega-3 fatty acid levels in red blood cell membranes of depressive patients. Biol Psychiatry 1998;43:315–319.
58 Freeman MP, Hibbeln JR, Wisner KL, et al: Omega-3 fatty acids: evidence basis for treatment and future research in psychiatry. J Clin Psychiatry 2006;67:1954–1967.
59 Hibbeln JR, Nieminen LR, Blasbalg TL, et al: Healthy intakes of n-3 and n-6 fatty acids: estimations considering worldwide diversity. Am J Clin Nutr 2006;83: 1483S–1493S.
60 Frasure-Smith N, Lespérance F, Julien P: Major depression is associated with lower omega-3 fatty acid levels in patients with recent acute coronary syndromes. Biol Psychiatry 2004;55:891–896.
61 Colangelo LA, He K, Whooley MA, et al: Higher dietary intake of long-chain omega-3 polyunsaturated fatty acids is inversely associated with depressive symptoms in women. Nutrition 2009;25: 1011–1109.
62 Sinn N, Milte CM, Street SJ, et al: Effects of n–3 fatty acids, EPA v. DHA, on depressive symptoms, quality of life, memory and executive function in older adults with mild cognitive impairment: a 6-month randomised controlled trial. Br J Nutr 2011;20:1–12.
63 Lespérance F, Frasure-Smith N, St-André E, et al: The efficacy of omega-3 supplementation for major depression: a randomized controlled trial. J Clin Psychiatry 2011;72:1054–1062.
64 Rondanelli M, Giacosa A, Opizzi A, et al: Effect of omega-3 fatty acids supplementation on depressive symptoms and on health-related quality of life in the treatment of elderly women with depression: a double-blind, placebo-controlled, randomized clinical trial. J Am Coll Nutr 2010;29:55–64.
65 Sublette ME, Ellis SP, Geant AL, Mann JJ: Meta-analysis of the effects of eicosapentaenoic acid (EPA) in clinical trials in depression. J Clin Psychiatry 2011;72: 1577–1584.
66 Martins JG: EPA but not DHA appears to be responsible for the efficacy of omega-3 long chain polyunsaturated fatty acid supplementation in depression: evidence from a meta-analysis of randomized controlled trials. J Am Coll Nutr 2009; 28:525–542.
67 Rondanelli M, Giacosa A, Opizzi A, Pelucchi C, La Vecchia C, Montorfano G, Negroni M, Berra B, Politi P, Rizzo AM: Long chain omega 3 polyunsaturated fatty acids supplementation in the treatment of elderly depression: effects on depressive symptoms, on phospholipids fatty acids profile and on health-related quality of life. J Nutr Health Aging 2011;15:37–44.
68 Hamazaki K, Choi KH, Kim HY: Phospholipid profile in the postmortem hippocampus of patients with schizophrenia and bipolar disorder: no changes in docosahexaenoic acid species. J Psychiatr Res 2010;44:688–693.
69 Balanzá- Martínez V, Fries GR, Colpo GD, et al: Therapeutic use of omega-3 fatty acids in bipolar disorder. Expert Rev Neurother 2011;11:1029–1047.
70 Osher Y, Bersudsky Y, Belmaker RH: Omega-3 eicosapentaenoic acid in bipolar depression: report of a small open-label study. J Clin Psychiatry 2005;66: 726–729.
71 Keck PE Jr, Mintz J, McElroy SL, Freeman MP, et al: Double-blind, randomized, placebo-controlled trials of ethyl-eicosapentanoate in the treatment of bipolar depression and rapid cycling bipolar disorder. Biol Psychiatry 2006; 60:1020–1022.
72 Evans SJ, Prossin AR, Harrington GJ, Kamali M, Ellingrod VL, Burant CF, McInnis MG: Fats and factors: lipid profiles associate with personality factors and suicidal history in bipolar subjects. PLoS One 2012;7:e29297.
73 Clayton EH, Hanstock TL, Hirneth SJ, Kable CJ, et al: Reduced mania and depression in juvenile bipolar disorder associated with long-chain omega-3 polyunsaturated fatty acid supplementation. Eur J Clin Nutr 2009;63:1037–1040.
74 Dinan T, Siggins L, Scully P, O'Brien S, et al: Investigating the inflammatory phenotype of major depression: focus on cytokines and polyunsaturated fatty acids. J Psychiatr Res 2009;43:471–476.
75 Jazayeri S, Keshavarz SA, Tehrani-Doost M, Djalali M, et al: Effects of eicosapentaenoic acid and fluoxetine on plasma cortisol, serum interleukin-1beta and interleukin-6 concentrations in patients with major depressive disorder. Psychiatry Res 2010;178:112–115.
76 Kiecolt-Glaser JK, Belury MA, Porter K, Beversdorf DQ, et al: Depressive symptoms, omega-6:omega-3 fatty acids, and inflammation in older adults. Psychosom Med 2007;69:217–224.
77 Kiecolt-Glaser JK, Belury MA, Andridge R, et al: Omega-3 supplementation lowers inflammation and anxiety in medical students: a randomized controlled trial. Brain Behav Immun 2011;25:1725–1734.

Cai Song, MD, PhD
Chinese Academy Engineering Institute for the Development of Endangered Medicinal Resources in Southwest China
Guangxi, 189 Changgang Road, Xing Ning District
Nanning, Guangxi 530023 (PR China)
E-Mail cai.song2011@gmail.com

Halaris A, Leonard BE (eds): Inflammation in Psychiatry.
Mod Trends Pharmacopsychiatry. Basel, Karger, 2013, vol 28, pp 90–99 (DOI: 10.1159/000343971)

The Brain-Gut Axis: A Target for Treating Stress-Related Disorders

Lucinda V. Scott · Gerard Clarke · Timothy G. Dinan

Department of Psychiatry and Alimentary Pharmabiotic Centre, University College Cork, Cork, Ireland

Abstract

The brain-gut axis provides a bidirectional means of communication between the microbiota within the gut and the brain. Stress acting via the brain can result in alteration of the microbial composition of the gut, but increasing evidence indicates that bacteria within the gut can influence brain neurochemistry and behaviour. It is clear that post-natal colonisation of the gut plays a key role in regulating the development of the hypothalamic-pituitary-adrenal axis and the development of pivotal neurotransmitter systems. Probiotics are defined as live bacteria which confer a health benefit. Recent studies in rodents have demonstrated the capacity of a probiotic, *Lactobacillus rhamnosus,* to alter the expression of GABA receptors centrally whilst producing anxiolytic type effects. Preliminary studies in humans are yielding encouraging findings. It may in the future be possible to use probiotic bacteria to treat depression and other stress-related disorder but we await the results of appropriately designed placebo-controlled trials.

Over the past decade there has been an exponential increase in research focusing on the brain-gut axis because of its potential role in a variety of disorders, not just relating to irritable bowel syndrome and inflammatory bowel disease, but also involvement in mood disorders. In this review we will explore the impact of stress on the brain-gut axis and the possibility of targeting the axis for the treatment of stress related disorders such as depression.

Studies of this brain-gut axis now of necessity include the gut microbiota, which in many respects is a metabolic organ in its own right. The general scaffolding of the brain-gut-enteric microbiota axis includes the central nervous system (CNS), the neuroendocrine and neuroimmune systems, the sympathetic and parasympathetic arms of the autonomic nervous system (ANS), the enteric nervous system (ENS) and the intestinal microbiota. These components interact to form a complex reflex network with afferent fibres that project to integrative CNS structures and efferent projections to the smooth muscle [1]. Through this bidirectional communication network, signals from the brain can influence the motor, sensory and secretory modalities of the gastrointestinal tract (GIT) and, conversely, visceral messages from the gut can influence brain function [1]. Less well studied but increasingly appreciated is the potential impact of the enteric microbiota on the relationships within the construct [2, 3] and the realisation it may play a part in CNS disorders such as major depression.

Microbiota Composition and Development

The GIT is inhabited with 10^{13}–10^{14} microorganisms, a figure thought to be ten times that of the number of human cells in our bodies and 150 times as many genes as our genome [4, 5]. The estimated species number varies greatly but it is generally accepted that the adult microbiome consists of greater than 1,000 species [6] and more than 7,000 strains [7]. It is an environment dominated by bacteria, mainly strict anaerobes, but also including viruses, protozoa, archae and fungi [8]. The microbiome is largely defined by 2 bacterial phylotypes, *Bacteroidetes* and *Firmicutes* with *Proteobacteria*, *Actinobacteria*, *Fusobacteria*, and *Verrucomicrobia* phyla present in relatively low abundance [9].

There are major methodological difficulties encountered while trying to delineate and enumerate the constituents of a normal microbiota. Culture-based analysis, the mainstay of traditional microbiological attempts to define the enteric flora, is only adequate for the minority of the gut microbiota that is amenable to cultivation [10, 11]. This problem has largely been circumvented by the use of culture-independent techniques, a portfolio consisting of sequencing-based methods, genetic fingerprinting, fluorescently labelled oligonucleotide probes (FISH), quantitative PCR as well as metagenomic approaches [12, 13]. The realisation that the secretory and metabolic capability of the microbiome was likely as important as phylotype composition has also led to the use of metabolomic and metaproteomic approaches to improve our understanding of what has been described as the forgotten organ [14]. Unfortunately, advances in culture methods have not kept pace with the rise of these alternative technologies and a dual-pronged line of attack may be required to complete the circle, a not inconsiderable logistical challenge [15]. Moreover, deficits remain to be redressed in relation to the microbial content of the small intestine due to an over-reliance on the analysis of faecal microbial composition [16].

As our knowledge of the previously unimagined scale of microbial diversity within the gut has expanded, so too has an understanding of its development and longitudinal variation. Colonisation of the infant gut commences at birth when delivery exposes the infant to a complex microflora and its initial microbiome has a maternal signature [17, 18]. The microbiome of unweaned infants is simple with high interindividual variability. The numbers and diversity of strict anaerobes increase as a result of diet and environment, and after 1 year of age a complex adult-like microbiome is evident [17, 18]. Despite a significant interpersonal variation in the enteric microbiota, there seems to be a balance that confers health benefits and an alteration in beneficial bacteria can negatively influence the wellbeing of the individual [3]. Several factors may alter the microbiome such as infection, disease, diet and antibiotics. However, it tends to revert to the stable diversity established in infancy once the threat of the initial distorting factor has subsided. Interestingly, it has been demonstrated that the core microbiota of an aged individual is distinct from that of younger adults [17] and that age related shifts in the composition of the intestinal microbiota are linked to adverse health effects in the elderly host [19, 20].

Evaluating the Role of the Microbiota

Of the varied strategies employed to study the function of the microbiota, perhaps the use of germ-free (GF) animals has offered the most revealing insights. Their use is based on the sterile uterine environment present during prenatal development [21] with surgical delivery replacing a normal vaginal birth, thus eliminating the opportunity for postnatal colonisation of the gut.

Subsequent comparison with their conventionally colonised counterparts allows inferences to be drawn regarding the morphological and physiological parameters that may be under the influence of the developing microbiota. However useful the complete absence of an intestinal microbiota may be in proof of principle studies, it is not reflective of real life scenarios. Other approaches which have more validity in this regard involve inducement of a dysbiosis of the enteric flora, either through administration of antibiotics or deliberate infection in preclinical studies. Broad-spectrum antibiotics in particular are known to perturb the microbiome by reducing biodiversity and delaying colonisation and are widely used as a method to intentionally The use of GF animals (with no bacterial exposure) has provided the most significant insights into the role of the microbiota in regulating the development of the hypothalamic-pituitary-adrenal axis (HPA).

Studies by Sudo et al. [21] provide insight into the role of the intestinal microbiota in the development of the HPA. In germ free mice a mild restraint stress induces an exaggerated release of corticosterone and ACTH compared to the specific pathogen-free (SPF) controls. The stress response in the GF mice is partially reversed by colonisation with faecal matter from SPF animals and fully reversed by monoassociation with *Bifidobacterium infantis* in a time-dependant manner [21]. This study clearly demonstrated that the microbial content of the gut is critical to the development of an appropriate stress response later in life and also that there is a narrow window in early life where colonisation must occur to ensure normal development of the HPA axis. Sudo et al. [21] also reported a decrease in brain-derived neurotrophic factor (BDNF), a key neurotrophin involved in neuronal growth and survival, and expression of the NMDA receptor subunit 2a (NR2a) in the cortex and hippocampus of GF animals compared to SPF controls. Dinan and Cryan [22] provide a comprehensive review of the microbiota and the regulation of the HPA.

Can the gut flora influence the neurochemistry and behaviour associated with the stress response? Neufeld et al. [23], despite reporting an unexplained increase in BDNF mRNA that was contrary to the protein decreases observed in the earlier study, found a less anxious phenotype for germ-free animals in the elevated plus maze. They also found alterations in NMDA and 5HT1a receptor expression. Both of these receptors are known to influence CRH release from the hypothalamus and changes in expression may explain altered HPA function in germ-free animals. This behavioural phenotype was replicated in another study with germ-free animals in both the elevated plus maze and the light dark box (where germ free animals spent more time in the light compartment) and associated with an altered gene expression profile in relevant brain regions.

Maternal separation, an early life stressor which can result in long-term HPA changes [1] has been shown to, cause a significant decrease in faecal lactobacilli on day 3 after separation, which returns to baseline by day 7 as assessed by enumeration of total and Gram-negative aerobic and facultative anaerobic bacterial species [24]. However, early life stress can also have long-term effects on the microbiome. Analysis of the 16S rRNA diversity in adult rats exposed to maternal separation for 3 h per day from post-natal days 2–12 revealed a significantly altered faecal microbiome when compared to the non-separated control animals [24]. A recent study found that chronic treatment with the probiotic *Lactobacillus rhamnosus* over 28 days produced animals with lower levels of corticosterone and reduced depressive behaviours in the forced swim test in addition to a less anxious phenotype in the elevated plus maze [25]. Alterations in GABA the main inhibitory neurotransmitter system were also observed. *L. rhamnosus*-treated animals showed alterations of $GABA_{B1b}$ mRNA in the brain with increased expression in cortical regions and decreased expression in the hippocampus, amygdala, and locus coeruleus as well as reduced $GABA_{A\alpha2}$ mRNA expression in the prefrontal cortex and amygdala and increased $GABA_{A\alpha2}$ in the hippocampus. The mechanism behind these changes was partially elucidated by studies in vagotomised animals. The neurochemical and behavioural effects of this bacterium do not occur following vagotomy, indicating that the vagus is a key route of communication between probiotic bacteria and the brain. A role for the gut microbiota in pain perception has also been indicated with one study demonstrating that specific *Lactobacillus* strains could induce the expression of μ-opioid and cannabinoid receptors in intestinal epithelial cells and mimic the effects of morphine in promoting analgesia [26].

Other potential mechanisms through which probiotics may influence the HPA and stress responsivity include neurotransmitter modulation. *B. infantis* 35624, for example, has been shown in Sprague-Dawley rats to induce an elevation in plasma tryptophan levels, a precursor to serotonin (5-HT) which is a key neurotransmitter within the brain-gut axis [27]. Since CNS tryptophan concentrations are largely dependent on peripheral availability and the enzymatic machinery responsible for the production of 5-HT is not saturated at normal tryptophan concentrations [28], the implication here is that the microbiota might play some role in the regulation of CNS as well as enteric nervous system 5-HT synthesis. This effect is potentially mediated by the impact of the microbiota on the expression of indoleamine-2,3-dioxygenase, a key enzyme in the physiologically dominant kynurenine pathway of tryptophan degradation but of course multiple mechanisms are possible and indeed likely, given the strain specific effects that have been observed in many probiotic studies to date [29].

Probiotics and Glucocorticoids

Demonstrating changes in glucocorticoid levels in either rodents or man based on single serum samples is notoriously unreliable. In a maternal separation model, Desbonnet et al. [27] found behavioural changes with bifidobacteria treatment but no reduction in corticosterone while, using a similar model, Gareau et al. [30] found that a lactobacillus reduced corticosterone levels and McKernan et al. [31] found that *B. infantis* reduced corticosterone levels, though the reduction did not reach statistical significance.

In a recent clinical study, healthy volunteers were given *Lactobacillus helveticus* R0052 and *Bifidobacterium longum R0175* in combination or placebo in a double-blind, randomised parallel group study for 30 days and assessed with the Hopkins Symptom Checklist (HSCL-90), the Hospital Anxiety and Depression Scale (HADS), the Perceived Stress Scale, the Coping Checklist (CCL) and 24-hour urinary free cortisol (UFC) [32]. Overall, positive psychological benefit from the probiotics was observed and the median UFC level in ng/ml of probiotic-treated subjects decreased between baseline and follow-up, whereas that of controls did not. These data support the view that probiotics can influence HPA activity in healthy volunteers.

The best evidence to date for the involvement of the microbiota in disease states comes from irritable bowel syndrome (IBS), a prototypical stress-related brain-gut axis disorder. IBS is the most common functional gastrointestinal disorder (FGID) encountered in clinical settings affecting an estimated 10–15% of the general population in Western Europe and North America [33]. It is characterised by the presence of abdominal pain or discomfort, an alteration in bowel habit, and the absence of reliable biomarker [34]. At least 50% of patients with IBS have a co-morbid psychiatric disorder, most commonly major depression or anxiety. A number of strands of evidence support a role for the microbiota in the pathophysiology of IBS and chief among these are the supporting data for post-infectious IBS (PI-IBS), a term which describes the development of IBS following an episode of bacteriologically confirmed gastroenteritis [35]. This phenomenon is now well documented and prospective studies showing that the risk of developing post-infectious IBS (PI-IBS) is in the order of 3–36% following an enteric infection with the precise incidence depending on the infecting organism [36]. The highest reported incidence of PI-IBS was associated with the Walkerton outbreak with 36% of individuals showing symptoms of IBS 2 years after infection with both *Campylobacter jejuni* and *Escherichia coli* 0157:H7 [35]. A mechanistic perspective has been provided by a number of studies which found a persistent elevation in rectal mucosal enteroendocrine cells, T lymphocytes and gut permeability following the infectious insult in subjects who went on to develop IBS [36]. These studies are regarded as important indicators of a link between alterations in the microbiota and mucosal inflammation in IBS.

On the basis that immunological changes in IBS might result from exposure to exogenous bacterial challenge, studies supporting the low-grade inflammation that has so frequently been reported in IBS are also taken as evidence of a role for the microbial perturbations in the disorder [34, 37]. Persistent low grade inflammation is a characteristic of PI-IBS and these patients exhibit greater IL-1β mRNA expression, both during and after the infection, compared with individuals who do not develop PI-IBS [38]. IBS patients with normal histology had increased intraepithelial lymphocytes and CD3+ and CD25+ cells in the lamina propria [39]. Increased CD25+ cells in IBS suggests an antigen challenge and these cells are preventing 'a more florid inflammatory response' [39]. There are increased numbers of activated mast cells in colonic biopsies from IBS patients and a direct correlation between proximity of the mast cells to neurons and pain severity [41]. Cenac et al. [42] demonstrated increased proteolytic activity in colonic washes from IBS compared with control patients while the analysis of TLR expression in colonic biopsies has demonstrated increased expression of TLR4 and TLR5 and decreased levels of TLR7 and TLR8 in IBS patients.

Others have investigated the involvement of the systemic immune system on the basis that it might reflect mucosal disturbances [40]. In a study of 78 IBS patients, O'Mahony et al. [43] demonstrated an abnormal IL-10/IL-12 ratio in stimulated peripheral blood mononuclear cells (PBMCs), indicating a proinflammatory, Th-1 state. An enhanced proinflammatory cytokine release (TNFα, IL-1β, IL-6) in lipopolysaccharide-stimulated peripheral blood mononuclear cells from IBS patients has also been observed. In another study, baseline plasma proinflammatory cytokines IL-6, IL-6R and IL-8 were elevated in IBS and accompanied by overactivation of the HPA axis [33, 47]. Indeed elevations in systemic proinflammatory mediators have repeatedly being demonstrated in both clinical datasets and reports from animal models of the disorder [44, 45]. What is not clear is whether they truly represent a disturbance at mucosal sites and this remains a subject in need of further investigation. Interestingly, it has recently been reported that IBS patients have a distinct pattern of exaggerated peripheral TLR activity as indicated by measurements of cytokine production following whole blood stimulations [46].

Microbiota and Obesity

The link between mood disorder and obesity has long been recognised and the microbiota may play an important role in connecting both disorders. As might be expected given the importance of the microbiota in supporting host digestion and metabolism, obesity has been considered as an illness with a potential microbial basis. It has been defined as a medical condition in which excess body fat has accumulated to produce adverse effects on health, leading to increased health problems and potentially a reduced life expectancy [48, 49]. The incidence of obesity is increasing in both developed and developing countries and has become a worldwide burden upon health care systems. Some of the most convincing evidence comes from studies in GF animals which have shown them to be naturally leaner than their colonised counterparts and colonisation of adult GF mice with a normal microbiota from the cecum of conventionally raised animals produces a 60% increase in body fat content and insulin resistance within 14 days despite reduced food intake [50]. GF mice are resistant to diet induced obesity but when colonised with faecal flora they gained weight due to altered fatty acid metabolism. Other studies have shown that the *ob/ob* mouse, a strain of mouse genetically predisposed to obesity, had a 50% decrease in the abundance of *Bacteroidetes* and a proportional increase in *Firmicutes* when compared to the lean wild-type and heterozygotic controls [51]. In humans, when the faecal microbiota of obese individuals was compared to lean controls there were fewer *Bacteroides* and more *Firmicutes* in the obese group. However, over time on a calorie restricted diet the relative abundance of *Bacteroidetes* increased and the abundance of *Firmicutes* decreased [52]. While these studies do offer some support for a perturbation of the resident flora in the accumulation of excess fat, microbial influences should not be considered in isolation. However, it is increasingly recognised that obesity is frequently associated with a pro-inflammatory phenotype, as in the case of major depression.

Microbiome, Stress and Depression

Although the bulk of research to date has focused on the impact of the microbiota on CNS function, there is also research to suggest that the brain can alter the microbiome [53]. Signalling molecules released into the gut lumen from cells in the lamina propria that are under the control of the CNS can result in changes in gastrointestinal motility and secretion as well as intestinal permeability, thus altering the gastrointestinal environment in which the bacteria reside. Under normal circumstances, non-pathogenic bacteria in the colon are restrained by the intestinal epithelium and it has long been established that conditions such as intestinal obstruction or haemorrhagic shock can alter gut permeability and thus translocation of bacteria across the epithelium. Accumulating evidence supports the view that psychological stress can also increase permeability of the gut allowing bacteria and bacterial antigens to cross the epithelial barrier and this can activate a mucosal immune response which in turn alters pro-inflammatory cytokines and perhaps activate the HPA. Acute stress was shown to cause an increase in colonic paracellular permeability [54] which involved mast cells and overproduction of IFN-γ with decreased expression of ZO-2 and occludin mRNA. The psychological component of social stress was shown to facilitate the translocation of indigenous bacteria into the host. Other studies have demonstrated that stress hormones promoted the growth of non-pathogenic isolates of *E. coli* as well as the pathogenic *E. coli* 0157:H7 strain via interactions with host catecholamines such as adrenaline and noradrenaline [55, 56]. Different psychological stressors are known to alter the composition of the microbiome by modulating the composition of total biomass in infants. Prenatal stressors have been shown to alter the microbiome in rhesus monkeys by reducing the overall numbers of bifidobacteria and lactobacilli [57]. Female monkeys were left undisturbed or were stressed during pregnancy using an acoustical startle paradigm for 6 weeks either early or late in their 24th week of gestation. Several types of intestinal microflora were repeatedly enumerated by faecal culture while infants were reared normally by their mothers. Significant changes in microflora concentrations occurred during the first 6 months of life. The profile of total aerobes and facultative anaerobes was biphasic, with peak concentrations occurring between 2 and 16 weeks of age. The numbers of bifidobacteria and lactobacilli were low at 2 days after birth but rapidly increased to a peak between 8 and 16 weeks of age. Although similar temporal patterns were evident in all infants, prenatal stress reduced the overall numbers of bifidobacteria and lactobacilli. Moderate disturbance during pregnancy leading to activation of the HPA was sufficient to alter the intestinal microflora in the newborn infant. These alterations could result in enhanced susceptibility to infection and suggest a mechanism for some effects of maternal pregnancy conditions on infant health.

Through what mechanism can stress impact on gut permeability? In a pivotal study, Clark et al. [58] showed that a rise in the pro-inflammatory cytokine interferon-γ can exploit lipid raft-mediated translocation pathways to cross the epithelium prior to cytokine-induced disruption of tight junctions. Meddings and Swain [59] found that restraint stress resulted in increased intestinal permeability and an associated increase in corticosterone. The time course of the HPA response to the psychological stress was similar to that of an ulcerogeneic stress model. Ait-Belgnaoui et al. [60] have found that prevention of gut leakiness by intestinal microbiota modulation leads to an attenuated HPA response to an acute psychological stress in rats; the probiotic *Lactobacillus farciminis* suppressed stress-induced hyperpermeability and endotoxemia and prevented HPA activation. To date in humans few studies have examined the microbiota in stress-related disorders with the obvious exception of irritable bowel syndrome.

Maes et al. [61] suggest stress-induced intestinal mucosal dysfunction with an increased translocation from Gram-negative bacteria plays an important role in the pathophysiology of depression and may account for HPA over-activity in some patients. The study examined the serum concentrations of IgM and IgA against LPS of the Gram-negative enterobacteria, *Hafnia alvei*, *Pseudomonas aeruginosa*, *Morganella morganii*, *Pseudomonas putida*, *Citrobacter koseri* and *Klebsielle pneumoniae* in patients with major depression and healthy controls. The prevalences and median values for serum IgM and IgA against LPS of enterobacteria were significantly greater in patients with major depression than in normal volunteers. These differences were significant to the extent that a diagnostic performance was obtained. The symptom profiles of increased IgM and IgA levels were reported as fatigue, autonomic and gastrointestinal symptoms and a subjective feeling of infection. The results suggest that intestinal mucosal dysfunction characterized by an increased translocation of Gram-negative bacteria (leaky gut) plays a role in the inflammatory pathophysiology of depression. It is possible that increased LPS translocation may mount an immune response and thus innate response system activation in some patients with major depression and specific 'sickness behaviour' symptoms. This 'leaky gut' effect of psychological stress may explain the HPA overactivity in some cases of depression. However, a detailed analysis of gut permeability and faecal microbiota in major depression needs to be undertaken before firm conclusions can be made.

Probiotics as Antidepressants?

Probiotics are defined as live bacteria which confer a health benefit. In a recent double-blind, placebo-controlled, randomised parallel group study volunteers received either the probiotic combination *L. helveticus* R0052 and *B. longum* or placebo for 30 days and were assessed by the Hopkins Symptom Checklist (HSCL-90), the Hospital Anxiety and Depression Scale (HADS), the Perceived Stress Scale, the Coping Checklist (CCL) and the measurement of 24-hour urinary free cortisol levels [32]. Daily administration of probiotic combination significantly reduced psychological distress in volunteers, as measured by the HSCL-90 scale, the HADS and by the CCL. Furthermore, urinary free cortisol levels were significantly reduced by the probiotics. This is the first demonstration of a probiotic influencing the HPA in humans.

Another study by Benton et al. [62] found that the consumption of a probiotic-containing yoghurt improved mood. One hundred and thirty-two physically healthy subjects with a mean age 61.8 years, volunteered in response to local media coverage. One hundred and twenty-four completed the trial. For a 3-week period, either a probiotic containing milk drink, or a placebo, were consumed daily. Mood and cognition were measured at baseline, and after 10 and 20 days of consumption. When the third with the lowest baseline mood were considered, they selectively responded by reporting themselves as happy rather than depressed after taking the probiotic.

In a study of patients with chronic fatigue syndrome, subjects were treated three times daily with *Lactobacillus casei* strain Shirota (LcS) or a placebo with identical taste and appearance [63]. Overall, there was a significant improvement in anxiety among those taking the active LcS compared to the placebo, providing further support for the view that a probiotic may have psychotropic effects.

Logan and Katzman [64] first proposed the use of probiotics as adjunct therapy in the management of depression. Since then there has been an accumulation of data from both clinical and preclinical studies supporting the view that probiotics may have a role in the treatment of depression. It is, however, certain that not all probiotics are the same and while some may have an impact on neurotransmitters and behaviour others will not. After all, they each have a unique genome with varying capacities to produce neuroactive substances. Certainly, for some patients probiotics would be far more acceptable than the current use of SSRIs and other antidepressants. However, before any firm conclusions can be drawn properly powered studies comparing an appropriate probiotic with an antidepressant and placebo and rigorously assessing the impact of a probiotic as adjunctive therapy to conventional antidepressants need to be conducted [65].

Conclusion

Studies of the gut microbiota are in their infancy. What is clear is that the microbiota is a highly complex metabolic structure with a significant potential to influence brain and gut activity. Likewise, environmental stress acting through the brain can influence the microbiota. An accumulating volume of evidence supports a role for the microbiota not just in conditions such as irritable bowel syndrome but also in obesity and major depression where a dysfunctional brain-gut-microbiota axis may account for the pro-inflammatory phenotype observed. It may be possible to use probiotic bacteria to treat depression and other stress-related disorder but we await the results of appropriately designed placebo-controlled trials.

References

1 O'Mahony SM, Hyland NP, Dinan TG, Cryan JF: Maternal separation as a model of brain-gut axis dysfunction. Psychopharmacology (Berl) 2011;214: 71–88.
2 Rhee SH, Pothoulakis C, Mayer EA: Principles and clinical implications of the brain–gut–enteric microbiota axis. Nat Rev Gastroenterol Hepatol 2009;6: 306–314.
3 Cryan JF, O'Mahony SM: The microbiome-gut-brain axis: from bowel to behavior. Neurogastroenterol Motility 2011;23:187–192.
4 Grenham S, Clarke G, Cryan JF, Dinan TG: Brain-gut-microbe communication in health and disease. Front Physiol 2011;2:94.
5 Gill SR, Pop M, Deboy RT, Eckburg PB, Turnbaugh PJ, Samuel BS, Gordon JI, Relman DA, Fraser-Liggett CM, Nelson KE: Metagenomic analysis of the human distal gut microbiome. Science 2006;312: 1355–1359.
6 Qin J, Li R, Raes J, Arumugam M, Burgdorf KS, Manichanh C, et al: A human gut microbial gene catalogue established by metagenomic sequencing. Nature 2010;464:59–65.
7 Ley RE, Peterson DA, Gordon J: Ecological and evolutionary forces shaping microbial diversity in the human intestine. Cell 2006;124:837–848.
8 Xu J, Mahowald MA, Ley RE, Lozupone CA, Hamady M, Martens EC, Henrissat B, Coutinho PM, et al: Evolution of symbiotic bacteria in the distal human intestine. PLoS Biol 2007;5:e156.
9 Eckburg PB, Bik EM, Bernstein CN, Purdom E, Dethlefsen L, Sargent M, Gill SR, Nelson KE, Relman DA: Diversity of the human intestinal microbial flora. Science 2005;308:1635–1638.

10 Shanahan F: Gut microbes: from bugs to drugs. Am J Gastroenterol 2010;105: 275–279.
11 Sekirov I, Russell SL, Antunes LCM, Finlay BB: Gut microbiota in health and disease. Physiol Rev 2010;90:859–904.
12 Archie E, Theis K: Animal behaviour meets microbial ecology. Anim Behav 2012;in press.
13 O'Hara AM, Shanahan F: The gut flora as a forgotten organ. EMBO Rep 2006;7: 688–693.
14 Forsythe P, Sudo N, Dinan T, Taylor VH, Bienenstock J: Mood and gut feelings. Brain Behav Immun 2010;24:9–16.
15 Mackie RI, Sghir A, Gaskins HR: Developmental microbial ecology of the neonatal gastrointestinal tract. Am J Clin Nutrit 1999;69:1035S–1045S.
16 Palmer C, Bik EM, Digiulio DB, Relman DA, Brown PO: Development of the human infant intestinal microbiota. PLoS Biol 2007;5:e177.
17 Kurokawa K, Itoh T, Kuwahara T, Oshima K, Toh H, Toyoda A, Takami H, Morita H, Sharma VK, Srivastava TP: Comparative metagenomics revealed commonly enriched gene sets in human gut microbiomes. DNA Res 2007;14:169.
18 Adlerberth I, Wold AE: Establishment of the gut microbiota in Western infants. Acta Paediatr 2009;98:229–238.
19 Claesson MJ, Cusack S, O'Sullivan O, Greene-Diniz R, de Weerd H, Flannery E, et al: Composition, variability, and temporal stability of the intestinal microbiota of the elderly. Proc Natl Acad Sci USA 2011;108(suppl 1): 4586–4591.
20 Woodmansey EJ: Intestinal bacteria and ageing. J Appl Microbiol 2007;102: 1178–1186.
21 Sudo N, Chida Y, Aiba Y, Sonoda J, Oyama N, Yu XN, et al: Postnatal microbial colonization programs the hypothalamic-pituitary-adrenal system for stress response in mice. J Physiol 2004;558:263–275.
22 Dinan TG, Cryan J: Regulation of the stress response by the gut microbiota: implications for psychoneuroendocrinology. Psychoneuroendocrinology 2012;in press.
23 Neufeld K, Kang N, Bienenstock J, Foster J: Reduced anxiety like behavior and central neurochemical change in germ free mice. Neurogastroenterol Motil 2011;23:255–264.
24 O'Mahony SM, Marchesi JR, Scully P, Codling C, Ceolho AM, Quigley EMM, Cryan JF, Dinan TG: Early life stress alters behavior, immunity, and microbiota in rats: implications for irritable bowel syndrome and psychiatric illnesses. Biol Psychiatr 2009;65:263–267.
25 Bravo JA, Forsythe P, Chew MV, Escaravage E, Savignac HM, Dinan TG, Bienenstock J, Cryan JF: Ingestion of *Lactobacillus* strain regulates emotional behavior and central GABA receptor expression in a mouse via the vagus nerve. Proc Natl Acad Sci USA 2011;108: 16050–16055.
26 Rousseaux C, Thuru X, Gelot A, Barnich N, Neut C, Dubuquoy, et al: *Lactobacillus acidophilus* modulates intestinal pain and induces opioid and cannabinoid receptors. Nat Med 2007;13:35–37.
27 Desbonnet L, Garrett L, Clarke G, Bienenstock J, Dinan TG: The probiotic *Bifidobacteria infantis*: an assessment of potential antidepressant properties in the rat. J Psychiatr Res 2008;43:164–174.
28 Ruddick JP, Evans AK, Nutt DJ, Lightman SL, Rook GA, Lowry CA: Tryptophan metabolism in the central nervous system: medical implications. Expert Rev Mol Med 2006;8:1–27.
29 Bercik P, Park AJ, Sinclair D, Khoshdel A, Lu J, Huang X, et al: The anxiolytic effect of *Bifidobacterium longum* NCC3001 involves vagal pathways for gut-brain communication. Neurogastroenterol Motil 2011;23:1132–1139.
30 Gareau MG, Wine E, Rodrigues DM, Cho JH, Whary MT, Philpott DJ, MacQueen G, Sherman PM: Bacterial infection causes stress-induced memory dysfunction in mice. Gut 2011;60:307–317.
31 McKernan DP, Fitzgerald P, Dinan TG, Cryan JF: The probiotic *Bifidobacterium infantis* 35624 displays visceral antinociceptive effects in the rat. Neurogastroenterol Motil 2010;22:1029–1035.
32 Messaoudi M, Lalonde R, Violle N, Javelot H, Desor D, Nejdi A, et al: Assessment of psychotropic-like properties of a probiotic formulation (*Lactobacillus helveticus* R0052 and *Bifidobacterium longum* R0175) in rats and human subjects. Br J Nutr 2011;105:755–764.
33 Dinan TG, Quigley E, O'Mahony S, O'Mahony L, Scully P, Shanahan F, Keeling PWN: Hypothalamic-pituitary-gut axis dysregulation in irritable bowel syndrome: plasma cytokines as a potential biomarker? Gastroenterology 2006;130: 304–311.
34 Quigley EM: Do patients with functional gastrointestinal disorders have an altered gut flora? Therap Adv Gastroenterol 2009;2:23–30.
35 Marshall JK, Thabane M, Garg AX, Clark WF, Salvadori M, Collins SM: Incidence and epidemiology of irritable bowel syndrome after a large waterborne outbreak of bacterial dysentery. Gastroenterology 2006;131:445–450.
36 Spiller RC, Jenkins D, Thornley JP, Hebden JM, Wright T, Skinner M, Neal KR: Increased rectal mucosal enteroendocrine cells, T lymphocytes, and increased gut permeability following acute *Campylobacter enteritis* and in post-dysenteric irritable bowel syndrome. Gut 2000;47:804–811.
37 Dunlop SP, Jenkins D, Neal KR, Spiller RC: Relative importance of enterochromaffin cell hyperplasia, anxiety, and depression in postinfectious IBS. Gastroenterol 2003;125:1651–1659.
38 Gwee K, Collins S, Read N, Rajnakova A, Deng Y, Graham J, Mckendrick M, Moochhala S: Increased rectal mucosal expression of interleukin 1 in recently acquired post-infectious irritable bowel syndrome. Gut 2003;52:523.
39 Chadwick VS, Chen W, Shu D, Paulus B, Bethwaite P, Tie A, Wilson I: Activation of the mucosal immune system in irritable bowel syndrome. Gastroenterology 2002;122:1778–1783.
40 Collins SM, Denou E, Verdu EF, Bercik P: The putative role of the intestinal microbiota in the irritable bowel syndrome. Dig Liver Dis 2009;41:850–853.
41 Barbara G, Stanghellini V, De Giorgio R, Cremon C, Cottrell GS, Santini D, Pasquinelli G, Morselli-Labate AM, Grady EF, Bunnett NW, Collins SM, Corinaldesi R: Activated mast cells in proximity to colonic nerves correlate with abdominal pain in irritable bowel syndrome. Gastroenterology 2004;126:693–702.
42 Cenac N, Andrews CN, Holzhausen M, Chapman K, Cottrell G, Andrade-Gordon P, Steinhoff M, Barbara G, Beck P, Bunnett NW, Sharkey KA, Ferraz JGP, Shaffer E, Vergnolle N: Role for protease activity in visceral pain in irritable bowel syndrome. J Clin Invest 2007;117:636–647.
43 O'Mahony L, McCarthy J, Kelly P, Shanahan F, Quigley EMM: Lactobacillus and bifidobacterium in irritable bowel syndrome: symptom responses and relationship to cytokine profiles. Gastroenterology 2005;28:541–551.

44 Dinan TG, Cryan J, Shanahan F, Keeling PW, Quigley EM: IBS: An epigenetic perspective. Nat Rev Gastroenterol Hepatol 2010;7:465–471.
45 Scully P, Mckernan DP, Keohane J, Groeger D, Shanahan F, Dinan TG, Quigley EM: Plasma cytokine profiles in females with irritable bowel syndrome and extra-intestinal co-morbidity. Am J Gastroenterol 2010;105:2235–2243.
46 McKernan DP, Gaszner G, Quigley EM, Cryan JF, Dinan TG: Altered peripheral Toll-like receptor responses in the irritable bowel syndrome. Alimenty Phar Therapeut 2011;33:1045–1052.
47 Dinan TG, Clarke G, Quigley EM, Scott LV, Shanahan F, Cryan J, Cooney J, Keeling PW: Enhanced cholinergic-mediated increase in the pro-inflammatory cytokine IL-6 in irritable bowel syndrome: role of muscarinic receptors. Am J Gastroenterol 2008;103:2570–2576.
48 Haslam DW, James WPT: Obesity. Lancet 2005;366:1197–1209.
49 Bloom SR, Kuhajda FP, Laher I, Pi-Sunyer X, Ronnett GV, Tan TMM, Weigle DS: The obesity epidemic pharmacological challenges. Mol Intervent 2008;8: 82–98.
50 Bäckhed F, Ding H, Wang T, Hooper LV, Koh GY, Nagy A, Semenkovich CF, Gordon JI: The gut microbiota as an environmental factor that regulates fat storage. Proc Natl Acad Sci USA 2004;101:15718–15723.
51 Ley RE, Bäckhed F, Turnbaugh P, Lozupone CA, Knight RD, Gordon JI: Obesity alters gut microbial ecology. Proc Natl Acadf Sci USA 2005;102:11070–11075.
52 Ley RE, Peterson DA, Gordon JI: Ecological and evolutionary forces shaping microbial diversity in the human intestine. Cell 2006;124:837–848.
53 Bercik P, Collins SM, Verdu EF: Microbes and the brain-gut axis. Neurogastroenterol Motil 2012;24:405–413.
54 Kiliaan AJ, Saunders PR, Bijlsma PB, Berin MC, Taminiau JA, Groot JA, Perdue MH: Stress stimulates transepithelial macromolecular uptake in rat jejunum. Am J Physiol Gastrointest Liver Physiol 1998;275:G1037–G1044.
55 Demaude J, Salvador-Cartier C, Fioramonti J, Ferrier L, Bueno L: Phenotypic changes in colonocytes following acute stress or activation of mast cells in mice: implications for delayed epithelial barrier dysfunction. Gut 2006;55:655–661.
56 Freestone PP, Williams PH, Haigh RD, Maggs AF, Neal CP, Lyte M: Growth stimulation of intestinal commensal *Escherichia* coli by catecholamines: a possible contributory factor in trauma-induced sepsis. Shock 2002;18:465–470.
57 Bailey MT, Lubach GR, Coe CL: Prenatal stress alters bacterial colonization of the gut in infant monkeys. J Pediatric Gastroenterol Nutrit 2004;38:414–421.
58 Clark E, Hoare C, Tanianis-Hughes J, Carlson GL, Warhurst G: Interferon gamma induces translocation of commensal *Escherichia coli* across gut epithelial cells via a lipid raft-mediated process. Gastroenterology 2005;12:1258–1267.
59 Meddings JB, Swain MG: Environmental stress-induced gastrointestinal permeability is mediated by endogenous glucocorticoids in the rat. Gastroenterology 2000;119:1019–1028.
60 Ait-Belgnaoui A, Durand H, Cartier C, Chaumaz G, Eutamene H, Ferrier L, Houdreau E, Fioramonti J, Bueno L, Theodorou V: Prevention of gut leakiness by intestinal microbiota modulation leads to attenuated HPA response to an acute psychological stress in rats. Psychoneuroendocrinology 2012;37: 1885–1895.
61 Maes M, Kubera M, Leunis JC: The gut-brain barrier in major depression: intestinal mucosal dysfunction with an increased translocation of LPS from Gram negative enterobacteria (leaky gut) plays a role in the inflammatory pathophysiology of depression. Neuro Endocrinol Lett 2008;29:117–124.
62 Benton D, Williams C, Brown A: Impact of consuming a milk drink containing a probiotic on mood and cognition. Eur J Clin Nutr 2007;61:355–361.
63 Rao AV, Bested AC, Beaulne TM, Katzman MA, Iorio C, Berardi JM, Logan AC: A randomized, double-blind, placebo-controlled pilot study of a probiotic in emotional symptoms of chronic fatigue syndrome. Gut Pathog 2009;1:6.
64 Logan AC, Katzman M: Major depressive disorder: probiotics may be an adjuvant therapy. Med Hypotheses 2005; 64:533–538.
65 Dinan TG, Quigley EM: Probiotics in the treatment of depression: science or science fiction? Aust NZ J Psychiatr 2011; 45:1023–1025.

Prof. Ted G. Dinan
Department of Psychiatry
Cork University Hospital
Wilton, Cork (Ireland)
E-Mail t.dinan@ucc.ie

Halaris A, Leonard BE (eds): Inflammation in Psychiatry.
Mod Trends Pharmacopsychiatry. Basel, Karger, 2013, vol 28, pp 100–116 (DOI: 10.1159/000343972)

The Question of Pro-Inflammatory Immune Activity in Schizophrenia and the Potential Importance of Anti-Inflammatory Drugs

Volker Arolt · Oliver Ambrée

Department of Psychiatry, University of Münster, Münster, Germany

Abstract

Throughout the last 20 years, numerous studies have indicated pro-inflammatory activity in patients with an acute schizophrenic episode. In this report, the most relevant findings concerning the cytokine systems in schizophrenia are demonstrated, together with recent studies on gene expression. Findings in humans are supplemented by observations from rodent models of schizophrenia. Furthermore, the current state of both neuroleptic and anti-inflammatory compounds on the immune system is reported, together with clinical data on the effects of anti-inflammatory medications on the course of the acute illness in patients with schizophrenia.

Schizophrenia is a devastating disease that can be described with sufficient reliability but is still not well understood, neither in its differing clinical presentations (e.g. clinical 'subtypes') nor its underlying biological and psychological mechanisms. However, it is now clear that about 60–70% of the risk variance is explained by genetic factors and an increasing number of candidate genes have been found during the last 20 years [1–3]. Nevertheless, there are also environmental causes that contribute to the pathophysiology of the disorder. Numerous different lines of research point to the possible importance of inflammation (or pro-inflammatory immunological activity, resp.) in the etiology of the disease.

As early as 1939, Lehmann-Facius from Frankfurt/Main carried out the first systematic research on the immunology of schizophrenia, then focusing on possible autoimmunity against brain lipid structures of patients with schizophrenia. Although his methods were criticized, his finding led to the notion of an autoimmune process in schizophrenia that dominated the field until modern methods were used. Today, there are a number of more global, as well as system-specific indicators that point to immune-mediated pathological mechanisms in schizophrenia. The relapsing/recurrent course of some subtypes of schizophrenia can be globally recognized under this perspective and compared with chronic inflammatory diseases such as rheumatoid arthritis and multiple sclerosis. Another global indicator is the association between maternal infections and the risk for schizophrenia in the offspring, a finding which can be recognized as epidemiologically proven. Maternal infection has long been regarded as a major influential factor in the pathophysiology of schizophrenia [4, 5], probably due to its profound influence on the developing physiology of the fetal/neonatal central nervous system. Furthermore, early infection or immune stimulation may influence the development of the immature immune system in the offspring and, by still largely unknown mechanisms, may result in what is later recognized as immunological abnormality or even inflammation in the schizophrenic adolescent and adult. Also clinically, the slow progression of grey matter loss in the brains of some patients with schizophrenia can be seen as the result of chronic mild inflammation. Furthermore, there are prominent side effects of the treatment with interferons and interleukins that resemble the psychopathological symptoms of schizophrenia. Global but comparatively strong indicators are enhanced serum levels of proteins that serve as common markers of ongoing inflammation. An often repeated and relatively strong finding is the enhanced levels of the soluble IL-2 receptor (sIL-2R) in humans with acute schizophrenia, an observation that is also made in the early stages of acute episodes of multiple sclerosis, rheumatoid arthritis, and other autoimmune disorders.

There are also system-specific indicators that point to an involvement of the immune system in schizophrenia. Single lines of research have focused on the possible actions of retroviruses but also of brain-specific antigens. A much larger body of research is concerned with certain subsystems of the peripheral immune system, particularly the amount and (dys-)functions of certain cellular populations. Throughout the last 20 years, these lines of research have dominated the field and there are now a substantial number of clinical findings in patients with schizophrenia that point to abnormalities in different parts of the immune system that are interpreted as pro-inflammatory. These findings are understood as signs of chronic mild inflammation by some researchers, although we know that the usual overt signs of inflammation, such as gliosis, are not found in the brains of subjects with schizophrenia.

Although numerous methodological problems such as different laboratory techniques, different disease stages, different ages and numerous intervening variables have hampered progress in the field [6–8], it must nevertheless be noticed that some findings have been constantly reproduced and even two meta-analyses have lately been carried out that underscore the increasing validity of those. Furthermore, due to novel technologies in the last 10 years, monocyte gene expression profiles have been compiled that supply us with information about the up- or downregulation of inflammation-relevant genes. Since the end of the 1990s, the field is also substantially stimulated by research on animal models that is based on rodents born under the influence of maternal (artificial) inflammation in utero. These models resemble psycho-physiological abnormalities that are found in humans with schizophrenia and have provided a possibility to study neuron pathological abnormalities that are similar to the subtle pathology in the brains of patients with schizophrenia. In these animal models, although much less studied, immunological abnormalities have been found that point to an early immaturity but also to pro-inflammatory activity.

If all these indicators of pro-inflammatory dysregulation of the immune system in men and mice have a significant biological meaning, it should be expected that anti-inflammatory compounds might be useful in the treatment of schizophrenia. Indeed, this seems to be the case, as studies with acetylsalicylate, celecoxib, minocyclin, and omega-3 fatty acids have shown. However, this proof-of-principle assumption only holds true if the anti-inflammatory properties of these compounds are mainly responsible for their antipsychotic effect, and not one of the numerous other biological mechanisms that are actuated along with the principal therapeutic action.

Schizophrenia and Inflammation – An Update on Clinical Findings

There are several lines of research on different cytokine systems in the CSF and in the blood of patients that point to immune dysfunction in schizophrenia, mostly in the direction of pro-inflammatory activity. In short, the primate immune system consists of two large subdivisions that are closely linked to each other: the innate and the acquired (or adaptive) system. The innate system consists of one humoral arm (the complement system with proteins that are needed to battle the invasion of micro-organisms) and one cellular arm, consisting of different cell types such as macrophages, granulocytes and natural killer cells. These cells communicate with each other and with the adaptive system by the means of cytokine secretion and receptor binding on the cell surfaces. The adaptive system consists of T cells which are differentiated into different cytotoxic (killer-T cells) or memory (T-helper cells) cells, and B cells that are responsible for the specific antibody production. There are also gamma-delta T cells which seem to play an intermediate role. The regular functioning of both the innate and the adaptive immune systems depends on a subtle balance of pro- and anti-inflammatory activities. So the main cytokines with pro-inflammatory properties in the innate immune system are IL-1β, TNF-α, and IFN-α, whereas IL-10 (produced by macrophages but also by T-helper cells) acts as an anti-inflammatory protein. Macrophages also produce IL-8 which is an important cytokine for chemotaxis, in order to attract neutrophils to the inflammatory/infectious site. The main pro-inflammatory cytokines of the adaptive immune system are IL-2 and IFN-γ (produced together with TNF-β by TH-1 cells that are upregulated by IL-12 which is mainly produced by monocytes), and the main anti-inflammatory cytokine is IL-4 that increases the differentiation of TH-0 cells into TH-2 cells (producing IL-4, IL-5, IL-6, IL-10 and IL-13) that represent the anti-inflammatory counterpart for TH-1 cells. The complexity of these regulatory mechanisms is far greater than hinted in this description and it must be of note that the same cytokine are produced by numerous cell types and also have numerous different effects that are still not fully understood. However, the amount of cytokines found in blood sera of patients or after certain cell types are stimulated may indicate the regulatory state of certain pathways. The upregulation of pro-inflammatory cytokines, e.g., has been proven to be useful to detect early inflammatory processes in autoimmune disorders. Along with the increasing knowledge on these messenger systems in the 1980s, the interest of researches was stimulated in order to gain closer information on the immune functions in the major psychiatric disorders.

In 1995, the group of van Kammen published a landmark study [9] that directly pointed to the possibility of an association between pro-inflammatory activity and onset of a schizophrenic episode. They used an intriguing experimental design that, however, could as yet not be reproduced (possibly also due to the special experimental setting they used). Seventy-nine male patients with schizophrenia were withdrawn from haloperidol treatment for up to 6 weeks and patients were evaluated for symptom recurrence. CSF and plasma levels were obtained by established procedures before haloperidol withdrawal and afterwards. Only the CSF levels of IL-2 were associated with the recurrence of psychotic symptoms. Relapse-prone patients, examined at two time points (medicated and after drug withdrawal) had significantly higher levels of CSF IL-2 than patients who did not relapse. The CSF IL-2 level during haloperidol treatment was a significant predictor of worsening in psychosis. Such an association could not be observed for IL-1, dihydroxyphenylacetic acid (the dopamine breakdown metabolite which could be expected to be increased due to an increased dopamine turnover) or the typical early psychopathological symptoms that go along with relapse, such as tension and sleep disturbances. Other groups have also investigated cytokine levels in the CSF, however with ambiguous results. Some show elevated IL-2 levels and others did not find a difference [7]. The results of the seven studies that have been carried out were included in a recent meta-analysis [10], showing that there were no significant differences between (in-)patients with schizophrenia and controls regarding IL-2, IL-1α and IL-6, but IL-1β levels were decreased. These results should be regarded with caution because there are only very few studies, and furthermore it should be taken into account that only the study of McAllister et al. [9] is based on an explicitly experimental follow-up procedure, designed to reflect the dynamics of the onset of an acute episode in schizophrenia, and that was different from the merely cross-sectional approach of other studies.

In the peripheral blood of patients with schizophrenia, numerous cytokines and related proteins have been measured throughout the last 20 years, now providing us an abundance of studies that are increasingly hard to evaluate, in contrast to the situation ten years ago [7, 11]. It is therefore highly beneficial that Potvin et al. [12] and Miller et al. [10] recently conducted the first meta-analyses in the field, also carefully regarding the clinical details of single studies.

With respect to in vivo cytokine levels, Potvin and co-workers analyzed data from 62 studies. Significant results were obtained for the IL-1 receptor antagonist (anti-inflammatory), IL-6, and sIL-2R, however, not for IL-1β, sIL-6R, TNF-α and IL-2. The authors interpret their results as further evidence for immune activation and an inflammatory syndrome in schizophrenia. They also point out that these results (together with their finding on the in vitro expressions of cytokines) do not support the hypothesis of an abnormal balance between the pro-inflammatory TH-1 (downregulation) and the anti-inflammatory TH-2 (upregulation) response ('TH-2 shift') that has been discussed in recent years [13]. Only recently, Miller and associates conducted a second meta-analysis. One of their main findings is also an increase of pro-inflammatory cytokines mainly produced by cells from the innate immune system, namely IL-1β, IL-6, IL-12, TNF-α, TGF-β that seems to be associated with the onset of a psychotic episode. During neuroleptic treatment, IL-1β, IL-6 and TGF-β tend to normalize. IFN-γ, a cytokine that is, along with IL-2 mainly produced by T cells was found to be increased and remained stable throughout neuroleptic treatment, whereas IL-2 was found normal. Interestingly, the soluble IL-2 receptor was also found to be continuously increased, as this has been repeatedly demonstrated, particularly in the earlier studies of Rapaport and colleagues [14–17]. When T cells are activated, they shed sIL-2R from their membrane surface. Hence, an increase in the blood level of sIL-2R can be regarded as a global sign of immune activation and is found in many diseases with ongoing pro-inflammatory activity. Another study [18] focused on the role of the transforming growth factor beta1 (TGF-β_1), a protein that is (together with IL-10) produced by TH-3 cells and suppresses the function of TH-1 cells, particularly the production of IFN-γ and IL-2, but also TH-2 cells. It is not clear if this cell type differs from T-regulatory cells. In schizophrenia, TGF-β_1 levels were found to be significantly increased, together with an increased IFN-y/IL-4 ratio. This finding points to the possibility that a pro-inflammatory activity with a preponderance of TH-1 cell activity over TH-2 functions is counterbalanced by TH-3 activity that downregulates both. This is an interesting observation as it indicates the possibility of an upregulated set point in immune activation in schizophrenia, similar to a recent study that combined gene expression with cellular cytokine production activity [19, see below].

The interpretation of different cytokine levels in the sera of humans can be obscured by numerous factors that are not easily controlled, such as different cellular sources, of which some do not belong to the humoral immune system, but also intervening variables such as body fat, smoking, and possibly psychotic medications [20–23]. For these reasons, research on the cytokine secretion functions of different cell populations seems to be a more valid way to unravel immunological abnormalities in patients. There is now a considerable amount of studies that have utilized different methods of in vitro stimulation of immune-competent cells of patients with schizophrenia, in order to detect such irregularities.

The first studies pointed to findings that were intensively discussed later on: a diminished production of the pro-inflammatory cytokine IL-2 which is produced by numerous cells [24–27]. This finding is the most often reproduced result in the field, however not unanimously, mostly due to methodological heterogeneity [7, 28]. In their recent meta-analysis, Potvin et al. [12] included all 20 studies that are available today and found a significant effect size. It seemed to be irrelevant whether the patients were treated with an antipsychotic or not. Interestingly, they found no significant result for the production of IFN-γ, another pro-inflammatory cytokine that is, along with IL-2, also produced by TH-1 cells. In a series of systematic studies, our group has redundantly reported significant decreases in IFN-y as well as in IL-2, abnormalities that were neither due to medication or other unknown serum factors, nor to an increased production of counter-regulatory TH-2 cytokines, like IL-4 or IL-10 [8, 29–32]. However, it may be the case that methodological differences (e.g. the type of assays that were used) account for at least some of the divergent results. It is also of note that Potvin and co-workers did not find differences between patients and controls regarding the in-vitro production of the (anti-inflammatory) cytokines IL-4 and IL-10 [12]. Hence the studies that reflected the activity of T cells point to a downregulation of TH-1-based immunological functions, together with normal TH-2 functioning suggesting that there is no overactivation of TH-2-based immunity that may account for the depleted TH-1 functions.

Taken together, there is now some evidence for an activation of pro-inflammatory monocyte-based immunological activity from in vivo studies on cytokine levels, together with a downregulated production of T cell-based pro-inflammatory activity. Although there are some interesting additional findings [13], the above mentioned form the core lines of evidence. Nevertheless, confounding factors have to be regarded that cannot be completely ruled out by the meta-analyses, although attempts were made to differentiate between recent onset schizophrenia and relapse in chronic schizophrenia, as well as medicated and unmedicated patients. Also, there is one trial [33] using a criss-cross technique, by which the cells and the sera of patients and controls where replaced with each other in order to control for (unknown) factors in the sera, such as medication, that might be responsible for deviant TH-1 (or TH-2) cell activity. No significant influence of such replacements was detected. Today, it seems appropriate to focus on those findings in the field that have been found to be reproducible and to act with caution when it comes to an interpretation of these findings which are not easily understood. It is, however, interesting to keep in mind findings in rheumatoid arthritis or multiple sclerosis that show an increased pro-inflammatory monocyte-based activity during an active inflammation in the joints or the CNS together with a deficit in the production of TH-1 cytokines. This finding seems to be typical for autoimmune disorders and may reflect the dynamics in the process of inflammation, in so far as a postonset acute state may be reflected. It is intriguing in this respect (paralleling autoimmune disorders) to recognize recent findings on microglia activation in the brains of patients with acute schizophrenia in postmortem studies [34, 35, for a review, see 36], and, more convincing, in two studies with PET technology [37, 38], both using the (11)-PK11195 marker. Some of these observations have been summoned and put into the context of a 'microglia theory' of schizophrenia [39, 40].

In recent years, DNA/RNA chip technology has unfolded its potential in detecting gene expression of known putative risk genes but for transcriptome screening. Interestingly, these methods have hardly been used in the neuroimmunology of schizophrenia, although it can be expected that these methods substantially add to the evidence that has already been outlined by research in the blood and PBMCs of patients but also in animal models (see below). As an exception and coming from research into inflammatory mechanisms in autoimmune disorders and, lately bipolar disorders, the group of H. and R. Drexhage in Rotterdam has used a respective approach in schizophrenia. As in the case of patients with bipolar disorders [41] they used a quantitative polymerase chain reaction assay (qPCR) in blood monocytes to detect the mRNA expression levels of a series of genes that code for proteins that are involved in inflammatory processes. In 27 patients with recent-onset schizophrenia [42], they demonstrated a substantial overlap with bipolar patients concerning the up-regulation of gene activity in two clusters of pro-inflammatory genes (subset 1a, characterized by *ATF3* and *DUSP2* and subset 1b, characterized by *EGR3* and *MXD1*). In contrast, in another cluster (subset 2, characterized by *PTPN7* and *NAB2*) there was an upregulation in 62 of the bipolar patients but a downregulation in 48% of the schizophrenia patients. In total, about 60% of the patients with schizophrenia showed an aberrant expression of a set of 34 genes coding for pro-inflammatory proteins. In another study [19], they investigated the T cell functions in a slightly younger cohort of 26 patients, partially overlapping with their first cohort. Using modern FACS technology they found on the one hand a higher percentage of circulating monocytes with a pro-inflammatory gene expression signature together with activated CD3+CD25+ T cells (including increased levels of sIL-2R) and higher levels of pro-inflammatory IL-17 T cells. On the other hand they detected higher numbers of IL-4 containing TH-2 cells and increases in CD4+CD25highFoxP3+ T cells (that counter-regulate the pro-inflammatory activity of macrophages/monocytes). Specifically, IL-4 containing cells were increased in those patients in whom also pro-inflammatory monocytes were found, indicating a situation in which anti-inflammatory cells are activated at the same time as pro-inflammatory activity occurs, perhaps in order to control the latter (as the authors interpret). This study, as some others, point to the possibility that, instead of a dysbalance between the TH-1 and the TH-2 system, both systems are upregulated, at least in the beginning of the (mostly life-long) disease.

Another observation that fits into the possibility of pro-inflammatory activity in schizophrenia refers to the tryptophan catabolism. Tryptophan is metabolized to kynurenine, and in the next step to 3-hydroxykynurenine (3-HK, which is neurotoxic) and to kynurenic acid (KYNA, which acts neuron-protective). Under the influence of IFN-γ (IL-4 seems to exert the opposite influence), that increases the activities of indolamin-2,3-dioxygenase (IDO) and the kynurenin-3-monooxygenase, the balanced degradation of tryptophan is shifted onto the neurotoxic pathway with an excess of 3-HK and a diminished share of KYNA. This seems to be exactly the situation in schizophrenia [43, 44]. These authors have shown not only an increase of 3-HK and an decrease of KYNA in (untreated and drug-naïve) patients with schizophrenia, but also a dramatic decline of 3-HT during pharmacotherapy, together with a slight increase in KYNA which means a restabilization of the 'healthy' balance between those molecules and points to possible anti-inflammatory properties of (atypical) neuroleptic compounds (see below). However, an activation of the IDO leads to a blocking of the conversion of T-regulatory cells into TH-17 cells; hence, correlation between the numbers of these cell types should be expected, but this was, as yet, not the case [19].

Another line of research links findings in humans with animal models that are based on prenatal immune stimulation or infection. In a large cohort of mothers, Brown et al. [45] were able to show that increases in maternal levels of the proinflammatory cytokine IL-8 (due to infections) during the second and third trimester were associated with an increased risk for schizophrenia in the offspring. Recently, this group demonstrated results from the same cohort that show an association between the blood levels of IL-8 in the mothers and structural brain abnormalities in the schizophrenic offspring [46] that have often been described in clinical cases of schizophrenia [47–49], such as increases in ventricular size, volume decreases in left entorhinal cortex and right posterior cingulate, and volumetric decreases that approached significance in the right caudate, the putamen bilaterally, and the right superior temporal gyrus. Such associations were only found in the schizophrenic sample, but not in the controls. These observations strongly resemble morphological peculiarities that have also been found in the rodent prenatal infection models [50].

Rodent Models of Prenatal Immune Challenge

Based on the epidemiological studies which suggest that viral or bacterial infections during the first or second trimester of pregnancy increase the offspring's risk to develop schizophrenia in later life [4] during the last 15 years, rodent models of maternal infection or maternal immune activation have been generated. Pregnant females were either exposed to influenza virus or treated with compounds such as the synthetic double-stranded RNA polyinosinic:polycytidilic acid (poly IC) or lipopolysaccharides (LPS), a component of the cell wall of Gram-negative bacteria [51]. Poly IC mimics a viral infection activating the innate immune response by binding to toll-like receptor 3, LPS imitates a bacterial infection and binds to Toll-like receptor 4 differing in their downstream signaling [52]. While infectious pathogens or immune activating substances do not pass the placenta, cytokines that are released by the mother's immune cells enter the fetal bloodstream and may influence fetal development [51, 53]. The exact pattern of cytokine changes depend on the time point of immune stimulation and the immunogenic agent.

The offspring of prenatally immune stimulated mothers show a number of behavioral abnormalities in adulthood which are thought to resemble phenotypes that also occur in schizophrenia patients. Deficits in the prepulse inhibition of the acoustic startle response is one of the most robust findings which could be replicated in a number of studies in rats and mice after prenatal influenza infection as well as after prenatal polyIC or LPS driven immune stimulation [54–58]. The behavioral abnormalities also include increased sensitivity to psychostimulants, e.g. locomotion induced by the indirect dopamine agonist amphetamine or the noncompetitive NMDA receptor antagonist dizocilpine (MK-801), decreased latent inhibition, reduced social exploration, and deficits in spatial and working memory versions of the Morris water-maze [56, 58–65]. Comparable to the onset of schizophrenia in humans it has been shown that behavioral effects with regard to prepulse and latent inhibition as well as amphetamine-induced locomotion occur only after puberty in response to prenatal poly IC administration [65–68]. So the type of behavioral alterations as well as their ontogeny in models of prenatal immune activation is similar to that of schizophrenia in humans.

Therapeutic and preventive pharmacological treatments with different substances ameliorate some of these behavioral abnormalities. Clozapine treatment in adult animals that were prenatally stimulated with poly IC improves object recognition memory after 2 weeks of treatment [57]. Three weeks of clozapine administration is sufficient to attenuate working memory deficits without affecting reduced adult hippocampal neurogenesis [62]. The preventive treatment of adolescent animals from postnatal day 35 to PND 70 with either haloperidol, the atypical neuroleptic clozapine or the selective serotonin reuptake inhibitor fluoxetine also significantly affects the behavioral abnormalities in adult animals. While haloperidol treatment during adolescence normalizes deficits in latent inhibition and reduced the hypersensitivity to amphetamine in adult mice after prenatal poly IC treatment, it also increases spontaneous locomotor activity in adult control animals as a side effect. Clozapine ameliorates the deficits in prepulse inhibition as well as latent inhibition, and fluoxetine was effective in improving prepulse inhibition deficits and reducing amphetamine-induced hyperlocomotion without side effects in the control group [69]. This efficacy of established drugs suggests that prenatal immune challenge in rodents leads to at least partly comparable molecular changes that appear in schizophrenia.

The underlying biological mechanisms of how the immune activation during pregnancy leads to the behavioral abnormalities in the adult offspring are far from being understood. There are several morphological, cellular and biochemical alterations that have been observed in the offspring of prenatally immune challenged mothers. Some of them are directly linked to the immune system while others are not. A closer link to the immune system is provided by findings on astrocytes, microglia and central cytokine expression. An increased number of astrocytes was reported in the cortex of young mice at postnatal day 14 and in the hippocampus at PND 35 after prenatal influenza viral infection [70]. LPS treatment of pregnant rats at gestational day 18 induces reactive astrogliosis in the corpus callosum at postnatal day 23 [71]. And rats whose mothers were treated with IL-6 early in pregnancy at days 8, 10, 12 or later at days 16, 18, 20 show increased astrogliosis as well as larger and more highly branched astrocytes in the hippocampus at 24 weeks of age [63]. Microglial activation in the hippocampus of newborn rats was found after LPS exposure at embryonic day 15 [72]. LPS administration at embryonic day 10.5 increased microglia density in substantia nigra of adult rats at 4, 14 and 17 months of age [73]. And in mice exposed to poly IC at embryonic day 9 an increased density of microglial cells was found in the hippocampus and striatum while there was no difference in frontal and occipital cortex [74]. With regard to cytokines different effects of prenatal LPS stimulation were reported. High doses of LPS at embryonic day 10.5 lead to elevated levels of TNF-α in the striatum and mesencephalon of rats at postnatal day 21 and in the striatum at PND 120 [75, 76]. High doses of LPS at gestational days 18 and 19 also increase levels of TNF-α, IL-1β, and IL-6 in the rat brain at postnatal day 7 [77, 78]. In contrast, reduced levels of TNF-α have been found in the frontal cortex of rats at postnatal day 7 following exposure to LPS at gestational days 14, 15 and 16 [79]. And after LPS administration at embryonic days 19 and 20 reduced brain levels of TNF-α mRNA are found in rats at postnatal day 7 while IL-1β mRNA is increased one day after birth [80].

In addition to central effects prenatal immune challenge also influences the peripheral immune response of the offspring. At postnatal day 21 both, serum levels as well as brain mRNA levels of IL-1β, IL-6, and TNF-α after acute LPS stimulation are significantly lower in rats that were prenatally treated with LPS at embryonic day 18 [81]. However, this effect seems to depend on the age of the animals. Prior to weaning at 19 days of age rats show a reduced, nearly absent response in serum TNF-α levels to acute LPS exposure when they had been treated with LPS at embryonic days 16, 18 and 20 compared to saline-treated controls. During adolescence at 50 days of age there is no difference between the groups that differ in prenatal LPS exposure. And in adulthood at postnatal day 100 prenatally LPS-exposed rats react to acute LPS stimulation with significantly increased serum TNF-α levels [82]. So the innate immune response of rats seems to develop delayed with a missing response in prepubertal animals that normalizes during adolescence and becomes hyperresponsive in adulthood. In general, the effects of prenatal immune stimulation on immune system function and integrity are far from being understood. Nevertheless, these numerous findings strongly support the notion that immune function is markedly influenced by prenatal immunogen exposure.

Antipsychotic Drugs in Schizophrenia: Impact on Immunological Functions

The anti-inflammatory properties of 1st generation ('traditional') neuroleptics as well as 2nd generation ('atypical') neuroleptics have gained some interest throughout the years for two reasons: one was the need to exclude a substantial effect of neuroleptics on the cytokine levels or the cellular production of cytokines in the sera of schizophrenic patients, in order to differentiate between genuine signs of inflammatory immune activity or medication effects. Still today, the influence of medication is not completely clear, although there seem to be substances which are more (clozapine) or less (risperidone, haloperidol) influential.

The second reason was based on the speculation that not only the main effects of neuroleptics on neurotransmitter systems may be responsible for their therapeutic action, but also their anti-inflammatory side effects, particularly as there seem to be multiple cross-links between cytokines and neurotransmitters in the brain.

Neuroleptics have experimentally been used as an adjunct to treat diseases with an inflammatory background, e.g. there is a published trial with haloperidol in rheumatoid arthritis that resulted in a positive, namely anti-inflammatory, effect in the verum group [83]. In patients with schizophrenia there are now numerous clinical studies that have indicated the influence of neuroleptics, particularly atypical neuroleptics on cytokine networks. These results have been reviewed in greater detail elsewhere [23, 84]. Although the heterogeneity of the data obscures a definite understanding of the influence of different antipsychotics on different cytokine actions, the global conclusion that can be drawn from a number of studies is that antipsychotics have an immune-modulatory if not an anti-inflammatory effect. It has repeatedly been shown that the levels of IL-6 as well as IL-2 seem to decrease during treatment. Furthermore, the the anti-inflammatory sIL-1RA and IL-10 levels seem to increase [85–90]. Consequently, it has been proposed that the anti-inflammatory properties of neuroleptics may also be responsible for their antipsychotic efficacy [13, 91].

An interesting finding was presented by Zhang et al. [92]. They showed in a clinical sample that both risperidone and haloperidol reduced elevated serum IL-2 levels during the therapy of a schizophrenic episode. However, neither risperidone nor haloperidol showed a significant influence on the higher serum IL-6 or IL-8 concentrations. The levels of IL-2 and IL-8 were correlated at baseline and during therapy and it was shown that patients presenting with low concentrations of serum IL-2 or IL-8 at baseline showed greater improvement and patients presenting with higher serum IL-2 or IL-8 concentrations at baseline showed less improvement after treatment. Also, Lin et al. [93] found increased levels of IL-6 in treatment-resistant patients, which were negatively correlated with the anti-inflammatory endogenous protein CC16.

However, it should be kept in mind that the biological mechanisms for changes in cytokine serum levels and production in vitro are far from fully understood. Furthermore, as these changes parallel the clinical improvement they may be direct effects of certain antipsychotic compounds, but it may also be the case that these changes just co-occur with the improvement of the symptoms of schizophrenia due to effects of other, perhaps unknown neurobiological mechanisms that are involved in the biological basis of the disease.

Anti-Inflammatory Drugs in Schizophrenia: Impact on Antipsychotic Action

Acetylsalicylate

Acetylsalicylate (ASA) is a classic drug that has analgesic, fever-lowering and anti-inflammatory properties. ASA (like e.g. ibuprofen) is a nonselective cyclooxygenase-1 (COX-1) and cyclooxygenase-2 (COX-2) inhibitor. It suppresses thrombocyte aggregation, due to COX-1 activity that is the only isoenzyme which is found in platelets. COX-2 (now also called prostaglandin endoperoxide synthase-2) is expressed in immune cells and is upregulated by bacterial lipopolysaccharides as well as by different cytokines. COX-2 inhibitors have anti-inflammatory properties. By their (additional) inhibition of the prostaglandin synthesis they reduce pain. In the light of the findings on the involvement of pro-inflammatory activity in acute schizophrenia, it was tempting to explore the possible impact of ASA on the treatment of the disease, namely since there had already been successful studies with the selective COX-2 inhibitor celecoxib (see below). There is now one randomized, double-blind, placebo-controlled trial with ASA as an adjunct to antipsychotic medication for the therapy of schizophrenia [94]. Seventy antipsychotic-treated patients were randomized to an add-on treatment with ASA (1,000 mg/day) or placebo and followed for 3 months. There was a significant effect on the decrease of the global and the positive symptoms. Interestingly, the treatment effect of aspirin was significantly larger in patients with a lowered serum IFN-y/IL-4 ratio. The authors interpreted this finding in the way that the greater the TH1/TH2 dysbalance, the better was the effect of ASA. Due to the fact that ASA is an old drug that has been used extremely frequently and for a long time these findings came up somewhat unexpected but, from a biological viewpoint, fit into the context and directly suggest that targeting immune functions in patients with schizophrenia is responsible for the (additional) beneficial effect of an anti-inflammatory agent.

Celecoxib

Celecoxib is a modern, nonsteroidal anti-inflammatory drug that is used in the treatment of inflammatory diseases like osteoarthritis, rheumatoid arthritis but also in acute pain. It is a highly selective COX-2 inhibitor. In 2002, Müller et al. [95] carried out the first prospective, placebo-controlled double-blind study with 50 patients with an acute exacerbation of schizophrenia. They were assigned at random to either risperidone plus an add-on of celecoxib or risperidone plus placebo. Whereas both groups showed significant improvement in scores on the Positive and Negative Syndrome Scale, the celecoxib group showed significantly greater improvement in the total score. The additional treatment with celecoxib had significant positive effects, especially during the first weeks of the therapy. Furthermore, this trial indicated that targeting pro-inflammatory mechanisms might produce an antipsychotic effect. In their discussion, the authors also pointed out the possibility that a nonimmunological therapeutic effect of celecoxib mediated by the N-methyl-D-aspartic acid receptor might have taken place. Nevertheless, the main result of this study was reproduced twice, in neuroleptic-naive, first-episode patients [96] as well as in more chronic patients [97]. However, when an additional treatment celecoxib was used in continuously ill, chronic patients, no superiority over placebo was found [98]. Consequently, in an analysis of these results it appears plausible that an anti-inflammatory treatment may be useful particularly in earlier stages of schizophrenia [99, 100]. And indeed, a recent, second study of these group with first-episode patients [101] showed a beneficial effect of an add-on of celecoxib (to amisulpride). Accordingly, in a recent meta-analysis, Sommer et al. [102] found a significant positive effect of celecoxib on the treatment of schizophrenia. However, the biological mechanisms by which celecoxib exert its influence on symptom reduction of schizophrenia are not well understood. One possibility may be a rebalancing of pro- and anti-inflammatory mechanisms [99] as this has been shown for celecoxib earlier [103]. Furthermore, it may also influence the dysbalanced tryptophan to KYNA ratio [104]. This therapeutic pathway may be due to the notion that KYNA acts as an NMDA receptor antagonist and therefore, like PCP, has pro-psychotic properties [105].

Minocycline

Minocycline is a second-generation broad-spectrum tetracycline that has been shown to exert a positive effect in some neurologic disorders with underlying inflammatory activity, namely multiple sclerosis, amyotrophic lateral sclerosis, Huntington's disease, and Parkinson's disease. Minocycline has anti-inflammatory but also neuroprotective properties. It inhibits apoptosis by a downregulation of pro-inflammatory cytokine production, particularly via TNF-α. It seems to have a direct effect on activated T cells that leads to a reduced ability of T cells to contact microglia by which the cytokine production in T cell-microglia signal transduction is downregulated [106]. Minocycline possibly also inhibits the activation of microglia by a blockade of NF-kappa B nuclear translocation. Also, by inhibition of the nitric oxide synthase it blocks nitric oxide-induced neurotoxicity and may thus influence the glutamatergic neurotransmission.

In animal models of schizophrenia, minocycline was shown to be efficacious in preventing or reducing psychosis-like behavior. As a schizophrenia model, treatment with dizocilpine, a noncompetitive NMDA antagonist, was used in two studies. Levcovitz et al. [107] demonstrated that, in rats, the intraperitoneal injection of dizocilpine led to changes in the acoustic startle response and the prepulse inhibition, as well as changes in visuospatial memory that have also been described in other animal models for schizophrenia (see above). Treatment with minocycline before the administration of dizocilpine prevented these neurophysiological deficits. Furthermore, these researchers demonstrated that, in reducing the relevant abnormalities, minocycline was as efficacious as haloperidol. By using a similar design in mice, Zhang et al. [108] demonstrated that reduced prepulse inhibition and hyperlocomotion (both induced by dizocilpine) could be prevented by pretreatment with minocycline. Furthermore, by in vivo microdialysis, they demonstrated that pretreatment with minocycline attenuated the dizocilpine-induced dopamine increase in the frontal cortex and the striatum, suggesting that this might be at least one important mechanism by which minocycline exerts a beneficial effect on psychosis-like behavior.

Other researchers used minocycline as treatment in order to reverse schizophrenia-resembling abnormalities. It is known that phencyclidine (PCP) induces a variety of behavioral changes in rodents (see above). Using a 'novel object recognition test' that was designed to detect explorative behavior, Fujita et al. [109] did not only demonstrate a downregulated preference for exploration in the phencyclidine-treated mice, but also a significant improvement of these deficits after treatment with minocycline. In a similar experiment in which methamphetamine was used, the subsequent treatment with minocycline led to a significant reduction of induced memory deficits. Hence, there is some evidence now that minocycline, by different neurochemical pathways, might also attenuate negative and cognitive symptoms in humans.

Miyaoka et al. [110] presented 2 cases of patients with catatonic schizophrenia in whom the (sole!) treatment with minocycline was efficacious. As the next step, Miyaoka [111] administered minocycline (150 mg/day) for 4 weeks as an add-on to antipsychotic medication to 22 patients with schizophrenia in an open-label design. There was a significant clinical improvement in the patients who were treated in the minocycline arm. The effect maintained throughout 4 weeks after the end of minocycline treatment, and no adverse events occurred. The first double-blind, randomized, placebo-controlled study appeared only recently: Levkovitz et al. [107] recruited 70 patients with 'early-phase' schizophrenia and randomly allocated 54 of them in a 2:1 ratio to minocycline 200 mg/day. All patients were put on antipsychotic treatment with an atypical antipsychotic only shortly before the study entry and were followed for 6 months. Minocycline had a significant beneficial effect on negative symptoms, but also on cognitive functioning particularly executive ('frontal') functions. Only few adverse side effects occurred.

Omega-3 Fatty Acids

Omega-3 fatty acids, specifically eicosapentaenoic acid (EPA) has a beneficial effect on a number of diseases, including arteriosclerosis and survival after myocardial infarction, but also immune-mediated diseases like rheumatoid arthritis [112, 113]. It has recently also been shown in a meta-analysis [114] that the administration of EPA as an adjunct to antidepressive pharmacotherapy is beneficial in major depression and that EPA may also have mood-stabilizing properties in bipolar disorder [115]. The biological mechanisms which underlie these effects are not completely understood, although there are lines of evidence for cell-membrane stabilizing properties, modulation of neurotransmission, anti-inflammatory effects, but also neuroprotection [116–118]. All these mechanisms may be of interest in the light of theories that contribute to the understanding of the biological basis of schizophrenia.

Particularly EPA has been tried in the prevention and treatment of schizophrenia; however, with ambiguous results. In a sample of young individuals at high risk for schizophrenia, a preventive effect on the transition to psychosis was shown in a placebo-controlled study with polyunsaturated fatty acids [119]. There are now 6 more randomized, double-blind, placebo-controlled trials in which mainly EPA was used as an adjunct to antipsychotic medication in patients with schizophrenia, and one trial in which EPA was used alone. A positive effect was shown in four studies, while no such effect could be found in three. A recent meta-analysis of these latter trials [120] could not confirm a beneficial effect on the symptoms of schizophrenia.

Taking together the outcomes of anti-inflammatory intervention studies in schizophrenia together, it appears that mainly those drugs proved to be useful in the reduction of psychotic symptoms (and psychosis-related cognitive dysfunction) that have clear anti-inflammatory properties. It is also remarkable that the evidence for beneficial effects has been proven by a substantial number of placebo-controlled, randomized, double-blind trails. Although as yet it cannot be ruled out that other effects that could not be controlled (e.g. a direct modulation of NMDA neurotransmission) may play a role, these findings seem to be concordant with the many replicated results that have pointed to a pro-inflammatory activity, at least in a subpopulation of patients with schizophrenia. For a proof-of-principle strategy it would be useful to monitor the putative immunological abnormalities on different levels (sera, cells, gene expression, microglia) during an anti-inflammatory intervention. Although such a strategy will claim substantial efforts, it will provide us with highly relevant information (1) whether the abnormalities are regularly found in the same patients, (2) whether it is this group of patients which profits from anti-inflammatory strategies, and (3) whether this profit is accompanied by a down-regulation of pro-inflammatory activity.

References

1 Gejman PV, Sanders AR, Kendler KS: Genetics of schizophrenia: new findings and challenges. Annu Rev Genomics Hum Genet 2011;12:121–144.

2 Pickard B: Progress in defining the biological causes of schizophrenia. Expert Rev Mol Med 2011;13:e25.

3 Rujescu D, Genius J, Benninghoff J, Giegling I: Current progress in the genetic research of schizophrenia: relevance for drug discovery? Curr Pharm Biotechnol 2012;in press.

4 Brown AS, Derkits EJ: Prenatal infection and schizophrenia: a review of epidemiologic and translational studies. Am J Psychiatry 2010;167:261–280.

5 Markham JA, Koenig JI: Prenatal stress: role in psychotic and depressive diseases. Psychopharmacology (Berl) 2011;214:89–106.

6 Arolt V, Rothermundt M, Peters M, Leonard B: Immunological research in clinical psychiatry: report on the consensus debate during the 7th Expert Meeting on Psychiatry and Immunology. Mol Psychiatry 2002;7:822–826.

7 Hinze-Selch D, Pollmächer T: In vitro cytokine secretion in individuals with schizophrenia: results, confounding factors, and implications for further research. Brain Behav Immun 2001;15:282–318.

8 Rothermundt M, Arolt V, Leadbeater J, Peters M, Rudolf S, Kirchner H: Cytokine production in unmedicated and treated schizophrenic patients. Neuroreport 2000;11:3385–3388.

9 McAllister CG, van Kammen DP, Rehn TJ, Miller AL, Gurklis J, Kelley ME, Yao J, Peters JL: Increases in CSF levels of interleukin-2 in schizophrenia: effects of recurrence of psychosis and medication status. Am J Psychiatry 1995;152:1291–1297.

10 Miller BJ, Buckley P, Seabolt W, Mellor A, Kirkpatrick B: Meta-analysis of cytokine alterations in schizophrenia: clinical status and anti-psychotic effects. Biol Psychiatry 2011;70:663–671.

11 Rothermundt M, Arolt V, Bayer TA: Review of immunological and immunopathological findings in schizophrenia. Brain Behav Immun 2001;15:319–339.

12 Potvin S, Stip E, Sepehry AA, Gendron A, Bah R, Kouassi E: Inflammatory cytokine alterations in schizophrenia: a systematic quantitative review. Biol Psychiatry 2008;63:801–808.

13 Müller N, Schwarz MJ: Immune system and schizophrenia. Curr Immunol Rev 2010;6:213–220.

14 Rapaport MH, McAllister CG, Pickar D, Nelson DL, Paul SM: Elevated levels of soluble interleukin 2 receptors in schizophrenia. Arch Gen Psychiatry 1989;46:291–292.

15 Rapaport MH, Doran AR, Nelson DL, McAllister C, Magliozzi JR, Paul SM: Haloperidol and soluble interleukin-2 receptors. Biol Psychiatry 1991;30:1063–1064.

16 Rapaport MH, McAllister CG, Kim YS, Han JH, Pickar D, Nelson DL, et al: Increased serum soluble interleukin-2 receptors in Caucasian and Korean schizophrenic patients. Biol Psychiatry 1994;35:767–771.

17 Rapaport MH, Lohr JB: Serum-soluble interleukin-2 receptors in neuroleptic-naive schizophrenic subjects and in medicated schizophrenic subjects with and without tardive dyskinesia. Acta Psychiatr Scand 1991;90:311–315.

18 Kim YK, Myint AM, Lee BH, Han CS, Lee HJ, Kim DJ, et al: Th1, Th2 and Th3 cytokine alteration in schizophrenia. Prog Neuropsychopharmacol Biol Psychiatry 2004;28:1129–1134.

19 Drexhage RC, Hoogenboezem TA, Cohen D, Versnel MA, Nolen WA, van Beveren NJ, Drexhage HA: An activated set point of T-cell and monocyte inflammatory networks in recent-onset schizophrenia patients involves both pro- and anti-inflammatory forces. Int J Neuropsychopharmacol 2011;14:746–755.

20 Beumer W, Drexhage RC, De Wit H, Versnel MA, Drexhage HA, Cohen D: Increasedlevel of serum cytokines, chemokines and adipokines in patients with schizophrenia is associated with disease and metabolic syndrome. Psychoneuroendocrinology 2012;in press.

21 Pollmächer T, Hinze-Selch D, Mullington J: Effects of clozapine on plasma cytokine and soluble cytokine receptor levels. J Clin Psychopharmacol 1996;16:403–409.

22 Pollmächer T, Hinze-Selch D, Fenzel T, Kraus T, Schuld A, Mullington J: Plasma levels of cytokines and soluble cytokine receptors during treatment with haloperidol. Am J Psychiatry 1997;154:1763–1765.

23 Pollmächer T, Haack M, Schuld A, Kraus T, Hinze-Selch D: Effects of antipsychotic drugs on cytokine networks. J Psychiatr Res 2000;34:369–382.

24 Bessler H, Levental Z, Karp L, Modai I, Djaldetti M, Weizman A: Cytokine production in drug-free and neuroleptic-treated schizophrenic patients. Biol Psychiatry 1995;38:297–302.

25 Ganguli R, Rabin BS, Belle SH: Decreased interleukin-2 production in schizophrenic patients. Biol Psychiatry 1989;26:427–430.

26 Ganguli R, Brar JS, Chengappa KR, DeLeo M, Yang ZW, Shurin G, et al: Mitogen-stimulated interleukin-2 production in never-medicated, first-episode schizophrenic patients. The influence of age at onset and negative symptoms. Arch Gen Psychiatry 1995;52:668–672.

27 Villemain F, Chatenoud L, Guillibert E, Pelicier Y, Bach JF: Decreased production of interleukin-2 in schizophrenia. Ann NY Acad Sci 1987;496:669–675.

28 Haack M, Hinze-Selch D, Fenzel T, Kraus T, Kuhn M, Schuld A, et al: Plasma levels of cytokines and soluble cytokine receptors in psychiatric patients upon hospital admission: effects of confounding factors and diagnosis. J Psychiatr Res 1989;33:407–418.

29 Hornberg M, Arolt V, Wilke I, Kruse A, Kirchner H: Production of interferons and lymphokines in leukocyte cultures of patients with schizophrenia. Schizophr Res 1995;15:237–242.

30 Wilke I, Arolt V, Rothermundt M, Weitzsch C, Hornberg M, Kirchner H: Investigations of cytokine production in whole blood cultures of paranoid and residual schizophrenic patients. Eur Arch Psychiatry Clin Neurosci 1996;246:279–284.

31 Rothermundt M, Arolt V, Leadbeater J, Peters M, Rudolf S, Kirchner H: Cytokine production in unmedicated and treated schizophrenic patients. Neuroreport 2000;11:3385–3388.

32 Arolt V, Rothermundt M, Wandinger KP, Kirchner H: Decreased in vitro production of interferon-gamma and interleukin-2 in whole blood of patients with schizophrenia during treatment. Mol Psychiatry 2000;5:150–158.

33 Rothermundt M, Arolt V, Weitzsch C, Eckhoff D, Kirchner H: Immunological dysfunction in schizophrenia: a systematic approach. Neuropsychobiology 1998;37:186–193.

34 Bayer TA, Buslei R, Havas L, Falkai P: Evidence for activation of microglia in patients with psychiatric illnesses. Neurosci Lett 1999;271:126–128.

35 Foster R, Kandanearatchi A, Beasley C, Williams B, Khan N, Fagerhol MK, Everall IP: Calprotectin in microglia from frontal cortex is up-regulated in schizophrenia: evidence for an inflammatory process? Eur J Neurosci 2006;24:3561–3566.

36 Bernstein HG, Steiner J, Bogerts B: Glial cells in schizophrenia: pathophysiological significance and possible consequences for therapy. Expert Rev Neurother 2009;9:1059–1071.

37 van Berckel BN, Bossong MG, Boellaard R, Kloet R, Schuitemaker A, Caspers E, Luurtsema G, Windhorst AD, Cahn W, Lammertsma AA, Kahn RS: Microglia activation in recent-onset schizophrenia: a quantitative [R]-11C]PK11195 positron emission tomography study. Biol Psychiatry 2008;64:820–822.

38 Doorduin J, de Vries EF, Willemsen AT, de Groot JC, Dierckx RA, Klein HC: Neuroinflammation in schizophrenia-related psychosis: a PET study. J Nucl Med 2009;50:1801–1817.

39 Monji A, Kato T, Kanba S: Cytokines and schizophrenia: microglia hypothesis of schizophrenia. Psychiatry Clin Neurosci 2009;63:257–65.

40 Monji A, Kato TA, Mizoguchi Y, Horikawa H, Seki Y, Kasai M, Yamauchi Y, Yamada S, Kanba S: Neuroinflammation in schizophrenia especially focused on the role of microglia. Prog Neuropsychopharmacol Biol Psychiatry 2011;in press.

41 Padmos RC, Hillegers MH, Knijff EM, Vonk R, Bouvy A, Staal FJ, de Ridder D, Kupka RW, Nolen WA, Drexhage HA: A discriminating messenger RNA signature for bipolar disorder formed by an aberrant expression of inflammatory genes inmonocytes. Arch Gen Psychiatry 2008;65:395–407.

42 Drexhage RC, van der Heul-Nieuwenhuijsen L, Padmos RC, van Beveren N, Cohen D, Versnel MA, Nolen WA, Drexhage HA: Inflammatory gene expression in monocytes of patients with schizophrenia: overlap and difference with bipolar disorder. A study in naturalistically treated patients. Int J Neuropsychopharmacol 2010;13:1369–1381.

43 Myint AM, Schwarz MJ, Verkerk R, Mueller HH, Zach J, Scharpé S, Steinbusch HW, Leonard BE, Kim YK: Reversal of imbalance between kynurenic acid and 3-hydroxykynurenine by antipsychotics in medication-naïve and medication-free schizophrenic patients. Brain Behav Immun 2011;25:1576–1581.

44 Myint AM: Kynurenines: From the perspective of major psychiatric disorders. FEBS J 2012;in press.

45 Brown AS, Hooton J, Schaefer CA, Zhang H, Petkova E, Babulas V, Perrin M, Gorman JM, Susser ES: Elevated maternal interleukin-8 levels and risk of schizophrenia in adult offspring. Am J Psychiatry 2004;161:889–895.

46 Ellman LM, Deicken RF, Vinogradov S, Kremen WS, Poole JH, Kern DM, Tsai WY, Schaefer CA, Brown AS: Structural brain alterations in schizophrenia following fetal exposure to the inflammatory cytokine interleukin-8. Schizophr Res 2010;121:46–54.

47 Borgwardt SJ, Riecher-Rossler A, Dazzan P, Chitnis X, Aston J, Drewe M, et al: Regional gray matter volume abnormalities in the at risk mental state. Biol Psychiatry 2007;61:1148–1156.

48 Pantelis C, Velakoulis D, McGorry PD, Wood SJ, Suckling J, Phillips LJ, et al: Neuroanatomical abnormalities before and after onset of psychosis: a cross-sectional and longitudinal MRI comparison. Lancet 2003;361:281–288.

49 Wright IC, Rabe-Hesketh S, Woodruff PW, David AS, Murray RM, Bullmore ET: Meta-analysis of regional brain volumes in schizophrenia. Am J Psychiatry 2000;157:16–25.

50 Patterson PH: Immune involvement in schizophrenia and autism. Etiology, pathology and animal models. Behav Brain Res 2009;204:313–321.

51 Boksa P: Effects of prenatal infection on brain development and behavior: a review of findings from animal models. Brain Behav Immun 2010;24:881–897.

52 Yang IV, Jiang W, Rutledge HR, Lackford B, Warg LA, De Arras L, Alper S, Schwartz DA, Pisetsky DS: Identification of novel innate immune genes by transcriptional profiling of macrophages stimulated with TLR ligands. Mol Immunol 2011;48:1886–1895.

53 Meyer U, Yee BK, Feldon J: The neurodevelopmental impact of prenatal infections at different times of pregnancy: the earlier the worse? Neuroscientist 2007;13:241–256.

54 Fortier ME, Luheshi GN, Boksa P: Effects of prenatal infection on prepulse inhibition in the rat depend on the nature of the infectious agent and the stage of pregnancy. Behav Brain Res 2007;181:270–277.

55 Makinodan M, Tatsumi K, Manabe T, Yamauchi T, Makinodan E, Matsuyoshi H, Shimoda S, Noriyama Y, Kishimoto T, Wanaka A: Maternal immune activation in mice delays myelination and axonal development in the hippocampus of the offspring. J Neurosci Res 2008;86:2190–2200.

56 Meyer U, Feldon J, Schedlowski M, Yee BK: Towards an immuno-precipitated neurodevelopmental animal model of schizophrenia. Neurosci Biobehav Rev 2005;29:913–947.

57 Ozawa K, Hashimoto K, Kishimoto T, Shimizu E, Ishikura H, Iyo M: Immune activation during pregnancy in mice leads to dopaminergic hyperfunction and cognitive impairment in the offspring: a neurodevelopmental animal model of schizophrenia. Biol Psychiatry 2006;59:546– 554.

58 Shi L, Fatemi SH, Sidwell RW, Patterson PH: Maternal influenza infection causes marked behavioral and pharmacological changes in the offspring. J Neurosci 2003;23:297–302.

59 Fortier ME, Joober R, Luheshi GN, Boksa P: Maternal exposure to bacterial endotoxin during pregnancy enhances amphetamine-induced locomotion and startle responses in adult rat offspring. J Psychiatr Res 2004;38:335–345.

60 Lanté F, Meunier J, Guiramand J, Maurice T, Cavalier M, de Jesus Ferreira MC, Aimar R, Cohen-Solal C, Vignes M, Barbanel G: Neurodevelopmental damage after prenatal infection: role of oxidative stress in the fetal brain. Free Radic Biol Med 2007;42:1231–1245.

61 Meyer U, Nyffeler M, Yee BK, Knuesel I, Feldon J: Adult brain and behavioral pathological markers of prenatal immune challenge during early/middle and late fetal development in mice. Brain Behav Immun 2008;22:469–486.
62 Meyer U, Knuesel I, Nyffeler M, Feldon J: Chronic clozapine treatment improves prenatal infection-induced working memory deficits without influencing adult hippocampal neurogenesis. Psychopharmacology (Berl) 2010;208:531–543.
63 Samuelsson AM, Jennische E, Hansson HA, Holmäng A: Prenatal exposure to interleukin-6 results in inflammatory neurodegeneration in hippocampus with NMDA/GABA(A) dysregulation and impaired spatial learning. Am J Physiol Regul Integr Comp Physiol 2006;290:R1345–R1356.
64 Smith SEP, Li J, Garbett K, Mirnics K, Patterson PH: Maternal immune activation alters fetal brain development through interleukin-6. J Neurosci 2007;27:10695–10702.
65 Zuckerman L, Rehavi M, Nachman R, Weiner I: Immune activation during pregnancy in rats leads to a postpubertal emergence of disrupted latent inhibition, dopaminergic hyperfunction, and altered limbic morphology in the offspring: a novel neurodevelopmental model of schizophrenia. Neuropsychopharmacology 2003;28:1778–1789.
66 Meyer U, Nyffeler M, Schwendener S, Knuesel I, Yee BK, Feldon J: Relative prenatal and postnatal maternal contributions to schizophrenia related neurochemical dysfunction after in utero immune challenge. Neuropsychopharmacology 2008;33:441–456.
67 Meyer U, Schwendener S, Feldon J, Yee BK: Prenatal and postnatal maternal contributions in the infection model of schizophrenia. Exp Brain Res 2006;173:243–257.
68 Vuillermot S, Weber L, Feldon J, Meyer U: A longitudinal examination of the neurodevelopmental impact of prenatal immune activation in mice reveals primary defects in dopaminergic development relevant to schizophrenia. J Neurosci 2010;30:1270–1287.
69 Meyer U, Spoerri E, Yee BK, Schwarz MJ, Feldon J: Evaluating early preventive antipsychotic and antidepressant drug treatment in an infection-based neurodevelopmental mouse model of schizophrenia. Schizophr Bull 2010;36:607–623.
70 Fatemi SH, Emamian ES, Sidwell RW, Kist DA, Stary JM, Earle JA, Thuras P: Human influenza viral infection in utero alters glial fibrillary acidic protein immunoreactivity in the developing brains of neonatal mice. Mol Psychiatry 2002;7:633–640.
71 Paintlia MK, Paintlia AS, Contreras MA, Singh I, Singh AK: Lipopolysaccharide-induced peroxysomal dysfunction exacerbates cerebral white matter injury: attenuation by N-acetyl cysteine. Exp Neurol 2008;210:560–576.
72 Roumier A, Pascual O, Béchade C, Wakselman S, Poncer JC, Réal E, Triller A, Bessis A: Prenatal activation of microglia induces delayed impairment of glutamatergic synaptic function. PLoSOne 2008;3:e2595.
73 Ling Z, Zhu Y, Tong CW, Snyder JA, Lipton JW, Carvey PM: Prenatal lipopolysaccharide does not accelerate progressive dopamine neuron loss in the rat as a result of normal aging. Exp Neurol 2009;216:312–320.
74 Juckel G, Manitz MP, Brüne M, Friebe A, Heneka MT, Wolf RJ: Microglial activation in a neuroinflammational animal model of schizophrenia – a pilot study. Schizophr Res 2011;131:96–100.
75 Ling Z, Gayle DA, Ma SY, Lipton JW, Tong CW, Hong JS, Carvey PM: In utero bacterial endotoxin exposure causes loss of tyrosine hydroxylase neurons in the postnatal rat midbrain. Mov Disord 2002;17:116–124.
76 Ling ZD, Chang Q, Lipton JW, Tong CW, Landers TM, Carvey PM: Combined toxicity of prenatal bacterial endotoxin exposure and postnatal 6-hydroxydopamine in the adult rat midbrain. Neuroscience 2004;124:619–628.
77 Kumral A, Baskin H, Yesilirmak DC, Ergur BU, Aykan S, Genc S, Genc K, Yilmaz O, Tugyan K, Giray O, Duman N, Ozkan H: Erythropoietin attenuates lipopolysaccharide-induced white matter injury in the neonatal rat brain. Neonatology 2007;92:269–278.
78 Yesilirmak DC, Kumral A, Baskin H, Ergur BU, Aykan S, Genc S, Genc K, Yilmaz O, Tugyan K, Giray O, Duman N, Ozkan H: Activated protein C reduces endotoxin-induced white matter injury in the developing rat brain. Brain Res 2007;1164:14–23.
79 Gilmore JH, Jarskog LF, Vadlamudi S: Maternal poly I: C exposure during pregnancy regulates TNF-alpha, BDNF, and NGF expression in neonatal brain and the maternal-fetal unit of the rat. J Neuroimmunol 2005;159:106–112.
80 Rousset CI, Chalon S, Cantagrel S, Bodard S, Andres C, Gressens P, Saliba E: Maternal exposure to LPS induces hypomyelination in the internal capsule and programmed cell death in the deep gray matter in newborn rats. Pediatr Res 2006;59:428–433.
81 Lasala N, Zhou H: Effects of maternal exposure to LPS on the inflammatory response in the offspring. J Neuroimmunol 2007;189:95–101.
82 Hodyl NA, Krivanek KM, Lawrence E, Clifton VL, Hodgson DM: Prenatal exposure to a pro-inflammatory stimulus causes delays in the development of the innate immune response to LPS in the offspring. J Neuroimmunol 2007;190:61–71.
83 Moots RJ, Al-Saffar Z, Hutchinson D, Golding SP, Young SP, Bacon PA, McLaughlin PJ: Old drug, new tricks: haloperidol inhibits secretion of proinflammatory cytokines. Ann Rheum Dis 1999;58:585–758.
84 Drzyzga L, Obuchowicz E, Marcinowska A, Herman ZS: Cytokines in schizophrenia and the effects of antipsychotic drugs. Brain Behav Immun 2006;20:532–545.
85 Müller N, Empl M, Riedel M, Schwarz M, Ackenheil M: Neuroleptic treatment increases soluble IL-2 receptors and decreases soluble IL-6 receptors in schizophrenia. Eur Arch Psychiatry Clin Neurosci 1997;247:308–313.
86 Sirota P, Meiman M, Herschko R, Bessler H: Effect of neuroleptic administration on serum levels of soluble IL-2 receptor-alpha and IL-1 receptor antagonist in schizophrenic patients. Psychiatry Res 2005;134:151–159.

87 Sugino H, Futamura T, Mitsumoto Y, Maeda K, Marunaka Y: Atypical antipsychotics suppress production of proinflammatory cytokines and up-regulate interleukin-10 in lipopolysaccharide-treated mice. Prog Neuropsychopharmacol Biol Psychiatry 2009;33:303–307.
88 Kim YK, Myint AM, Verkerk R, Scharpe S, Steinbusch H, Leonard B: Cytokine changes and tryptophan metabolites in medication-naïve and medication-free schizophrenic patients. Neuropsychobiology 2009;59:123–129.
89 Song C, Lin A, Kenis G, Bosmans E, Maes M: Immunosuppressive effects of clozapine and haloperidol: enhanced production of the interleukin-1 receptor antagonist. Schizophr Res 2000;42:157–164.
90 Song XQ, Lv LX, Li WQ, Hao YH, Zhao JP: The interaction of nuclear factor-kappa B and cytokines is associated with schizophrenia. Biol Psychiatry 2009;65:481–488.
91 Müller N, Schwarz MJ: COX-2 inhibition in schizophrenia and major depression. Curr Pharm Des 2008;14:1452–1465.
92 Zhang XY, Zhou DF, Cao LY, Zhang PY, Wu GY, Shen YC: Changes in serum interleukin-2, -6, and -8 levels before and during treatment with risperidone and haloperidol: relationship to outcome in schizophrenia. J Clin Psychiatry 2004;65:940–947.
93 Lin A, Kenis G, Bignotti S, Tura GJ, De Jong R, Bosmans E, Pioli R, Altamura C, Scharpé S, Maes M: The inflammatory response system in treatment-resistant schizophrenia: increased serum interleukin-6. Schizophr Res 1998;32:9–15.
94 Laan W, Grobbee DE, Selten JP, Heijnen CJ, Kahn RS, Burger H: Adjuvant aspirin therapy reduces symptoms of schizophrenia spectrum disorders: results from a randomized, double-blind, placebo-controlled trial. J Clin Psychiatry 2010;71:520–527.
95 Müller N, Riedel M, Scheppach C, Brandstätter B, Sokullu S, Krampe K, Ulmschneider M, Engel RR, Möller HJ, Schwarz MJ: Beneficial antipsychotic effects of celecoxib add-on therapy compared to risperidone alone in schizophrenia. Am J Psychiatry 2002;159:1029–1034.
96 Zhang Y, Chun Chen D, Long Tan Y, Zhou DF: A double blind, placebo controlled trial of celecoxib added to risperidone in first-episode and drug-naive patients with schizophrenia (abstract). Eur Arch Psychiatry Clin Neurosci 2006;256 (suppl 2):II/50.
97 Akhondzadeh S, Tabatabaee M, Amini H, AhmadiAbhari SA, Abbasi SH, Behnam B: Celecoxib as adjunctive therapy in schizophrenia: a double-blind, randomized and placebo-controlled trial. Schizophr Res 2007;90:179–185.
98 Rapaport MH, Delrahim KK, Bresee CJ, Maddux RE, Ahmad- pour O, Dolnak D: Celecoxib augmentation of continuously ill patients with schizophrenia. Biol Psychiatry 2005;57:1594–1596.
99 Müller N, Ulmschneider M, Scheppach C, Schwarz MJ, Ackenheil M, Möller HJ, Gruber R, Riedel M: COX-2 inhibition as a treatment approach in schizophrenia: immunological considerations and clinical effects of celecoxib add-on therapy. Eur Arch Psychiatry Clin Neurosci 2004;254:14–22.
100 Meyer U, Schwarz MJ, Müller N: Inflammatory processes in schizophrenia: a promising neuroimmunological target for the treatment of negative/cognitive symptoms and beyond. Pharmacol Ther 2011;132:96–110.
101 Müller N, Krause D, Dehning S, Musil R, Schennach-Wolff R, Obermeier M, Möller HJ, Klauss V, Schwarz MJ, Riedel M: Celecoxib treatment in an early stage of schizophrenia: results of a randomized, double-blind, placebo-controlled trial of celecoxib augmentation of amisulpride treatment. Schizophr Res 2010;121:118–124.
102 Sommer IE, de Witte L, Begemann M, Kahn RS: Nonsteroidal anti-inflammatory drugs in schizophrenia: ready for practice or a good start? A meta-analysis. J Clin Psychiatry 2011; in press.
103 Stolina M, Sharma S, Lin Y, Dohadwala M, Gardner B, Luo J, et al: Specific inhibition of cyclooxygenase 2 restores antitumor reactivity by altering the balance of IL-10 and IL-12 synthesis. J Immunol 2000;164:361–370.
104 Müller N, Myint AM, Schwarz MJ: Kynurenine pathway in schizophrenia: pathophysiological and therapeutic aspects. Curr Pharm Des 2011;17:130–136.
105 Steiner J, Bogerts B, Sarnyai Z, Walter M, Gos T, Bernstein HG, Myint AM: Bridging the gap between the immune and glutamate hypotheses of schizophrenia and major depression: Potential role of glial NMDA receptor modulators and impaired blood-brain barrier integrity. World J Biol Psychiatry 2011; in press.
106 Giuliani F, Hader W, Yong VW: Minocycline attenuates T cell and microgliaactivity to impair cytokine production in T cell-microglia interaction. J Leukoc Biol 2005;78:135–143.
107 Levkovitz Y, Mendlovich S, Riwkes S, Braw Y, Levkovitch-Verbin H, Gal G, Fennig S, Treves I, Kron S: A double-blind, randomized study of minocycline for the treatment of negative and cognitive symptoms in early-phase schizophrenia. J Clin Psychiatry 2010;71:138–149.
108 Zhang L, Shirayama Y, Iyo M, Hashimoto K: Minocycline attenuateshyperlocomotion and prepulse inhibition deficits in mice after administration of the NMDA receptor antagonist dizocilpine. Neuropsychopharmacology 2007;32:2004–2010.
109 Fujita Y, Ishima T, Kunitachi S, Hagiwara H, Zhang L, Iyo M, Hashimoto K: Phencyclidine-induced cognitive deficits in mice are improved by subsequent subchronic administration of the antibiotic drug minocycline. Prog Neuropsychopharmacol Biol Psychiatry 2008;32:336–339.
110 Miyaoka T, Yasukawa R, Yasuda H, Hayashida M, Inagaki T, Horiguchi J: Possible antipsychotic effects of minocycline in patients with schizophrenia. Prog Neuropsychopharmacol Biol Psychiatry 2007;31:304–307.
111 Miyaoka T: Clinical potential of minocycline for schizophrenia. CNS Neurol Disord Drug Targets 2008;7:376–381.
112 Ruggiero C, Lattanzio F, Lauretani F, Gasperini B, Andres-Lacueva C, Cherubini A: Omega-3 polyunsaturated fatty acids and immune-mediated diseases: inflammatory bowel disease and rheumatoid arthritis. Curr Pharm Des 2009;15:4135–4148.
113 Wall R, Ross RP, Fitzgerald GF, Stanton C: Fatty acids from fish: the anti-inflammatory potential of long-chain omega-3 fatty acids. Nutr Rev 2010;68:280–289.

114 Sublette ME, Ellis SP, Geant AL, Mann JJ: Meta-analysis of the effects of eicosapentaenoic acid (EPA) in clinical trials in depression. J Clin Psychiatry 2011;72:1577–1584.
115 Turnbull T, Cullen-Drill M, Smaldone A: Efficacy of omega-3 fatty acid supplementation on improvement of bipolar symptoms: a systematic review. Arch Psychiatr Nurs 2008;22:305–311.
116 Palacios-Pelaez R, Lukiw WJ, Bazan NG: Omega-3 essential fatty acids modulate initiation and progression of neurodegenerative disease. Mol Neurobiol 2010;41:367–374.
117 Tull SP, Yates CM, Maskrey BH, O'Donnell VB, Madden J, Grimble RF, Calder PC, Nash GB, Rainger GE: Omega-3 Fatty acids and inflammation: novel interactions reveal a new step in neutrophil recruitment. PloS Biol 2009;7:e1000177.
118 Zhang W, Hu X, Yang W, Gao Y, Chen J: Omega-3 polyunsaturated fatty acid supplementation confers long-term neuroprotection against neonatal hypoxic-ischemic brain injury through anti-inflammatory actions. Stroke 2010;41:2341–2347.
119 Amminger GP, Schäfer MR, Papageorgiou K, Klier CM, Cotton SM, Harrigan SM, Mackinnon A, McGorry PD, Berger GE: Long-chain omega-3 fatty acids for indicated prevention of psychotic disorders: a randomized, placebo-controlled trial. Arch Gen Psychiatry 2010;67:146–154.
120 Fusar-Poli P, Berger G: Eicosapentaenoic acid interventions in schizophrenia: meta-analysis of randomized, placebo-controlled studies. J Clin Psychopharmacol 2012;32:179–185.

Volker Arolt, MD, PhD
Department of Psychiatry, University of Münster
Albert Schweitzer Campus 1/A9
DE-48149 Münster (Germany)
E-Mail arolt@uni-muenster.de

Halaris A, Leonard BE (eds): Inflammation in Psychiatry.
Mod Trends Pharmacopsychiatry. Basel, Karger, 2013, vol 28, pp 117–126 (DOI: 10.1159/000343974)

Inflammation as the Cause of the Metabolic Syndrome in Depression

Brian E. Leonard

Pharmacology Department, National University of Ireland, Galway, Ireland; Department of Psychiatry and Psychotherapy, Ludwig Maximilian University, Munich, Germany; European School of Neuroscience, Department of Psychiatry and Neuropsychology, Maastricht University, Maastricht, The Netherlands

Abstract

Physical ill-health is a common feature of depression. Cardiovascular disease, type 2 diabetes and cancer are, for example, common occurrences and lead to increased morbidity and mortality. The question arises whether these changes are attributable to an unhealthy life style or due to genetic or other causes that are exacerbated by stress and adverse environmental factors. Of the various factors which have been implicated, chronic low-grade inflammation has received much attention in recent years. For this reason, the following review assesses the evidence that implicates inflammation in the psychopathology of major depression and how this is linked to the increased incidence of physical illness in these patients.

Chronic ill health such as coronary heart disease, type-2 diabetes, cancer and neurodegenerative diseases associated with ageing, are some of the most costly health problems in the industrialised countries of the world. Psychological and psychiatric disorders are commonly associated with physical ill-health and the question arises if there is a pathological link between them [1, 2]. This is not an unreasonable assumption as there is good evidence that the frequency of coronary heart disease and type 2 diabetes are often associated with middle aged and elderly patients who are suffering from depression [3]. If depression is a trigger for such physical illnesses, could factors such as an unhealthy life style, which often accompanies chronic psychiatric disorders as depression and schizophrenia, be a contributing factor to the adverse physical and psychological state? Besides the unhealthy life-style and poor diet, there is also evidence that some psychotropic drugs can contribute to the metabolic syndrome and thereby contribute to physical illness [4].

While it seems reasonable to suggest that physical and mental ill-health can share a common pathological basis, the links between them are very complex. One possible approach is to consider the role that the hypothalamic-pituitary-adrenal (HPA) axis, together with aspects of the immune system, may play [5]. For example, it is known that the changes the concentrations of the glucocorticoids and the pro-inflammatory cytokines in coronary heart disease and depression are qualitatively similar. Furthermore, physical and psychological stress play a crucial role in

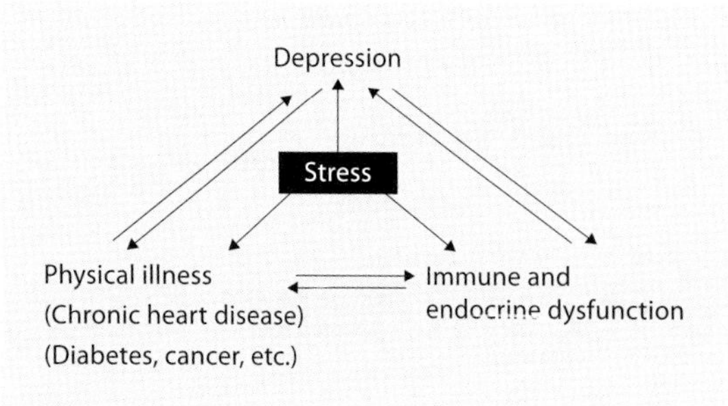

Fig. 1. Interaction between depression, physical illness and changes in the endocrine and immune systems associated with chronic stress.

both depression and chronic physical illness with which it is associated [6]. Thus, it may be hypothesised that a pathway develops in depression whereby the increased vulnerability to stress initiates changes in the HPA and immune axes that initiate the metabolic syndrome. The purpose of this review is to assess the evidence that links the triad of pathologies, namely depression, physical illness and dysfunction of the immune and endocrine system with stress. This is illustrated in figure 1.

Inflammation, the Immune System and Sickness Behaviour

Experimental studies in rodents have demonstrated that chronic stress, or the parenteral administration of a bacterial endotoxin, activates the innate immune system and thereby triggers a cascade of inflammatory changes that impact on the central nervous system [7]. The subsequent increase in pro-inflammatory cytokines from the stress activated macrophages and monocytes (IL-1, IL-6, TNFα and IFNγ) is reflected in similar changes in the brain due to the stress-induced activation of the microglia. The inflammatory changes in the brain are also the result of the impact of the peripheral inflammatory cytokines on the endothelial cells that line the blood-brain barrier and which signal changes to the microglia [8]. Once activated, the microglia release pro-inflammatory cytokines and thereby contribute to the central inflammatory response [9]. The peripheral cytokines can also trigger changes in the brain by activating the terminals of the vagus nerve [10]. Thus any inflammatory changes in the periphery that are initiated by infection, injury or stress impact on the brain.

While such experimental studies offer a plausible explanation for the mechanisms whereby inflammation in the periphery affect the brain, it is important to emphasise the clinical evidence that substantiates this view. Perhaps the most telling evidence is provided by the therapeutic treatment of non-psychiatric patients with hepatitis or leukaemia who were treated with IFNα. A substantial minority of these patients (30–50% in some studies) develop severe depressive-like episodes that are attenuated by appropriate antidepressant treatment [11, 12]. These symptoms are collectively associated sickness behaviour and are characterised by neurovegetative (fatigue, anorexia, disrupted sleep profile) as well as psychological symptoms of depressed mood, anhedonia and cognitive dysfunction. This raises the question if depression is a form of sickness behaviour or a component of depression that reflects the initial inflammatory state which is responsible for the neurovegetative symptoms? Clinical studies by Capuron, Raison and colleagues [11, 12] suggest that the effective response of the major depressive symptoms (particularly the depressed mood state) to antidepressant treatments that have little effect on the neurovegetative symptoms suggest that sickness behaviour is a component of, but separate from, depression. This has recently been critically assessed by Maes et al. [13]. The conceptual importance of the separation of sickness behaviour from depression has also been discussed recently by Janszky et al. [14] who suggest that the pro-inflammatory cytokine-induced sickness response is better considered as a form of vital exhaustion than depression.

Thus, both clinical and experimental evidence has established that activation of the innate immune system either by infection or stress results in profound peripheral and central changes. But what are the consequences of the inflammatory changes in the brain? Clearly, the changes in central neurotransmitters are major contributors to the behavioural symptoms and will be considered later in this review. Besides the impact of immune activation on the neurotransmitter function, the pro-inflammatory cytokines strongly activate the HPA axis and thereby enhance the synthesis and release of glucocorticoids from the adrenal glands [15, 16]. Thus, inflammation, whether caused by stress or some other vulnerability factor associated with depression, contributes to the hypercortisolaemia that is a common feature of the pathology of major depression [17].

Comment on the Hygiene Hypothesis of Depression

Infection and injury have always been a component of human existence. Physiological adaptation to the impact of such assaults on physical survival has therefore been an important factor in mitigating these challenges [18]. The HPA axis and the sympathetic nervous system (SNS) play an essential role in the protection against infection and the consequences of injury by changing the responsiveness of the immune cells so that the invading organisms can be destroyed [19]. The role of the HPA axis in the acute stress response is particularly relevant as the glucocorticoids by enhancing gluconeogenesis provide glucose for the increased metabolic demands of skeletal and cardiac muscles and the brain. Activation of the SNS results in the appropriate response of the vascular system to enhance the blood supply to the central organs involved in the 'fight or flight' response. In addition, noradrenaline and adrenaline increase the synthesis of pro-inflammatory cytokines which, together with the activated macrophages and monocytes, destroy any invading microorganism. These are highly adaptive processes for survival and it is evident that the lethargy and other cytokine-induced behavioural symptoms that accompany the inflammatory response are beneficial for recovery. Injured or disease infected humans living in a hostile environment would have a limited chance of survival unless they could find safe shelter for rest and recuperation. The symptoms of acute sickness behaviour are therefore beneficial under these conditions and the problems only arise when they become chronic, nonadaptive and potentially limit survival as occurs in major psychiatric disorders such as major depression and schizophrenia.

Adverse Effects of Hypercortisolaemia in Depression

Healthy individuals respond to stress with the activation of the sympatho-adrenal system and the HPA axis that result in an elevation in glucocorticoids and the sympathomimetic amines. As a consequence of these physiological changes, the resting heart rate increases and there is a reduction in vascular resistance. However, in individuals who are obese, or already show insulin resistance, endothelial dysfunction can reduce the normal compensatory reduction in blood pressure and thereby limit the distribution of glucose to essential organs. These events contribute to central obesity and the development of the metabolic syndrome [20].

Many of the features of the metabolic syndrome in depression overlap with Cushing's syndrome [21]. At the cellular level, the enzymes 11β-hydroxysteroid dehydrogenase type 1 converts metabolically inactive cortisone to active cortisol. The pro-inflammatory cytokines IL-6 and TNF-alpha increase the hepatocyte expression of 11β-hydroxysteroid dehydrogenase type 1 thereby increasing the intracellular concentration of cortisol. These cytokines also enhance the expression of glucocorticoids receptors thereby further potentiating the effects of cortisol, changes that contribute to the mobilisation of lipids and glucose that contribute to the metabolic syndrome.

Activation of the glucocorticoids receptors inhibits the transcription of many immune response genes. This occurs due to the suppression of the activation of the nuclear gene promoter by the glucocorticoids receptor, by the activation of the transcription induction of the anti-inflammatory genes such as NF-kβ, and the non-genomic antagonism of pro-inflammatory transcription factors, such as NF-kβ, via protein-protein interactions [22]. Thus, the usual response which occurs following injury or infection leads to the activation of the HPA axis, the release of glucocorticoids from the adrenal glands and an inhibition of the immune response [23]. However, this mechanism is compromised by chronic stress, in those suffering from a major mood disorder, or as a result of adverse social factors such as a bereavement, unemployment or chronic physical illness. Such individuals exhibit a high expression of pro-inflammatory cytokines in the blood even in the presence of high glucocorticoids levels [24]. As evidence that the glucocorticoids receptors are desensitised following chronic social stress or example, it has been shown that there is a decrease in the signalling of the glucocorticoids receptors which are located on monocyte membranes [25]. This is associated with shift in gene transcription towards an increase in NF-kβ and AP-1-mediated pro-inflammatory gene expression even under basal, non-stress conditions (fig. 2).

Depression and Comorbidity with Physical Illness

Physical illnesses, such as coronary heart disease, type 2 diabetes, cancer and autoimmune diseases such as asthma and psoriasis, are known to be more frequent in patients with chronic depression and vice versa [26]. Increased cardiovascular morbidity and mortality is associated with increased inflammatory changes as indicated by raised serum acute-phase proteins and C-reactive proteins in addition to pro-inflammatory cytokines such as IL-6 [27, 28]. Such inflammatory markers are suggestive of a poor prognosis among the survivors of acute coronary events [29] and commonly occur in depressed patients [30]. This raises the possibility that the metabolic syndrome, which commonly occurs in physical illness and in depression, is a disorder of the brain!

Stress, Early Life Trauma and Inflammation

There is a growing appreciation of the importance of epigenetic factors in early life that have adverse effects on the subsequent physical and mental development of the child [31]. Damage to the neural networks during early development may have an impact in later life by changing hormonal, immune and neurotransmitter signals and disrupt the adaptive response to stress. This has been shown in experimental animals [32] and also in adults reporting childhood maltreatment. The maladaptive consequences are thought to be related to a reduction in glucocorticoids receptor signalling [33]. As glucocorticoids normally inhibit inflammation, these changes suggest that childhood maltreatment may be connected to the increase in inflammation shown in the adult. In support of this hypothesis, Danese et al. [32] demonstrated that more than 10% of cases of low-grade inflammation in the population, as indicated by an increase in the serum concentration of hsCRP, may be attributed to childhood trauma. In the adults, the inflammatory changes were also associated with an increase in fibrinogen and the white cell count and linked to the increased frequency of such psychiatric disorders as depression [34], and physical illnesses such as coronary

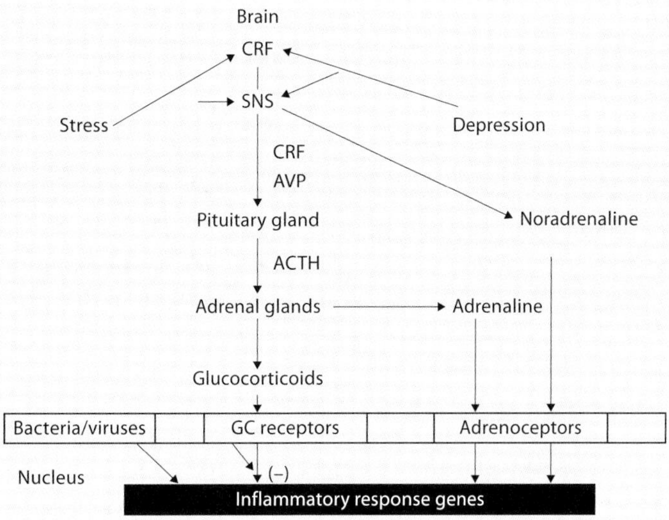

Fig. 2. Glucocorticorticoid receptors are widely distributed on tissues throughout the periphery and brain. Acute activation of glucocorticoid receptors on immune cells, e.g. leucocytes, inhibit the inflammatory networks in which NF-kβ mediated transcription of pro-inflammatory cytokine genes (IL-1β, IL-6, TNFα, IFNγ, etc.) occur. Antibacterial and viral genes are also activated by GC receptors. Pro-inflammatory cytokines activate the HPA axis and thereby contribute to hypercortisolaemia. After an acute stress response, the secretion of glucocorticoids decreases as a result of a negative feedback mechanism whereby glucocorticoids activate GC receptors in the hypothalamus and thereby decrease CRF release and the activation of the pituitary gland. Following chronic stress or depression; however, the hypersecretion of glucocorticoids is enhanced by the pro-inflammatory cytokines and by arginine vasopressin (AVP) release from the hypothalamus. These chronic changes result in the desensitisation of the GC receptors and therefore the continuing secretion of glucocorticoids in the presence of pro-inflammatory cytokines. Stress, whether acute or chronic, activates the sympathetic nervous system (SNS) as an essential component of the fight or flight response. The locus coeruleus is the main sympathetic centre in the brain. Noradrenaline stimulates the adrenal glands to release adrenaline. The catecholamines then activate α- and β-adrenoceptors on vascular tissue and immune cells in particular. The adrenoceptors on immune cells, via the activation of transcription factors such as CREB, enhance the expression of pro-inflammatory response genes and thereby contribute to the inflammatory response. Thus, both the positive and negative mediation of the inflammatory processes occur in response to stress, the nature of the response depending upon whether it is acute, and adaptive, or chronic and non-adaptive.

heart disease, diabetes and cancer [35]. Closely associated with hypercortisolaemia and increased peripheral sympathetic activity, there was also a reduction in cholinergic function [36]. The involvement of the cholinergic system in the maladaptive response to stress reflects the imbalance between the parasympathetic and sympathetic system; under normal physiological conditions, the parasympathetic system enhances the anti-inflammatory response, a situation which would appear to be defective in those subject to early life trauma.

While it is self-evident that arthritis and allergies are consequences of an underlying inflammatory process, it is less obvious that chronic inflammation plays a role in other physical disorders. Recent studies by Christian et al. [37] have shown that self-related health is an independent predictor after the objective health status, behavioural risk factors and sociodemographic characteristics of the patient are taken into account. These authors demonstrated that there was a direct correlation between the rise in serum IL-6 and CRP and the decrease in self-related general health status. The analysis was based on a sample of 250 relatively healthy older adults and the relationship between the health status and the inflammatory markers was not accounted for by the objective health diagnosis, BMI, age, gender, depressive symptoms, neuroticism score or health behaviours. As inflammation is linked to sickness behaviour, it is hypothesised that self-related health reflects an adaptive response by promoting behaviours such as rest that conserve metabolic energy and reduce the likelihood of social conflict [38]. These observations are important because they integrate the effects of chronic inflammation on the endocrine, immune and neurotransmitter systems with physical and psychiatric disorders and how these pathological changes that are linked to these disorders may arise.

Link between Inflammation, the Metabolic Syndrome and Depression

Over the past decade, overweight and obesity is recognised as a major health issue in most high income countries, particularly in North America and some Western European countries. Overweight is often linked to hyper-insulinaemia, increased angiotensin activity while central obesity is followed by glucose intolerance, the development of type 2 diabetes, hypertension, low high-density lipoprotein (HDL) levels and hypertriglyceridaemia. These changes characterise the metabolic syndrome [39].

It is self-evident that diet and changes in life-style contribute to the metabolic syndrome. Thus, the increased intake of saturated fats, salt, refined sugar and red meat, rather than fish and traditional grain based diets, combined with an increasing sedative life-style, have been major factors in the adverse health effects of obesity [40, 41].

The question arises to what extent these changes in life-style impact on mental ill-health. The vulnerability of depressed patients to stressful life events and the stress-activated HPA axis and sympathetic nervous system are essential components of the stress response. In depression, the increased sympathetic drive results in the activation of the NF-kβ-dependent inflammatory cascade involving both the peripheral macrophages and central microglia [42]. The cells of visceral white matter fat and the endothelial cells are also activated by the sympathomimetic amines and thereby contribute to the inflammatory cascade by activating the acute phase proteins in the liver and choroids plexus [43]. These changes in the peripheral and central immune systems are co-ordinated with the metabolic changes induced by hypercortisolaemia.

The stress-induced activation of visceral fat is often overlooked as an important source of pro-inflammatory cytokines such as IL-6 and TNFα, chemokines, adipokines, leptin, resistin as well as acute-phase proteins [44]. Of these, leptin occupies a central role in the regulation of food intake. The plasma concentration of leptin is correlated with the degree of adiposity and with the concentration of dopamine in the cerebrospinal fluid [45]. As dopamine release is associated with reward and also initiates and co-ordinates motor activity, it is hypothesised that dopamine facilitates the integration of sensory cues related to hunger possibly due to the anatomical link between the nucleus accumbens and the lateral hypothalamus, a region of the brain which regulates feeding behaviour [46]. As there is evidence that the dopaminergic system is malfunctioning in depression,

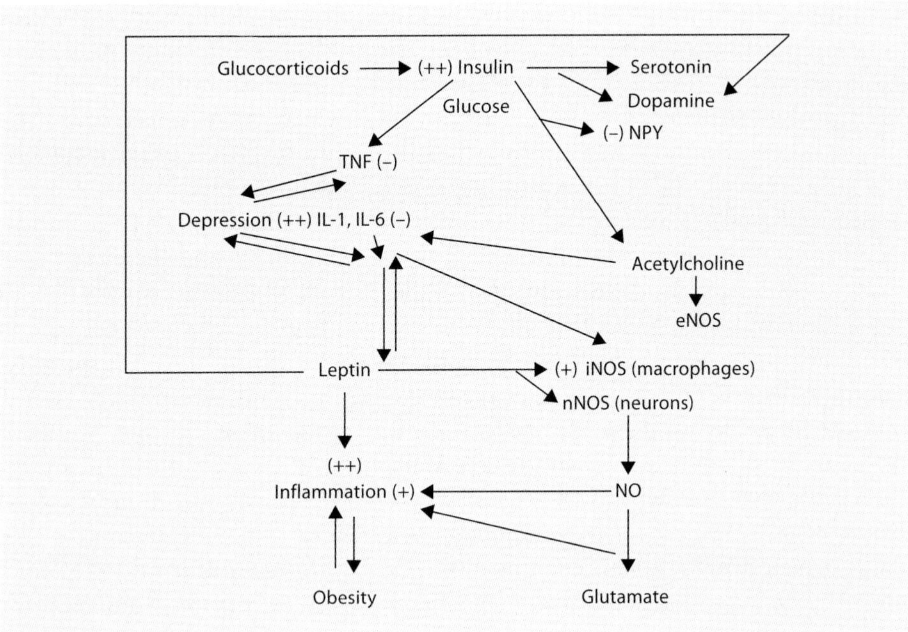

Fig. 3. Interaction between some central neurotransmitters, inflammatory mediators and the metabolic syndrome in major depression. Glucocorticoids (GCs) mobilise fatty acids and glucose, and increase the gluconeogenesis that contributes to insulin resistance. Hypercortisolaemia, a common feature of depression, is enhanced by the direct action of pro-inflammatory cytokines on the release of corticotrophin-releasing factor (CRF). Depression and stress are associated with increased macrophage and microglia activity that are the main source of the pro-inflammatory cytokines. These cytokines activate inducible and neuronal nitric oxide synthase (NOS) thereby increasing nitric oxide (NO), an activator of the glutamatergic system. The pro-inflammatory cytokines also enhance the activity of the glutamatergic system by reducing the re-uptake of glutamate into astrocytes. Leptin, derived primarily from white visceral fat, contributes to inflammation directly and also indirectly by activating NOS.

a possible link is provided between changes in metabolism and central neurotransmitter malfunction in depression. This link is summarised in figure 3.

The metabolic effects of the glucocorticoids are well known and involve gluconeogenesis, mobilisation of fatty acids and, in the brain, a reduction in the synthesis of neurotrophic factors such as brain-derived neurotrophic factor. As a consequence of hypercortisolaemia, insulin insensitivity that underlies type 2 diabetes, deposition of fat in the coronary circulation that contributes to myocardial infarction, and a reduction in neurotrophic factor synthesis that reduces neuronal repair occur.

A pre-diabetic status is frequently recorded in depressed patients and the first observation that suggested that depression was associated with diabetes was made in 1684 [47]. The point prevalence of diabetes in depression is 11% and may be as high as 28.5% [48]. From the clinical perspective, it is important to note that depression is associated with a worsening of the glycaemic control [49] and while the hypercortisolaemia was suggested to be a major pathophysiological mechanism, it now appears likely that the hyperglycaemia itself contributes to the worsening of the depressive symptoms [50]. The reason for the adverse effect of hyperglycaemia on the symptoms of depression are uncertain but one possibility is the activation of the sympatho-adrenal system which, as has already been discussed, activates the inflammatory cascade and thereby contributes indirectly to the pathological changes which underlie depression.

In addition to the physical ill-health associated with the metabolic effects of hypercortisolaemia, there is also epidemiological evidence that the frequency of dementia is significantly increased in elderly patients who have suffered from chronic major depression [51]. This is a possible consequence of the reduction in the synthesis of neurotrophic factors combined with the neurotoxic effects of the activated glutamatergic system. The pro-inflammatory cytokines activate the glutamatergic system indirectly by reducing the activity of the amino acid transporter for glutamate on the neuroglia [52] and by activating the tryptophan-kynurenine pathway in the brain [53]. This occurs as a consequence of the action of these cytokines on indoleamine dioxygenase, the rate-limiting enzyme in the tryptophan-kynurenine pathway, and by the activation of tryptophan dioxygenase by the glucocorticoids. In depression, the activation of this pathway leads to a reduction in the synthesis of serotonin, the amine neurotransmitter considered to play a crucial role in the mood state, and the synthesis of the neurotoxins quinolinic acid (an NMDA glutamate receptor agonist) and 3-hydroxykynurenine which enhances oxidative stress [54]. These neurotoxic effects cause apoptosis of neurons in the hippocampus, amygdala, pre-frontal cortex and to a lesser extent to sub-cortical regions of the brain [55] and therefore contribute to the neurodegenerative changes that are a prelude to dementia in later life [56].

Conclusion

While physical ill-health, commonly associated with major depression, is frequently attributed to an unhealthy life style associated with a nutritionally poor diet and lack of exercise, it is now apparent that there are underlying metabolic abnormalities that predispose depressed patients to both physical and mental ill-health. There is now substantial evidence to show that chronic low-grade inflammation plays a role in causing both the metabolic syndrome and the symptoms of depression. Recent insight into the links between the HPA axis, the immune system and central neurotransmitters which play a crucial role in the aetiology of depression may suggest ways in which novel antidepressants may be developed that improve both the physical and mental state of the patients.

References

1 Frasure-Smith N, Lesperance F, Talajic MD: Depression and 18 month prognosis of myocardial infarction. Circulation 1995;91:999–1005.
2 Anisman H, Merali Z: Cytokines, stress and depressive illness: brain-immune interactions. Ann Med 2003;35:2–11.
3 Dantzer R, Capuron L, Irwin M, et al: Identification and treatment of symptoms associated with inflammation in medically ill patients. Psychoneuroendocrinology 2008;33:18–29.
4 Heiskanen TH, Niskanen LK, Hintikka JJ, et al: Metabolic syndrome and depression: a cross sectional analysis. J Clin Psychiatry 2006;67:1422–1427.
5 Irwin MR, Cole SW: Reciprocal regulation of the neural and innate immune systems. Nat Rev Immunol 2011;11: 625–632.
6 Dantzer R, O'Connor JC, Freund GG, et al: From inflammation to sickness and depression: when the immune system subjugates the brain. Nat Rev Neurosci 2008;9:46–57.
7 Hart BL: Biological basis of the behaviour of sick animals. Neurosci Biobehav Rev 1988;12:1232–1237.

8 Haroon E, Raison CL, Miller AH: Psychoneuroimmunology meets neuropsychopharmacology: translational implications of the impact of inflammation on behavior. Neuropsychopharmacol Rev 2012;37:137–162.
9 Frank MG, Barattar MV, Springer DB, et al: Microglia serve as a neuroimmune substrate for stress-induced potentiation of CNS proinflammatory cytokine responses. Brain Behav Immunol 2007;21:47–59.
10 Bluthe RM, Walter V, Parnet P, et al: Lipopolysaccharide induces sickness behaviour in rats by vagal mediated mechanism. C R Acad Sci 1994;111:499–503.
11 Capuron L, Gumnick JF, Musselman DL, et al: Neurobehavioural effects of interferon alpha in cancer patients: phenomenology and paroxetine responsiveness of symptom dimensions. Neuropsychopharmacoology 2002;26:643–652.
12 Raison CL, Borisov AS, Majer M, et al: Activation of CNS inflammatory pathways by interferon alpha: relationship to monoamines and depression. Biol Psychiatry 2007;65:296–307.
13 Maes M, Berk M, Song C, et al: Major depression, a progressive disorder and sickness behaviour. Submitted, 2012.
14 Janszky I, Lekander M, Blom M, et al: Self-rated health and vital exhaustion, but not depression,is related to inflammation in women with coronary heart disease. Brain Behav Immunol 2005;19: 555–563.
15 Besedovsky HO, delRey A: Immune-neuro-endocrine interactions: facts and hypothesis. Endocr Rev 1996;17:64–102.
16 Pariante CM, Pearce BD, Pisell TL, et al: The proinflammatory cytokine, interleukin 1 alpha, reduces glucocorticoid receptor translocation and function. Endocrinology 1997;140:4359–4366.
17 Gold PW, Chrousos GP: Organisation of the stress system and its dysregulation in melancholic and atypical depression: high verses low CRH/ NE states. Mol Psychiatry 2002;7:254–275.
18 von Mutius E: Allergies, infections and the hygiene hypothesis: the epidemiological evidence. Immunobiology 2007;212:435–439.
19 Panina-Bordignan P: Beta-2 agonists prevent Th-1 development by selective inhibition of IL-12. J Clin Invest 1997; 100:1513–1519.
20 Das UN: Obesity, metabolic syndrome X and inflammation. Nutrition 2002;18: 430–438.
21 Farmer A, Korszun A, Owen MJ, et al: Medical disorders in people with recurrent depression. Br J Psychiatr 2008;192: 351–355.
22 Irwin MR, Cole SW: Reciprocal regulation of the neural and innate immune systems. Nat Rev Immunol 2011;11: 625–632.
23 Pariante CM, Lighman SL: The HPA axis in major depression: classic theories and new developments. Trends Neurosci 2008;31:464–468.
24 Christian LM, Glaser R, Porter K, et al: Poorer self-rated health is associated with elevated inflammatory markers among older adults. Psychoneuroendocrinology 2011;36:1495–1504.
25 Pariante CM, Miller AH: Glucocorticoid receptors in major depressionrelevance to pathophysiology and treatment. Biol Psychiatr 2001;49:391–404.
26 Black PH: The inflammatory consequences of psychological stress: relationship to insulin resistance, atherosclerosis and diabetes mellitus, type-2. Med Hypoth 2006;67:879–891.
27 Victor W, Vieweg R, Pandurangi AK: Editorial: The relation of stress and psychiatric illness to coronary heart disease. Acta Psychiatr Scand 2006;113:241–244.
28 Das UN: Is type 2 diabetes mellitus a disorder of the brain? Nutrition 2002;18: 667–672.
29 Davidson KW, Schwartz JE, Kirkland SA, et al: Relation of inflammation to depression and coronary heart disease (from the Canadian Nova Scotia Health Survey Prospective population study). Am J Cardiol 2009;103:755–761.
30 Miller AH, Maletic V, Raison CL: Inflammation and its discontents: the role of cytokines in the pathophysiology of major depression. Biol Psychiatr 2009; 65:732–741.
31 Danese A, Pariante CM, Caspi A, et al: Childhood maltreatment predicts adult inflammation in a life-course study. PNAS 2007;104:1319–1324.
32 Danese A, Moffitt TZ, Pariante CM, et al: Elevated inflammation levels in depressed adults with a history of childhood maltreatment. Arch Gen Psychiatr 2008;65:4405–4015.
33 Besedovsky H, del Rey A, Sorkin F, Dinarello CA: Immunoregulatory feedback between IL-1 and glucocorticoid hormones. Science 1986;233:652–654.
34 Ershler WB, Keller E: Age associated increased IL-6 gene expression, late life diseases and frailty. Ann Rev Med 2000;51:245–270.
35 Hamerman D, Berman JW, Albers GW, et al: Emerging evidence for inflammation in conditions frequently affects older adults: report of a symposium. J Am Geriatr Soc 1999;47:1016–1025.
36 Borovikova LV, Ivanova S, Zhang M, et al: Vagus nerve stimulation attenuates the systemic inflammatory response to endotoxin. Nature 2000;405:458.
37 Christian LM, Glaser R, Porter K, et al: Poorer self-rated health is associated with elevated inflammatory markers among older adults. Psychoneuroendocrinology 2011;36:1495–1504.
38 Dantzer R: Cytokine-induced sickness behaviour: mechanisms and implications. Ann NY Acad Sci 2001;933:222–234.
39 Das UN: Is obesity an inflammatory condition? Nutrition 2001;17:953–966.
40 Kelcolt-Glaser JK: Stress, food and inflammation: psychoneuroimmunology and nutrition at the cutting edge. Psychosom Med 2010;72:365–369.
41 Cunnane SC, McAdoo KR, Horrobin DF: n–3 essential fatty acids decrease weight gain in genetically obese mice. Br J Nutr 2986;56:87–91.
42 Grebe KM: Cutting edge: sympathetic nervous system increases proinflammatory cytokines and exacerbates influenza A virus pathogenesis. J Immunol 2009;184:540–544.
43 Nanee DM, Sandersw VM: Autoimmune innervation and regulation of the immune system (1987–2007). Brain Behav Immunol 2007;21:736–745.
44 Hotamisligil GS: Inflammation and metabolic disorders. Nature 2006;444: 860–867.
45 Muenzberg H: Differential leptin access into the brain: a hierarchical organization of hypothalamic leptin target sites? Physiol Behav 2008;94:664–669.
46 Martel P, Fantino M: Mesolimbic dopaminergic system activity as a function of food reward:a microdialysis study. Pharmacol Biochem Behav 1996;53:221–226.
47 Willis T: Diabetes: A Medical Odyssey. New York, Tuckahoe, 1971.

48 Lustman PJ, Clouse RE: Depression in diabetic patients: the relationship between mood and glycaemic control. J Diab Compl 2005;19:113–122.
49 Lustman PJ, Anderson PJ, Freedland KE, et al: Depression and poor glycaemic control: a meta analytic review of the literature. Diabetes Care 2000;23:934–942.
50 Lustman PJ, Anderson P, Freedland KE, et al: Depression and poor glycaemic control: a meta analytic review of the literature. Semin Clin Neuropsychiatr 1998;3:102–114.
51 Geerlings MI, Schoevers RA, Beckman, et al: Depression and the risk of cognitive decline and Alzheimer's disease. Results of two prospective community based studies in the Netherlands. Br J Psychiatr 2000;176:556–575.
52 Bezzi P, Domercq M, Brambilla L, et al: CXCR4 –activated astrocyte glutamate release via TNF-alpha: amplification of microglia triggers neurotoxicity. Nat Neurosci 2001;4:702–710.
53 Dantzer R, O'Connor JC, Freund GG, et al: From inflammation to sickness and depression when the immune system subjugates the brain. Nature Rev Neurosci 2008;9:46–57.
54 Myint A-M, Kim Y-K, Verkerk R, et al: Kynurenine pathway in major depression: evidence of impaired neuroprotection. J Affect Dis 2007;98:143–151.
55 Schwarcz RT: The kynurenine pathway of tryptophan degradation as a drug target. Curr Opin Pharmacol 2004;4:12–17.
56 Leonard BE, Myint A-M: Inflammation and depression: is there a causal relationship with dementia? Neurotox Res 2006;10:149–160.

Prof. Brian E. Leonard
Pharmacology Department, National University of Ireland
Galway (Ireland)
E-Mail belucg@iol.ie

Halaris A, Leonard BE (eds): Inflammation in Psychiatry.
Mod Trends Pharmacopsychiatry. Basel, Karger, 2013, vol 28, pp 127–143 (DOI: 10.1159/000343980)

Glucocorticoids and Inflammation: A Double-Headed Sword in Depression?

How Do Neuroendocrine and Inflammatory Pathways Interact during Stress to Contribute to the Pathogenesis of Depression?

M.A. Horowitz[a] · P.A. Zunszain[a,b] · C. Anacker[a–c] · K. Musaelyan[a] · C.M. Pariante[a,b]

[a]Department of Psychological Medicine, Section of Perinatal Psychiatry and Stress, Psychiatry and Immunology (SPI-lab), Institute of Psychiatry, King's College London, [b]Institute of Psychiatry and South London and Maudsley NHS Foundation Trust, National Institute for Health Research 'Biomedical Research Centre for Mental Health' and [c]Centre for the Cellular Basis of Behaviour, Institute of Psychiatry, King's College London, London, UK

Abstract

Both glucocorticoids and inflammation have been implicated in the pathogenesis of depression. There is a large body of literature indicating that hyperactivity of the hypothalamic-pituitary-adrenal (HPA) axis and glucocorticoid receptor (GR) dysfunction are present in a significant proportion of depressed patients. There is also evidence of increased inflammatory processes in depressed populations, with higher levels of cytokines being a prominent finding – including raised levels of IL-6, and IL-1. These findings appear difficult to reconcile given the well-recognised property of glucocorticoids as prominent anti-inflammatory molecules. There are three potential solutions posed to this dilemma. Firstly, it has been argued that the glucocorticoid system and the inflammatory system exist in balance with one another and chronic stress can disrupt this balance in favour of inflammatory processes at the expense of glucocorticoid signalling. It has also been suggested that glucocorticoids have more complex actions than typically thought, and, in low levels can actually be pro-inflammatory, rather than universally anti-inflammatory. Lastly, it is possible that inflammation and glucocorticoid signalling may act on the same processes and structures without direct interaction to give rise to cumulative damage. Improved understanding of this interaction will allow further progress in determining targets for treatment.

As seen previously in this book, there is compelling evidence for both a cytokine theory of depression and a glucocorticoid theory of depression (involving glucocorticoid resistance and a hyperactive hypothalamic-pituitary-adrenal (HPA) axis). At first glance, it seems deeply problematic that these two theories co-exist, in the context of the canonical understanding of glucocorticoids as potent anti-inflammatory agents. There are three plausible hypotheses for how both these groups of molecules can contribute to depression. The first argues that the presence of glucocorticoid receptor (GR) resistance and impaired glucocorticoid signalling provides a permissive environment for inflammation as the usual inhibitory influence it provides is removed [1]. A second position argues that glucocorticoids are not, as widely supposed, uniformly anti-inflammatory and can in specific circumstances actually be permissive towards and even potentiate inflammation [2]. Finally, inflammation (largely in the form of cytokines) and glucocorticoids have been shown to have overlapping effects on important processes implicated in the pathogenesis of depression – including their influence on the monoamine system [3, 4], neurodegeneration [5] and neurogenesis [6, 7], as well as brain region function and connectivity [8] – allowing for the possibility of cumulative effects on these processes. There are therefore a variety of ways in which inflammation and the neuroendocrine system can interact to contribute to the pathogenesis of depression.

Before embarking on a discussion of the possible modes of interaction between glucocorticoids and inflammation in the pathogenesis of depression, we briefly review the literature linking each to depression.

Cytokines Are Implicated as Pathogenic of Depression

A now robust literature implicates inflammation as integral to the pathogenesis of depression, with a particular emphasis on the role of cytokines. This link first came to clinical attention through the observation of increased rates of depression in the medically ill [9], particularly in those suffering from disorders with inflammatory components (e.g. neurodegenerative, infectious, and autoimmune diseases), above and beyond what could be explained by the psychological distress of having these illnesses [10]. The hypothesis generated from this finding that inflammation was a potential cause of depression was supported by evidence that patients with clinical depression in the absence of any medical co-morbidity also demonstrated increased levels of inflammation, as measured in blood, cerebrospinal fluid and in the prefrontal cortex of suicide victims [11, 12]. The validity of this finding has been strengthened through meta-analyses which confirm higher levels of IL-1, IL-6, CRP and TNF-α in depressed patients [13, 14]. Controversy concerning the direction of causation between inflammation and depression has been resolved in part by the quasi-experimental setting of interferon alpha treatment for various medical illnesses, including hepatitis C, which has a well-recognised side effect of inducing depression in 30–50% of patients (depending on the dosage used) [15–17], thus indicating that inflammation in this paradigm is causative of depression.

Inflammation, particularly in the form of cytokines, provides a highly plausible agent for the pathogenesis of depression. On administration of cytokines to humans, they elicit 'sickness behaviour' – depressed mood, fatigue, anxiety, sleep disturbances, anhedonia and cognitive dysfunction – which closely resembles many key symptoms of major depression [18]. For some time, debate focused on the means by which cytokine increase in the periphery could affect the brain. Several mechanisms were identified – entrance through leaky parts of the blood-brain barrier, active transport by cytokine specific transporters located on the brain endothelium, and activation of vagal afferent fibres which transmit cytokine signals to the brain [19]. However, more recently it has been shown that the brain itself produces cytokines [20] indicating an even more direct exposure of the brain to the potentially harmful effects of these molecules.

Cytokines affect many of the processes implicated in the development of depression. This includes their ability to affect the neurotransmitter metabolism of dopamine, noradrenaline and serotonin [19]; synthesis of dopamine in rats [21]; and, reuptake of serotonin, noradrenaline and dopamine [22–24]. Particular attention has focused on the ability of cytokines to induce indoleamine 2,3-dioxygenase (IDO), leading to increased metabolism and shunting of tryptophan, the direct metabolic precursor to serotonin [25]. Pertinently, patients who develop depression during interferon alpha therapy demonstrate particularly marked reductions in tryptophan and serotonin [26, 27]. Additionally, the cytokine-induced activation of IDO increases the neurotoxic products of the kynurenine pathway [5, 28] which has also been shown to decrease neurogenesis [29] another process which has recently been implicated in the pathogenesis of depression [30]. Cytokines can also contribute to neurodegeneration which has been implicated as potentially an important process in depression [31].

Furthermore, cytokines affect neuroendocrine function in a number of profound ways – from activation of the HPA axis by induction of CRH, ACTH and also cortisol [32] as well as by directly inducing impairment and resistance in the function of the GR [33]. These interactions will be explored in further detail below. Cytokines can also affect regional brain activity, particularly in cortico-striatal-limbic circuits [19], as well as neuroplasticity [34]. In particular, IL-1β inhibits LTP which is involved in learning and memory, known to be affected in MDD [35]. All of these processes have been identified as relevant to the development of depression.

Animal models have further offered evidence for causative effects of cytokines on depression. Rodents (and non-human primates) given cytokines show an increase in depression-like behaviours [36, 37] and they also demonstrate some of the characteristic findings in depression – such as typical neuroendocrine changes [36, 37]. The most widely used paradigm to induce depression-like behaviour in mice – chronic mild stress – increases IL-1β in the hippocampus. To emphasise the importance of cytokines in the behavioural and neuroendocrine changes that accompany this paradigm, IL-1 knockout mice show no such behavioural change in response to these stress paradigms [38]. This abolition of the behavioural and neuroendocrine effects of chronic stress has also been replicated with antagonism of IL-1 receptors [39]. These studies demonstrate the integral importance of inflammation to the induction of depression.

Furthermore, it has recently been shown that anti-inflammatory drugs demonstrate efficacy in the treatment of depression, particularly in those patients resistant to conventional pharmacotherapy. Aspirin (a COX-1, COX-2 and prostaglandin inhibitor) and celecoxib (a COX-2 inhibitor) both increase antidepressant efficacy when added onto conventional antidepressants [40, 41] and the TNF-α blocker, etanercept, significantly improved depressive symptoms in a group of psoriasis patients, independent of improvements to disease severity [42]. This contributes to the overall literature suggesting a pivotal role for inflammation in the pathogenesis of depression.

Role of the Neuroendocrine System in the Pathogenesis of Depression

The neuroendocrine theory of depression, were it not for the force with which the success of SSRIs has lodged serotonin in the popular discourse, would be viewed as the now canonical theory of depression. There is vast and growing evidence for the role of glucocorticoids, a dysfunctional HPA axis and an impaired glucocorticoid receptor (GR) as key elements in the pathogenesis of depression. Fifty percent of depressed patients (80% if severely depressed) show hyperactivity of the HPA axis [32, 43–45]. Evidence of a hyperactive HPA axis includes increased pituitary and adrenal gland volume and increased levels of glucocorticoids in saliva, CSF, blood and urine [46–48]. For example, in one study depressed patients were found to have twice the levels of salivary cortisol as control subjects [48]. A functional measure of HPA axis hyperactivity and impaired feedback mechanisms found in many depressed patients is the dexamethasone suppression test in which healthy controls show 85% suppression of endogenous cortisol production in response to administration of the synthetic glucocorticoid dexamethasone but depressed patients only show about 45% suppression [48].

There has been some question over the direction of causality: whether depression gives rise to high levels of cortisol or whether high levels of cortisol are causative of the illness. There are clinical situations and experimental paradigms that have contributed to the clarification of this question. Patients placed on glucocorticoids for a variety of immunological or infectious reasons have a greatly increased incidence of depression [49]. In rats, chronic glucocorticoid treatment induces depression and anxiety-like behaviour [50–52].

Glucocorticoids are also thought to represent a plausible causative agent for depression because they affect many of the structures and processes that are associated with depression. They have been shown to affect adult hippocampal neurogenesis, neuronal survival, neuronal excitability, learning, and memory amongst other processes [53]. For example, in rodents chronic stress and glucocorticoid administration induced neural cell death, atrophy of neuronal processes and a reduction of adult hippocampal neurogenesis [52, 54, 55]. The link between glucocorticoids and reduced hippocampal neurogenesis has been well established. Removal of the adrenal glands in rats lowers glucocorticoid levels and increases hippocampal neurogenesis which can be reversed by exogenous glucocorticoid administration [56]. Furthermore, it has been shown that chronically elevated glucocorticoids impair synaptic plasticity and cognition, decrease neurogenesis and spine density, and cause dendritic atrophy [57]. Indeed, the glucocorticoid receptor has been shown to be important in regulating the expression of neurotrophic factors, such as brain-derived neurotrophic factor (BDNF), which may also contribute to the changes in the depressed brain by increasing neuronal cell death and altering hippocampal neurogenesis [58].

These findings are supported by in vitro studies in which the proliferation of rodent stem cells has been shown to be reduced [59]. Even more pertinently, this effect has been demonstrated with human hippocampal stem cells in which dexamethasone and cortisol (in high levels) show dose-dependent effects on both the reduction of cell proliferation and neurogenesis [7]. In clinical samples the data is scarce – one study which examined human post mortem brain tissue showed no change in neurogenesis in depressed patients [60] and another found a decrease [61]. To add to the controversy, the reduction in hippocampal volume found in depressed patients [62, 63] has also been attributed to the effects of glucocorticoids as chronic treatment with glucocorticoids reduces hippocampal volume in animals [64].

Interaction of Glucocorticoids and Inflammation in Depression

In short, there is a vast body of evidence supporting both a cytokine and glucocorticoid theory of depression. The question then becomes: How do these two systems interact with each other in the pathogenesis of this disease? This issue can be separated into three logical categories – firstly, the ways in which inflammation can enhance the abnormalities in the neuroendocrine system characteristically found in depression; secondly, the way in which the neuroendocrine system can exacerbate the abnormalities found in the inflammatory system in depression, and, thirdly, those processes affected by both that may not necessitate direct interaction between them. There are debates in the literature in both the first and third categories but neither is as impassioned and exciting as that in the second category, namely how the neuroendocrine system can contribute to the pathogenesis of inflammation demonstrated in depression. This category will therefore receive special attention.

Inflammation Plays an Important Role in the Effect of Glucocorticoids in Inducing GR Resistance and HPA Axis Hyperactivity

We turn first to the ability of inflammation to modulate the neuroendocrine system. There are two broad mechanisms by which inflammation may contribute to neuroendocrine abnormalities – through direct effects on the glucocorticoid receptor (GR) or by activating the HPA axis. We consider their effects on the GR first.

Direct Effects of Inflammation on the GR

One of the principal mechanisms by which inflammation is thought to contribute to the action of glucocorticoids in the pathogenesis of depression is by the action of cytokines on the GR. GR dysfunction is thought by some to be a central feature of the pathology in the neuroendocrine system of depressed patients [47, 65]. There is evidence that cytokines can impair every aspect of GR function – GR translocation from the nucleus, GR protein-protein interactions, GR binding to response elements on the DNA – as well as affect the expression profile of GR isoforms, favouring the isoform that has relatively reduced actions compared to the dominant isoform [33].

The ability of cytokines to induce impaired GR function (GR resistance) in vitro has been widely demonstrated. For example, in a mouse fibroblast line, IL-1α administration was found to induce a decrease of dexamethasone (DEX)-induced GR translocation from cytoplasm to nucleus, which also caused decreased GR-DNA binding and decreased DEX-induced transcription of target genes [66]. The direct relationship between the actions of IL-1α and the induction of this GR dysfunction was demonstrated by the abolishment of this effect on co-administration of an IL-1 receptor antagonist [67]. There is now a large body of evidence suggesting that different cytokines (in particular the cytokines that have been implicated as pro-inflammatory) – IL-1β, IL-2, IL-4, IL-5, TNF-α, and IFN-α – can each induce all or some of the aspects of glucocorticoid resistance in vitro as represented by decreased DEX or prednisolone-induced GR translocation, decreased GR-mediated gene transcription, decreased GR-GRE (glucocorticoid response element) binding, and decreased GR nuclear ligand-binding affinity [68–70]. Activation of the p38 signalling pathway, NF-κβ signalling pathways, and STAT5 have been implicated in this process, involving phosphorylation of the GR or protein interactions involving the GR (such as with STAT5) [71]. Ongoing debate in this field focuses on the exact means by which cytokines disrupt GR functioning.

Animal Models

These in vitro findings have been replicated in animal models. The ability of inflammation to induce DEX non-suppression in rats was first observed after endotoxin treatment [72] and, more recently, after HSV-1 infection [73]. The mechanisms underlying these effects have been studied in mice subjected to social disruption. This paradigm reliably induces GR resistance – manifested as decreased GR translocation to the nucleus and a lack of glucocorticoid (GC)-induced suppression of NF-κβ – in splenic monocytes [74, 75]. It could be concluded from this that either the inflammation or glucocorticoid release induced by social stressors (both are recognised components of the response to such stress) could be responsible for this GR resistance. However, a critical study showed that in IL-1 receptor knockout mice subjected to the same paradigm, despite evidence of glucocorticoid overdrive (adrenal hypertrophy, thymic involution, elevated serum corticosterone levels) GR resistance was not induced [76]. This experiment provides evidence for the hypothesis that inflammation is a necessary precondition for stress/glucocorticoids to induce GR resistance. That is, the effect of stress on the GR may be mediated by inflammation.

Clinical Studies

Given the body of evidence in cell and animal models implicating inflammation as a cause of GR resistance, investigation of the relationship between these two phenomena has been conducted in clinical populations with major depression. These studies have yielded mixed results. An early report showed a correlation between the degree of GR resistance (as measured by post-DEX cortisol levels) and levels of IL-1β (assessed after mitogen stimulation of PBMCs), suggesting a potential relationship between cytokine levels and the induction of GR resistance [77]. In depressed patients increased blood concentration of TNF-α was also linked to GR resistance (as indicated by reduced skin sensitivity to topical glucocorticoids) [78]. However, evidence of reduced levels of cytokines has also been found in patients with GR resistance and GC excess [79]. Another recent study provides support for the hypothesis that inflammatory and neuroendocrine abnormalities arise in conjunction. Peripheral blood monocytes isolated from subjects exposed to the chronic stress of looking after a patient with brain cancer showed an increase in the expression of genes containing promoter response elements for NF-κβ and a decrease in genes containing promoter elements for the GR [80]. This represents a clear relationship between increased inflammatory signalling and decreased glucocorticoid signalling in the context of chronically stressed (but not depressed) subjects.

The problem these clinical studies face is determining the direction of causation. If increased inflammatory markers and GR resistance were found in a depressed population, two conclusions are possible – either that the increased inflammation caused the GR resistance, or that the presence of GR resistance caused the increased presence of inflammation (presumably by a release of the inhibition that glucocorticoids can apply to inflammation). While some evidence for the former direction of causation has been gained from the animal literature [38], evidence of this in clinical populations is difficult because of the careful longitudinal studies required.

One paradigm that has allowed some insight into the direction of causality between increased inflammation and impaired glucocorticoid signalling is IFN-α administration for the treatment of various medical illnesses, as mentioned previously. It has been shown in a group of patients undergoing IFN-α therapy that after 12 weeks of treatment, they experienced a flattening of their cortisol curve (a measure of HPA axis dysfunction) compared with control subjects and this flattening significantly correlated with the development of depressive symptoms in these subjects [81]. Interestingly, the degree of flattening identified was similar to that found in clinically depressed populations [81]. This study provides some evidence that inflammatory agents may be the cause of the HPA axis dysfunction found in depression. However, as cytokines are known to activate the HPA axis, it is still not clear whether the HPA axis dysfunction seen in these patients is the result of direct cytokine action on the GR or due to the increased HPA activity induced by cytokines. Indeed, the lack of a correlation between immune and HPA axis measures makes interpretation of causation more difficult [81]. One recent finding also confirms a role for cytokines on influencing the HPA axis – a relationship between polymorphisms of the IL-1β gene and post-DEX levels of cortisol in healthy subjects [82].

Non-GR Mechanisms by which Cytokines Can Affect GC Signalling

Cytokines, aside from their effects on GR function, can also cause dysfunctional glucocorticoid signalling by their effects on the transcription of GR isoforms and on enzymes in the metabolic pathways of glucocorticoids. IL-1 and TNF-α have been shown to induce the beta isoform of the GR which is both unable to bind cortisol and limits the activity of the functional alpha isoform as a transcription factor [83, 84]. Cytokines can also affect the availability of glucocorticoids by inhibiting the enzyme responsible for glucocorticoid metabolism in end organs, 11β-hydroxysteroid dehydrogenase [85]; by inhibiting corticosteroid-binding globulin which prevents diffusion of corticosteroids into cells (which has been demonstrated in septic shock and burn patients) [86] and by acutely inhibiting the *p*-glycoprotein pump responsible for removing glucocorticoids from a cell in high doses (though chronic, lower doses tend to increase expression of this protein) [87]. Through all these myriad processes, cytokines can impair GR function and increase the concentration of glucocorticoids available to cells, thus enhancing the abnormalities of the glucocorticoid system.

Cytokines also Activate the HPA Axis and Could Thereby Contribute to HPA Hyperactivity/GR Resistance

Apart from direct effects on the GR, cytokines can also contribute to abnormalities in the neuroendocrine system by activation of the HPA axis. Again, the paradigm of IFN-α treatment for medical illnesses has been used in studying these effects. IFN-α (like other cytokines) acutely activates the HPA axis in humans with increases in ACTH and cortisol seen within hours [88] and CRH increases observed as well [32, 89]. This fits with what has been observed in rats in which IFN-α induces the release of CRF from the hypothalamus (in vivo and ex vivo) and the amygdala [90]. It is of interest to note that patients who go on to develop depression following IFN-α treatment showed more pronounced increases in ACTH and cortisol following their first IFN-α treatment than those who did not [88]. Additionally, the degree of HPA axis hyperactivity (as indicated by the ACTH and cortisol rise in response to IFN-α therapy) on the first visit significantly predicted depressive symptoms at week 8 of therapy. There is the possibility that this hyper-responsiveness represents a pre-existing sensitivity of the HPA axis in these patients already predisposed towards depression. However, this work suggests a means by which cytokines could exacerbate a potentially pre-existing vulnerability to induce a depressive illness and thus represents another means by which inflammation can worsen abnormalities of the neuroendocrine system.

How Abnormalities of the Neuroendocrine System Can Contribute to the Damage Perpetrated by Cytokines

There have been two dominant perspectives on how abnormalities in the neuroendocrine system can lead to abnormalities in the inflammatory system. The first perspective argues that a dysfunctional neuroendocrine system is less able to fulfil its usual job of inhibiting the inflammatory response and in the absence of this inhibition, inflammation is increased. The other perspective argues that glucocorticoids, rather than being uniformly anti-inflammatory, in certain contexts (including in the brain) can actually be pro-inflammatory. Both of these perspectives are explored.

Less GC Signalling Means Less Inhibition of Inflammation and Therefore More Inflammation

There is evidence that a dysfunctional neuroendocrine system means that the usual inhibitory control it wields over inflammation is lessened allowing inflammation to increase in depressed patients. An interesting study, already mentioned, demonstrated that in the context of stress there is a shift towards inflammatory signalling and away from glucocorticoid signalling through changes in the relative expression of transcripts bearing response elements to each system [80]. This may be indicative more widely of processes occurring during chronic stress and depression where the balance of signalling moves from glucocorticoid to inflammatory signalling, explaining the dysfunctional glucocorticoid signalling and the increased inflammation found in depressed populations.

This shift in balance between these two systems has been observed in animal models of depression. One study used social disruption to induce depression-like symptoms [91]. Splenic monocytes removed from these animals demonstrated glucocorticoid resistance, along with an exaggerated pulmonary immune response to influenza infection (associated with increased mortality). However, it should be noted that while this fits a model of increased inflammatory signalling and decreased glucocorticoid signalling, the GR resistance was only evident in splenic monocytes and not in cells from other tissues [92]. Other studies have demonstrated that mice subjected to stress paradigms demonstrate both increased inflammation and glucocorticoid resistance [75]. Reduced GR function leads to a lack of transcriptional suppression of NF-κβ, suggesting the mechanism by which GR resistance is linked to an increased inflammatory response [75]. Indeed there are a number of interactions between the GR and NF-κβ signalling pathways suggesting they can be mutually inhibitory at a number of different levels – including protein-protein interactions and induction of inhibitory molecules [93], although the full complexity of their interactions remains to be elucidated. There is evidence that a balance exists between these two competing systems and that this balance may be impacted by chronic stress.

The relationship between decreased glucocorticoid signalling and increased inflammation has also been demonstrated in a dynamic system. It has been shown that rats which receive the GR antagonist RU-486 for 3 days prior to receiving a lipopolysaccharide (LPS) injection show increased induction of TNF-α, IL-1β and INF-α than those that receive just an LPS injection, and this is associated with more profound and prolonged depressive symptoms [94]. This provides direct evidence for an increased inflammatory response along with increased depressive symptoms in the context of reduced glucocorticoid signalling and provides a potential analogy to the clinical situation suggesting impaired glucocorticoid signalling could lead to increased inflammation. However, the short time period examined limits the conclusions that can be drawn. Indeed, a similar experimental set up showed a decrease or no change in TNF-α and IL-6 levels with RU486 pretreatment in a variety of rat strains, which the authors attributed to the increased corticosterone levels [95]. There is therefore some evidence that decreased glucocorticoid signalling in depression may be responsible for the increased inflammation detected in depressed patients.

Glucocorticoids in Some Situations Can Fail to Suppress Inflammation and Actually Potentiate It

Another perspective on the manner in which the neuroendocrine system can exacerbate the abnormalities in the inflammatory system is that glucocorticoids do not have a unipolar mode of action, but rather a bipolar one – they can be anti- or pro-inflammatory, depending on dosage and the temporal relationship to an inflammatory insult. This perspective thus differs strongly from the above interpretation, as it suggests that it is not the presence of inadequate glucocorticoid signalling that accounts for increased inflammation but rather a normal (but often overlooked) aspect of glucocorticoid signalling.

Contrary to the widely held idea that glucocorticoids always inhibit inflammation, there is evidence that acute stress can actually be associated with a stimulation of the immune response as part of its adaptive response. This evidence is represented below in a graph of glucocorticoid actions showing both anti-inflammatory and pro-inflammatory effects at different dosages (fig. 1).

This body of evidence suggests that glucocorticoids are pro-inflammatory at certain intermediate dosages while they are anti-inflammatory at very low dosages and at high dosages (such as those used in their well-known clinical applications). The ability of glucocorticoids to be pro-inflammatory has been demonstrated in cell and animal models and early evidence has been established in human subjects.

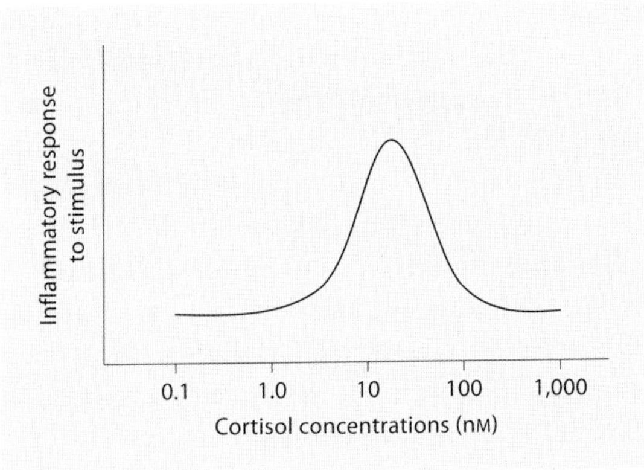

Fig. 1. Estimation of the biphasic nature of glucocorticoids. The well-established anti-inflammatory effects of glucocorticoids are present at high dosages, whereas at low-to-intermediate dosages, glucocorticoids are permissive and even stimulatory towards the inflammatory response mounted to a relevant stimulus. Timing (not represented in this graph) also plays an important role in determining the nature of the inflammatory effects of glucocorticoids. Adapted from Yeager et al. [96].

Cell Examples

Early studies examined the ability of low levels of glucocorticoids during pre-treatment or co-treatment to exacerbate an inflammatory response to inflammatory stimulation in cell cultures. One such study showed that peritoneal murine macrophages demonstrate a 2- to 4-fold enhancement of Fc receptor-mediated phagocytosis in response to IFN-γ stimulation when co-incubated with 10–100 nM hydrocortisone, but higher or lower concentrations were less effective [97]. It was shown in rat macrophages that cells treated with corticosterone demonstrated an increased response to subsequent stimulation with LPS and IFN-γ, as measured by increased levels of nitrite, IL-6, and TNF-α [98]. Increase in the expression of these cytokines was associated with increased activation of NF-κβ. More recently, it was shown that bone marrow-derived macrophages co-incubated with corticosterone demonstrate a dose-dependent increase in LPS-stimulated TNF-α and IL-6 release that is only observed at intermediate corticosterone concentrations of 10–20 nM [99].

Animal Examples

The ability of glucocorticoids to potentiate inflammation in animals has also been demonstrated. It has been shown that administration of exogenous corticosterone in rats before LPS stimulation increased inflammatory responses, as indicated by levels of TNF-α, IL-1β and IL-6, in both the hippocampus and the liver, while corticosterone administered after LPS stimulation inhibited this increase [100]. Of note, the levels of corticosterone used in this experiment were similar to those induced in rats subjected to inescapable shocks (a paradigm often used to induce depressive-like behaviour in rats). This effect has also been demonstrated in the hypothalamus, hippocampus, pituitary, cerebellum and the cortex of rats [101]. A more naturalistic iteration of this paradigm involving inescapable shock administered before ex vivo LPS stimulation of hippocampal microglia also increased the inflammatory response (as measured by IL-6, IL-1β, NF-κβ inhibitor alpha) [102]. This effect was abolished by pre-treatment with RU-486 or adrenalectomy, indicating the central role played by glucocorticoids in this paradigm. To emphasise the importance of timing in understanding the relationship between glucocorticoids and inflammation, acute stress after LPS administration decreases the same inflammatory response [103].

Human Evidence

Glucocorticoids given before an immune challenge can also prime the immune response in human subjects. Six hours of hydrocortisone treatment 12 or 144 h before LPS administration greatly increased the amount of circulating IL-6 and TNF-α, while hydrocortisone directly before and 6 h before LPS administration reduced the levels of these cytokines [104]. A more recent study showed both that baseline cortisol levels do not exert anti-inflammatory effects and that pre-treatment with an intermediate dose of cortisol increased the inflammatory response to an LPS injection in a patient population [96]. Again, the intermediate dose of cortisol that potentiated subsequent inflammatory responses was one that produced plasma levels normally observed after a major physical stress.

Mechanisms

There have been a number of proposed mechanisms by which glucocorticoid priming may have a pro-inflammatory effect. Glucocorticoids have been shown to increase the expression of receptors for IL-1 in human monocytes [105], and IL-6 in human epithelial and hepatoma cell lines [106], as well as other cytokine receptors. It has been shown that the pro-inflammatory effects of corticosterone are associated with an increased activation of NF-κβ, the MAPK signalling cascade and a reduced activation of the PI3K-Akt pathway following LPS stimulation [99]. NF-κβ, a transcription factor, and the MAPK signalling cascade are essential for the expression of chemokines and cytokines following inflammatory activation of a cell. Conversely, the PI3K-Akt pathway is an important component of the negative regulation of this expression. The increased activation of these pathways upon glucocorticoid treatment was thought to be due to an upregulation of SHIP1, a phosphatase that negatively regulates the PI3K signalling pathway, as knockdown of SHIP1 increased the PI3K-Akt signalling pathway and reduced cytokine expression [99].

This pro-inflammatory effect may be mediated by an increase in expression of Toll-like receptor 2 (TLR2) receptors which are upregulated in corticosterone pretreated rats [100]. It has been shown that elevated corticosterone before inflammatory insult increases basal activation of ERK1 and 2, p38, SAPK/JNK and AKT, all enzymes important in the expression of cytokines, in both the hippocampus and frontal cortex of rats, suggesting another plausible mechanism by which this priming may occur [107]. It has been shown that acute stress in animal models increases NF-κβ activity, TNF-α levels and the activity of its synthesising enzyme TNF-α convertase, as well as PGE2 and COX-2 levels – enzymes responsible for activating inflammatory responses – in the cortex [108–110]. Similarly, dexamethasone treatment induces NLRP3, a critical component of the inflammasome, mRNA and protein in human macrophages, significantly enhancing the release of IL-1β, TNF-α and IL-6 in a mechanism that was dexamethasone and GR dependent [111]. All of these pathways are likely to play a role in this effect.

Unresolved Questions

Nothwithstanding the above findings, there are a still a number of unanswered questions and complexities that need to be resolved in this field. For example, an explanation is required why this phenomenon of increased inflammation after priming with glucocorticoids was observed in some brain regions – the hippocampus and frontal cortex – and not others – the hypothalamus [107]. There has also been some debate as to whether these effects exist only in the context of acute stress or whether they are also present in chronic stress paradigms. Two studies have recently implemented chronic stress paradigms before inflammatory insult, yielding pro-inflammatory effects [107, 112]. Another question has also been posed as to whether the pro-inflammatory effects elicited by low dose glucocorticoid treatment are due to activation of the MR or GR with some authors suggesting the former [2] and other the latter [96].

Issues with Respect to Depression

A more general question which arises is: Are these models of the pro-inflammatory effects of glucocorticoids relevant to depression? There are several reasons to believe so. It has been noted that cortisol levels in depressed patients are not hugely increased – that is they do not demonstrate levels found in Cushing's patients – who have average plasma cortisol levels in the range of 220–2,800 nM [113] or those on exogenous glucocorticoid treatment but are rather more modestly elevated. For example, one study found that controls had 24-hour average cortisol levels of 177 nM, whereas patients with depression had an average of 223 nM [114]. Another study showed similar findings with morning sampling – controls had on average 223 nM of plasma cortisol at 8 a.m., whilst depressed patients had 345 nM on average [115]. These findings would fit into the pro-inflammatory model outlined above (and dosages given in the experiments outlined above) [96]. As pointed out in this chapter, such a model of pro-inflammatory effects mediated by glucocorticoids may be more applicable to repetitive stress and injury, rather than sustained chronic inflammation and stress, which may well model the sporadic instances of stress that have been implicated in the pathogenesis of depression [96].

GCs and Cytokines Act on the Same Processes that Have Been Implicated in the Pathogenesis of Depression

A number of processes that have been implicated in the pathogenesis of depression are affected by both cytokines and glucocorticoids. These groups of substances may exert their effects at different times or in co-ordination and, thereby, produce greater damage than each of them would exert alone. A number of processes fall into this category; the ability of both groups of compounds to contribute to GR resistance and HPA hyperactivity has already been discussed. Both groups have been implicated in a number of processes – including their effects on the monoamine systems, kynurenine metabolism, brain region connectivity, and neural plasticity. We will now focus on their contribution toward impairing neurogenesis and neurodegenerative processes.

Neurogenesis

Neurogenesis has been implicated as a key process in the pathogenesis of depression (potentially in its ability to modulate the activity of the HPA axis) [30]. Both cytokines and glucocorticoids have been shown to inhibit neurogenesis. It has been demonstrated that stress can cause alterations in dendrite and spine morphology in specific brain regions and eventually suppress adult neurogenesis in animal models [54, 55, 116, 117]. This effect has been demonstrated as being specific to glucocorticoids – both the endogenous glucocorticoid cortisol and the synthetic analogue dexamethasone – which reduce neurogenesis in human hippocampal progenitor cells [7].

Cytokines, particularly IL-1β can have similar effects. It is known that IL-1 decreases adult hippocampal neurogenesis in animal models [38, 39]. Previously, this has been demonstrated indirectly by known inducers of IL-1 – LPS and radiation – that produce reductions in neurogenesis in rats [118, 119]. Interestingly, both acute and chronic administration have this effect and, as mentioned previously, the specificity of this phenomenon to IL-1 was demonstrated by abolishment of the effect with an IL-1β receptor antagonist [38]. Indeed, overexpression of the IL-1 receptor in mice leads to a decrease in neurogenesis [120]. More recently, the ability of IL-1β to reduce neurogenesis in human hippocampal cells has been demonstrated [29]. It is possible that both of these effects on adult neurogenesis could be additive (or even synergistic) in clinical populations.

Neurodegeneration

There is evidence that neurodegeneration is also present in depression. The strongest evidence is the decrease in size of the hippocampus present in depressed populations [121]. Meta-analyses have shown this reduction to be associated with recurrent bouts of depression [63, 122] and longer duration of illness [123]. The cause of this reduction in volume is unclear – it could be induced by neurodegeneration or by decreased neurogenesis [31], though many other causes have been proposed. Markers of neurodegeneration have been detected in both depressed populations and in animal models of depression [124]. However, there have also been studies that find no evidence of neurodegeneration [125, 126].

Both cytokines and glucocorticoids can induce neurodegeneration. Glucocorticoids have been shown to induce neurodegeneration in animal models. Glucocorticoid signalling via the GR makes neurons (particularly in the hippocampus) less capable of surviving a variety of insults. Subacute corticosterone treatment worsens CA1 hippocampal neuron loss caused by global ischaemia [127]. In chronic stress paradigms elevated levels of glucocorticoids are associated with neuronal death along with suppressed adult neurogenesis and dendritic retraction [128]. However, there is also evidence suggesting that glucocorticoids can play a protective role to threatened neurons. One study demonstrated that GR inhibition (with RU486) caused greatly increased neurodegeneration in an animal model exposed to immune activation [129] suggesting that an intact glucocorticoid system might play an important suppressive role on inflammation in the brain and that this suppression prevents neurodegeneration.

Cytokines can also induce neurodegeneration, probably by their effects on the kynurenine pathway. As explored above, IFN-α induces the enzymes IDO and kynurenine 3-monooxygenase (KMO) which leads to an increase in the neurotoxic metabolites of the kynurenine pathway at the expense of the neuroprotective products. The importance of these neurotoxic chemicals to the pathogenesis of depression has been emphasised by experiments showing that blockade of IDO activation prevents the onset of depressive-like symptoms in cytokine induced depression in animal models [130]. Furthermore, administration of kynurenine dose dependently induces depressive-like behaviour in animal models [130]. It is possible that both cytokines and glucocorticoids can therefore contribute to the process of neurodegeneration and play an additive role. This possibility must be tempered by some evidence that glucocorticoids can play a protective role in some instances of inflammation induced neurodegeneration.

Conclusions

It is well established that both inflammation and dysfunction of the glucocorticoid system contribute to depression. What is unclear is how these two pathways interact in this process. A number of potential means by which these two systems might interact to cause depression has been explored. One explanation holds that chronic stress leads to GR resistance and decreased glucocorticoid signalling, which allows inflammation to increase in the relative absence of its usual inhibitory influence. This may involve chronic stress shifting the balance from glucocorticoid signalling to inflammatory signalling, as found in the study of chronically stressed carers for brain cancer sufferers [80]. One issue with this account is the finding of increased levels of cortisol in depressed patients. It would therefore need to be demonstrated that GR resistance in relevant cell populations would still lead to an overall decrease in glucocorticoid signalling. Moreover, overall decreased glucocorticoid signalling would need to be shown to be a necessary pre-condition for increased inflammation in depressed subjects. There are plausible mechanisms for such an interaction as the GR and NF-κβ signalling pathways have been shown to be mutually inhibitory at a number of different levels [93], suggesting a balance between these two competing systems. In animal models it has indeed been shown that increased inflammatory responses after stress paradigms are associated with glucocorticoid resistance [75]. Again, further evidence of changing ratios between these two systems in animal and clinical populations would strengthen this hypothesis.

On the other hand, it has also been argued that there are situations in which glucocorticoid and inflammatory signalling are not mutually inhibitory, under certain concentration and temporal conditions. There is evidence that this happens in key parts of the brain – including the hippocampus. What this hypothesis requires is to explain the clinical relevance of this – are stress-primed inflammatory responses analogous to conditions that give rise to depression in clinical populations? Are conditions sufficiently similar in the brain and the periphery that the investigation of peripheral cells offers a relevant model? If central and peripheral tissues respond differently, and even different parts of the brain respond differently to one another, what molecular mechanisms might explain these differences? This question equally hangs over the effects of temporal differences between glucocorticoid action and inflammatory insult.

A third explanation focuses on processes that are implicated in depression that are both affected by inflammation and glucocorticoids – processes like neurogenesis and neurodegeneration. It is indeed interesting that in these instances glucocorticoids and inflammation can both cause damage and suggests a possible way in which they may therefore cause damage in a cumulative manner. However, what this scenario begs is an explanation of how it is possible that these two systems could be acting on these target processes without interaction. Is it really plausible that they could exert their effects at different times or in different positions without direct interaction? If there is direct interaction then the explanation reverts to one of the above – increased inflammation is permitted because there is less glucocorticoid signalling or because glucocorticoid signalling can actually potentiate inflammation in certain circumstances.

Finally, a combination of these explanations is possible. The dominant cause may vary across individuals, or within individuals depending on the stage of illness, accounting for some of the heterogeneity in the literature. Elucidation of the relationship between inflammation and the glucocorticoid system in cell and animal models may inform the design of clinical studies leading to clarification of this relationship and a better understanding of the pathogenesis of depression.

Acknowledgements

PAZ and CMP are supported by a grant from the Commission of European Communities Seventh Framework Programme (Collaborative Project Grant Agreement no. 22963, Mood Inflame), the National Institute for Health Research Mental Health Biomedical Research Centre in Mental Health at South London and Maudsley NHS Foundation Trust and King's College London and the Medical Research Council (UK), grant MR/J002739/1. PAZ is also supported by a NARSAD Young Investigator Award. CMP is also supported by a grant from the Psychiatry Research Trust, UK (McGregor 97). MH is supported by a King's Overseas Research Studentship from the King's College London Graduate School.

References

1 Raison C: When not enough is too much: the role of insufficient glucocorticoid signalling in the pathophysiology of stress-related disorders. Am J Psychiatry 2003;160:1554–1565.

2 Sorrells SF, Caso JR, Munhoz CD, Sapolsky RM: The stressed CNS: when glucocorticoids aggravate inflammation. Neuron 2009;64:33–39.

3 Schiepers OJG, Wichers MC, Maes M: Cytokines and major depression. Prog Neuropsychopharmacol Biol Psychiatry 2005;29:201–217.

4 Miller AH: Mechanisms of cytokine-induced behavioral changes: psychoneuroimmunology at the translational interface. Brain Behav Immun 2009;23:149–158.

5 Zunszain PA, Anacker C, Cattaneo A, Carvalho LA, Pariante CM: Glucocorticoids, cytokines and brain abnormalities in depression. Prog Neuropsychopharmacol Biol Psychiatry 2011;35:722–729.

6 Lucassen PJ, Meerlo P, Naylor AS, Dam AMV, Dayer AG, Fuchs E, et al: Regulation of adult neurogenesis by stress, sleep disruption, exercise and inflammation: implications for depression and antidepressant action. Eur Neuropsychopharmacol 2010;20:1–17.
7 Anacker C, Zunszain PA, Cattaneo A, Carvalho LA, Garabedian MJ, Thuret S, et al: Antidepressants increase human hippocampal neurogenesis by activating the glucocorticoid receptor. Mol Psychiatry 2011;16:738–750.
8 Stone EA, Lin Y, Quartermain D: A final common pathway for depression? Progress toward a general conceptual framework. Neurosci Biobehav Rev 2008;32: 508–524.
9 Evans DL, Charney DS, Lewis L, Golden RN, Gorman JM, Krishnan KRR, et al: Mood disorders in the medically ill. Biol Psychiatry 2005;58:175–189.
10 Pollak Y, Yirmiya R: Cytokine-induced changes in mood and behaviour: implications for 'depression due to a general medical condition', immunotherapy and antidepressive treatment. Int J Neuropsychopharmacol 2002;5:389–399.
11 Raison CL, Capuron L, Miller AH: Cytokines sing the blues: inflammation and the pathogenesis of depression. Trends Immunol 2006;27:24–31.
12 Pandey GN, Rizavi HS, Ren X, Fareed J, Hoppensteadt DA, Roberts RC, et al: Proinflammatory cytokines in the prefrontal cortex of teenage suicide victims. J Psychiatric Res 2012;46:57–63.
13 Howren M, Lamkin D, Suls J: Associations of depression with C-reactive protein, IL-1, and IL-6: a meta-analysis. Psychosom Med 2009;186:171–186.
14 Dowlati Y, Herrmann N, Swardfager W, Liu H, Sham L, Reim EK, et al: A meta-analysis of cytokines in major depression. Biol Psychiatry 2010;67:446–457.
15 Asnis G, De La Garza R: Interferon-induced depression in chronic hepatitis C: a review of its prevalence, risk factors, biology, and treatment approaches. J Clin Gastroenterol 2006;40:322–335.
16 Capuron L, Miller AH: Cytokines and psychopathology: lessons from interferon-alpha. Biol Psychiatry 2004; 56:819–824.
17 Raison C, Borisov A, Broadwell S: Depression during pegylated interferon-alpha plus ribavirin therapy: prevalence and prediction. J Clin Psychiatry 2005; 66:41–48.
18 Capuron L, Gumnick J, Musselman D: Neurobehavioral effects of interferon-α in cancer patients: phenomenology and paroxetine responsiveness of symptom dimensions. Neuropsychopharmacology 2002;26:643–652.
19 Miller AH, Maletic V, Raison CL: Inflammation and its discontents: the role of cytokines in the pathophysiology of major depression. Biol Psychiatry 2009;65:732–741.
20 Szelenyi J: Cytokines and the central nervous system. Brain Res Bull 2001;54: 329–338.
21 Kitagami T, Yamada K, Miura H, Hashimoto R, Nabeshima T, Ohta T: Mechanism of systemically injected interferon-alpha impeding monoamine biosynthesis in rats: role of nitric oxide as a signal crossing the blood-brain barrier. Brain Res 2003;978:104–114.
22 Zhu C-B, Blakely RD, Hewlett WA: The proinflammatory cytokines interleukin-1beta and tumor necrosis factor-alpha activate serotonin transporters. Neuropsychopharmacology 2006;31:2121–2131.
23 Moron J, Zakharova I, Ferrer J, Al E: Mitogen-activated protein kinase regulates dopamine transporter surface expression and dopamine transport capacity. J Neurosci 2003;23:8480–8488.
24 Chang L, Karin M: Mammalian MAP kinase signalling cascades. Nature 2001; 410:37–40.
25 Fujigaki H, Saito K, Fujigaki S, Takemura M, Sudo K, Ishiguro H, et al: The signal transducer and activator of transcription 1alpha and interferon regulatory factor 1 are not essential for the induction of indoleamine 2,3-dioxygenase by lipopolysaccharide: involvement of p38 mitogen-activated protein kinase and nuclear factor. J Biochem 2006;139:655–662.
26 Bonaccorso S, Marino V, Biondi M, Grimaldi F, Ippoliti F, Maes M: Depression induced by treatment with interferon-alpha in patients affected by hepatitis C virus. J Affect Disord 2002;72:237–241.
27 Wichers MC, Koek GH, Robaeys G, Verkerk R, Scharpé S, Maes M: IDO and interferon-alpha-induced depressive symptoms: a shift in hypothesis from tryptophan depletion to neurotoxicity. Mol Psychiatry 2005;10:538–544.
28 Myint AM, O'Mahony S, Kubera M, Kim YK, Kenny C, Kaim-Basta A, et al: Role of paroxetine in interferon-alpha-induced immune and behavioural changes in male Wistar rats. J Psychopharmacol 2007;21:843–850.
29 Zunszain, Anacker C, Cattaneo A, Choudhury S, Musaelyan K, Myint AM, et al: Interleukin-1β: a new regulator of the kynurenine pathway affecting human hippocampal neurogenesis. Neuropsychopharmacology 2012;37:939–949.
30 Snyder JS, Soumier A, Brewer M, Pickel J, Cameron HA: Adult hippocampal neurogenesis buffers stress responses and depressive behaviour. Nature 2011;476:458–461.
31 Sapolsky RM: Is impaired neurogenesis relevant to the affective symptoms of depression? Biol Psychiatry 2004;56: 137–139.
32 Pariante C, Miller AH: Glucocorticoid receptors in major depression: relevance to pathophysiology and treatment. Biol Psychiatry 2001;49:391–404.
33 Pace TWW, Miller AH: Cytokines and glucocorticoid receptor signaling. Relevance to major depression. Ann NY Acad Sci 2009;1179:86–105.
34 Hayley S, Poulter MO, Merali Z, Anisman H: The pathogenesis of clinical depression: stressor- and cytokine-induced alterations of neuroplasticity. Neuroscience 2005;135:659–678.
35 Pickering M, O'Connor JJ: Pro-inflammatory cytokines and their effects in the dentate gyrus. Prog Brain Res 2007;163:339–354.
36 Felger J, Alagbe O, Hu F, Mook D: Effects of interferon-alpha on rhesus monkeys: a nonhuman primate model of cytokine-induced depression. Biol Psychiatry 2007;23:1324–1333.
37 Song C, Wang H: Cytokines mediated inflammation and decreased neurogenesis in animal models of depression. Prog Neuropsychopharmacol Biol Psychiatry 2011;35:760–768.
38 Goshen I, Kreisel T, Licht T, Weidenfeld J, Yirmiya R, Ben-Menachem-Zidon O, et al: Brain interleukin-1 mediates chronic stress-induced depression in mice via adrenocortical activation and hippocampal neurogenesis suppression. Mol Psychiatry 2008;13:717–728.
39 Koo JW, Duman RS: IL-1β is an essential mediator of the antineurogenic and anhedonic effects of stress. Proc Natl Acad Sci USA 2008;105:751–756.

40 Mendlewicz J, Kriwin P, Oswald P, Souery D, Alboni S, Brunello N: Shortened onset of action of antidepressants in major depression using acetylsalicylic acid augmentation: a pilot open-label study. Int Clin Psychopharmacol 2006;21:227–231.
41 Müller N, Schwarz MJ, Dehning S, Douhe A, Cerovecki A, Goldstein-Müller B, et al: The cyclooxygenase-2 inhibitor celecoxib has therapeutic effects in major depression: results of a double-blind, randomized, placebo controlled, add-on pilot study to reboxetine. Mol Psychiatry 2006;11:680–684.
42 Tyring S, Gottlieb A, Papp K, Gordon K, Leonardi C, Wang A, et al: Etanercept and clinical outcomes, fatigue, and depression in psoriasis: double-blind placebo-controlled randomised phase III trial. Lancet 2006;367:29–35.
43 Young EA, Haskett RF, Murphy-Weinberg V, Watson SJ, Akil H: Loss of glucocorticoid fast feedback in depression. Arch Gen Psychiatry 1991;48:693–699.
44 Holsboer F: The corticosteroid receptor hypothesis of depression. Neuropsychopharmacology 2000;23:477–501.
45 Pariante C: Depression, stress and the adrenal axis. J Neuroendocrinol 2003;15: 811.
46 Nemeroff CB, et al: Adrenal gland enlargement in major depression. Arch Gen Psychiatry 1992;49:384–347.
47 Pariante CM: Risk factors for development of depression and psychosis. Glucocorticoid receptors and pituitary implications for treatment with antidepressant and glucocorticoids. Ann NY Acad Sci 2009;1179:144–152.
48 Juruena MF, Cleare AJ, Papadopoulos AS, Poon L, Lightman S, Pariante C: Different responses to dexamethasone and prednisolone in the same depressed patients. Psychopharmacology 2006;189: 225–235.
49 Brown ES, Suppes T: Mood symptoms during corticosteroid therapy: a review. Harv Rev Psychiatry 1998;5:239–246.
50 Ardayfio P, Kim K-S: Anxiogenic-like effect of chronic corticosterone in the light-dark emergence task in mice. Behav Neurosci 2006;120:249–256.
51 Murray F, Smith DW, Hutson PH: Chronic low dose corticosterone exposure decreased hippocampal cell proliferation, volume and induced anxiety and depression like behaviours in mice. Eur J Pharmacol 2008;583:115–127.
52 David D, Samuels BA, Rainer Q, Wang J-W, Marsteller D, Mendez I, et al: Neurogenesis-dependent and -independent effects of fluoxetine in an animal model of anxiety/depression. Neuron 2009;161:479–493.
53 Herbert J, Goodyer IM, Grossman AB, Hastings MH, de Kloet ER, Lightman SL, et al: Do corticosteroids damage the brain? J Neuroendocrinol 2006;18: 393–411.
54 McEwen BS, Seeman T: Protective and damaging effects of mediators of stress. Elaborating and testing the concepts of allostasis and allostatic load. Ann NY Acad Sci 1999;896:30–47.
55 Sapolsky RM: The possibility of neurotoxicity in the hippocampus in major depression: a primer on neuron death. Biol Psychiatry 2000;48:755–765.
56 Gould E, Cameron H, Daniels DC, Woolley CS, McEwen BS: Adrenal hormones suppress cell division in the adult rat dentate gyrus. J Neurosci 1992;12:3642–3650.
57 McEwen BS, Magarinos AM: Stress and hippocampal plasticity: implications for the pathophysiology of affective disorders. Human Psychopharmacol 2001; 16(S1):S7–S19.
58 Sousa N, Cerqueira JJ, Almeida OFX: Corticosteroid receptors and neuroplasticity. Brain Res Rev 2008;57:561–570.
59 Kim JB, Ju JY, Kim JH, Kim T-Y, Yang B-H, Lee Y-S, et al: Dexamethasone inhibits proliferation of adult hippocampal neurogenesis in vivo and in vitro. Brain Res 2004;1027:1–10.
60 Reif A, Fritzen S, Finger M, Strobel A, Lauer M, Schmitt A, et al: Neural stem cell proliferation is decreased in schizophrenia, but not in depression. Mol Psychiatry 2006;11:514–522.
61 Boldrini M, Underwood MD, Hen R, Rosoklija GB, Dwork AJ, Mann JJ, et al: Antidepressants increase neural progenitor cells in the human hippocampus. Neuropsychopharmacology 2009;34: 2376–2389.
62 Sheline YI, Wang PW, Gado MH, Csernansky JG, Vannier MW: Hippocampal atrophy in recurrent major depression. Proc Natl Acad Sci USA 1996;93: 3908–3913.
63 Videbech P, Ravnkilde B: Hippocampal volume and depression: a meta-analysis of MRI studies. Am J Psychiatry 2004; 161:1957.
64 Sapolsky RM: Depression, antidepressants, and the shrinking hippocampus. Proc Natl Acad Sci USA 2001;98: 12320–12322.
65 Anacker C, Zunszain PA, Carvalho LA, Pariante C: The glucocorticoid receptor: Pivot of depression and of antidepressant treatment? Psychoneuroendocrinology 2010;36:415–425.
66 Pariante C, Pearce B, Pisell T, Sanchez C: The proinflammatory cytokine, interleukin-1α, reduces glucocorticoid receptor translocation and function. Endocrinology 1999;140:4359–4366.
67 Wang X, Wu H, Miller H: Interleukin 1alpha (IL-1alpha) induced activation of p38 mitogen-activated protein kinase inhibits glucocorticoid receptor function. Mol Psychiatry 2004;9:65–75.
68 Goleva E, Kisich KO, Leung DYM: A role for STAT5 in the pathogenesis of IL-2-induced glucocorticoid resistance. J Immunol 2002;169:5934–5940.
69 Onda K, Nagashima M, Kawakubo Y, Inoue S, Hirano T, Oka K: Mitogen-activated protein kinase kinase 1/extracellular signal-regulated kinase (MEK-1/ERK) inhibitors sensitize reduced glucocorticoid response mediated by TNFalpha in human epidermal keratinocytes (HaCaT). Biochem Biophys Res Commun 2006;351:266–272.
70 Xu Q, Goleva E, Ou L-S, Li L-B, Leung DYM: CD56+ cells induce steroid resistance in B cells exposed to IL-15. J Immunol 2004 ;172:7110–7115.
71 Pace TWW, Hu F, Miller AH: Activation of cAMP-protein kinase A abrogates STAT5-mediated inhibition of glucocorticoid receptor signaling by interferon-alpha. Brain Behav Immun 2011;25: 1716–1724.
72 Weidenfeld J, Yirmiya R: Effects of bacterial endotoxin on the glucocorticoid feedback regulation of adrenocortical response to stress. Neuroimmunomodulation 1996;3:352–357.
73 Bener D, Wohlman A, Itzik A, Yirmiya R, Ben-Hur T, Weidenfeld J: Glucocorticoid resistance following herpes simplex-1 infection: role of hippocampal glucocorticoid receptors. Neuroendocrinology 2007;85:207–215.
74 Quan N, Avitsur R, Stark JL, He L, Shah M: Social stress increases the susceptibility to endotoxic shock. J Neuroimmunol 2001;115:36–45.

75 Quan N, Avitsur R, Stark JL, He L, Lai W: Molecular mechanisms of glucocorticoid resistance in splenocytes of socially stressed male mice. J Neuroimmunol 2003;137:51–58.
76 Engler H, Bailey MT, Engler A, Stiner-jones LM, Quan N, Sheridan JF: Interleukin-1 receptor type 1-deficient mice fail to develop social stress-associated glucocorticoids resistance in the spleen. Psychoneuroendocrinology 2008;33: 108–117.
77 Maes M, Bosmans E, Meltzer H: Interleukin-1β: a putative mediator of HPA axis hyperactivity in major depression. Am J Psychiatry 1993;150: 1189–1193.
78 Fitzgerald P, O'Brien S, Scully P: Cutaneous glucocorticoid receptor sensitivity and pro-inflammatory cytokine levels in antidepressant-resistant depression. Psychol Med 2006;36:37–43.
79 Schuld A, Schmid D, Haack M, Holsboer F, Friess E, Pollmächer T: Hypothalamo-pituitary-adrenal function in patients with depressive disorders is correlated with baseline cytokine levels, but not with cytokine responses to hydrocortisone. J Psychiatr Res 2003;37:463–470.
80 Miller, Chen E, Sze J, Marin T, Arevalo JMG, Doll R, et al: A functional genomic fingerprint of chronic stress in humans: blunted glucocorticoid and increased NF-kB signaling. Biol Psychiatry 2008; 64:266–272.
81 Raison CL, Borisov AS, Woolwine BJ, Massung B, Vogt G, Miller AH: Interferon-a effects on diurnal hypothalamic-pituitary-adrenal axis activity: relationship with proinflammatory cytokines and behavior. Mol Psychiatry 2008;15: 535–547.
82 Sasayama D, Hori H, Iijima Y, Teraishi T, Hattori K, Ota M, et al: Modulation of cortisol responses to the DEX / CRH test by polymorphisms of the interleukin-1beta gene in healthy adults. Behav Brain Sci 2011;7:23.
83 Lewis-Tuffin LJ, Cidlowski JA: The physiology of human glucocorticoid receptor beta (hGRbeta) and glucocorticoid resistance. Ann NY Acad Sci 2006;1069:1–9.
84 Webster JC, Oakley RH, Jewell CM, Cidlowski JA: Proinflammatory cytokines regulate human glucocorticoid receptor gene expression and lead to the accumulation of the dominant negative beta isoform: a mechanism for the generation of glucocorticoid resistance. Proc Natl Acad Sci USA 2001;98:6865–6870.
85 Kossintseva I, Wong S, Johnstone E: Pro-inflammatory cytokines inhibit human placental 11beta-hydroxysteroid dehydrogenase type 2 activity through Ca^{2+} and cAMP pathways. Am J Physiol Endocrinol Metab 2006;2:282–288.
86 Emptoz-Bonneton A, Crave J: Corticosteroid-binding globulin synthesis regulation by cytokines and glucocorticoids in human hepatoblastoma-derived (HepG2) cells. J Clin Endocrinol Metab 1997;82:3758–3762.
87 Bauer B, Hartz A: Tumor necrosis factor α and endothelin-1 increase P-glycoprotein expression and transport activity at the blood-brain barrier. Mol Pharmacol 2007;71:667–675.
88 Capuron L, Raison C: Association of exaggerated HPA axis response to the initial injection of interferon-alpha with development of depression during interferon-alpha therapy. Am J Psychiatry 2003;160:1342–1345.
89 Besedovsky H: Immune-neuro-endocrine interactions: facts and hypotheses. Endocr Rev 1996;17:64–102.
90 Raber J, Koob GF, Bloom FE: Interferon-alpha and transforming growth factor-beta 1 regulate corticotropin-releasing factor release from the amygdala: comparison with the hypothalamic response. Neurochem Int 1997;30:455–463.
91 Sheridan J, Stark J, Avitsur R: Social disruption, immunity, and susceptibility to viral infection: role of glucocorticoid insensitivity and NGF. Ann NY Acad Sci 2000;917:894–905.
92 Avitsur R, Powell N, Padgett DA, Sheridan JF: Social interactions, stress, and immunity. Neurol Clin 2009;29:285–293.
93 Rhen T, Cidlowski J: Brain interleukin-1 mediates chronic stress-induced depression in mice via adrenocortical activation and hippocampal neurogenesis suppression. N Engl J Med 2005;353:1711–1723.
94 Wang D, Lin W, Pan Y, Kuang X, Qi X, Sun H: Brain, behavior, and immunity chronic blockade of glucocorticoid receptors by RU486 enhances lipopolysaccharide-induced depressive-like behaviour and cytokine production in rats. Brain Behav Immun 2011;25:706–714.
95 Yang B, Trump RP, Shen Y, Mcnulty JA, Clifton LG, Stimpson SA, et al: RU486 did not exacerbate cytokine release in mice challenged with LPS nor in db/db mice. BMC Pharmacol 2008;10:1–10.
96 Yeager M, Pioli P, Guyre P: Cortisol exerts bi-phasic regulation of inflammation in humans. Dose Response 2011; 1949:332–347.
97 Warren MK, Vogel SN: Opposing effects of glucocorticoids on interferon-gamma-induced murine macrophage Fc receptor and Ia antigen expression. J Immunol 1985;134:2462–2469.
98 Smyth GP, Stapleton PP, Freeman TA, Concannon EM, Mestre JR, et al: Glucocorticoid pretreatment induces cytokine overexpression and nuclear factor-B activation in macrophages. J Surg Res 2004;261:253–261.
99 Zhang T, Daynes RA: Macrophages from 11β-hydroxysteroid dehydrogenase type 1-deficient mice exhibit an increased sensitivity to lipopolysaccharide stimulation due to TGF-β-mediated up-regulation of SHIP1 expression. J Immunol 2007;179:6325–6335.
100 Frank MG, Miguel ZD, Watkins LR, Maier SF: Prior exposure to glucocorticoids sensitizes the neuroinflammatory and peripheral inflammatory responses to *E. coli* lipopolysaccharide. Brain Behav Immun 2010;24:19–30.
101 Johnson J, O'Connor K, Deak T, Stark M: Prior stressor exposure sensitizes LPS-induced cytokine production. Brain Behav Immun 2002;476:461–476.
102 Frank M, Thompson B, Watkins L: Glucocorticoids mediate stress-induced priming of microglial pro-inflammatory responses. Brain Behav Immun 2011;26: 337–345.
103 Goujon E, Parnet P, Laye S, Combe C: Stress downregulates lipopolysaccharide-induced expression of proinflammatory cytokines in the spleen, pituitary and brain. Brain Behav Immun 1995;9:292–303.
104 Barber A, Coyle S, Marano M: Glucocorticoid therapy alters hormonal and cytokine responses to endotoxin in man. J Immunol 1993;150:1999–2006.
105 Spriggs K, Cosmans D, Lioubins PJ, Dower K: Induction of an interleukin-1 receptor (IL-1R) on monocytic cells. J Biol Chem 1990;265:22499–22505.
106 Snyers L, De Wit L, Content J: Glucocorticoid up-regulation of high-affinity interleukin-6 receptors on human epithelial cells. Proc Natl Acad Sci USA 1990;87:2838–2842.

107 Munhoz, Lepsch LB, Kawamoto EM, Malta MB, Lima LDS, Avellar MCW, et al: Chronic unpredictable stress exacerbates lipopolysaccharide-induced activation of nuclear factor-kappaB in the frontal cortex and hippocampus via glucocorticoid secretion. J Neurochem 2006;26:3813–3820.
108 Madrigal JL, Moro MA, Lizasoain I, Lorenzo P, Castrillo A, Boscá L, et al: Inducible nitric oxide synthase expression in brain cortex after acute restraint stress is regulated by nuclear factor kappaB-mediated mechanisms. J Neurochem 2001;76:532–538.
109 Madrigal JLM, Hurtado O, Moro MA, Lizasoain I, Lorenzo P, Castrillo A, et al: The increase in TNF-alpha levels is implicated in NF-kappaB activation and inducible nitric oxide synthase expression in brain cortex after immobilization stress. Neuropsychopharmacology 2002; 26:155–63.
110 Madrigal JLM, Moro MA, Lizasoain I, Lorenzo P, Fernández AP, Rodrigo J, et al: Induction of cyclooxygenase-2 accounts for restraint stress-induced oxidative status in rat brain. Neuropsychopharmacology 2003;28:1579–1588.
111 Busillo JM, Azzam KM, Cidlowski JA: Glucocorticoids sensitize the innate immune system through regulation of the NLRP3 inflammasome. J Biol Chem 2011;286:38703–38713.
112 de Pablos RM, Villarán RF, Argüelles S, Herrera AJ, Venero JL, Ayala A, et al: Stress increases vulnerability to inflammation in the rat prefrontal cortex. J Neurosci 2006;26:5709–5719.
113 Starkman MN, Gebarski SS, Berent S, Schteingart DE: Hippocampal formation volume, memory dysfunction, and cortisol levels in patients with Cushing's syndrome. Biol Psychiatry 1992;32:756–765.
114 Halbreich U, Asnis G, Shindledecker R: Cortisol secretion in endogenous depression. I. Basal plasma levels. Arch Gen Psychiatry 1985;42:904–908.
115 Fang VS, et al: Plasma ACTH and cortisol levels in depressed patients: relation to dexamethasone suppression test. Life Sci 1981;29:931–938.
116 Mirescu C, Gould E: Stress and adult neurogenesis. Hippocampus 2006;16: 233–238.
117 Joëls M, Karst H, Krugers HJ, Lucassen PJ: Chronic stress: implications for neuronal morphology, function and neurogenesis. Front Neuroendocrinol 2007;28: 72–96.
118 Ekdahl CT, Claasen J-H, Bonde S, Kokaia Z, Lindvall O: Inflammation is detrimental for neurogenesis in adult brain. Proc Natl Acad Sci USA 2003;100: 13632–13637.
119 Monje ML, Toda H, Palmer TD: Inflammatory blockade restores adult hippocampal neurogenesis. Science 2003;302: 1760–1765.
120 Spulber S, Oprica M, Bartfai T, Winblad B, Schultzberg M: Blunted neurogenesis and gliosis due to transgenic overexpression of human soluble IL-1ra in the mouse. Eur J Neurosci 2008;27:549–558.
121 Sheline YI: Neuroimaging studies of mood disorder effects on the brain. Biol Psychiatry 2003;54:338–352.
122 Campbell S, Marriott M, Nahmias C, MacQueen GM: Lower hippocampal volume in patients suffering from depression: a meta-analysis. Am J Psychiatry 2004;161:598.
123 McKinnon MC, Yucel K, Nazarov A, MacQueen GM: A meta-analysis examining clinical predictors of hippocampal volume in patients with major depressive disorder. J Psychiatry Neurosci 2009;34:41–54.
124 Maes M, Yirmyia R, Noraberg J, Brene S, Hibbeln J, Perini G, et al: The inflammatory & neurodegenerative (I&ND) hypothesis of depression: leads for future research and new drug developments in depression. Metab Brain Dis 2009;24: 27–53.
125 Müller MB, Lucassen PJ, Yassouridis A, Hoogendijk WJ, Holsboer F, Swaab DF: Neither major depression nor glucocorticoid treatment affects the cellular integrity of the human hippocampus. Eur J Neurosci 2001;14:1603–1612.
126 Lucassen P, Holsboer F, Bauer J: Hippocampal apoptosis in major depression is a minor event and absent from subareas at risk for glucocorticoid overexposure. Am J Pathol 2001;158:453–468.
127 Sapolsky R, Pulsinelli WA: Glucocorticoids potentiate ischemic injury to neurons: therapeutic implications. Science 1985;229:1397–1400.
128 Czéh B, Lucassen PJ: What causes the hippocampal volume decrease in depression? Are neurogenesis, glial changes and apoptosis implicated? Eur Arch Psychiatry Clin Neurosci 2007;257:250–260.
129 Nadeau S, Rivest S: Glucocorticoids play a fundamental role in protecting the brain during innate immune response. J Neurosci 2003;23:5536–5544.
130 O'Connor JC, Lawson MA, André C, Briley EM, Szegedi SS, Lestage J, et al: Induction of IDO by Bacille Calmette-Guérin is responsible for development of murine depressive-like behavior. J Immunol 2009;182:3202–3212.

Dr. C.M. Pariante
Institute of Psychiatry, The James Black Centre, King's College London
125 Coldharbour Lane
London SE5 9NU (UK)
E-Mail carmine.pariante@kcl.ac.uk

Halaris A, Leonard BE (eds): Inflammation in Psychiatry.
Mod Trends Pharmacopsychiatry. Basel, Karger, 2013, vol 28, pp 144–161 (DOI: 10.1159/000343981)

Co-Morbidity between Cardiovascular Pathology and Depression: Role of Inflammation

Angelos Halaris

Department of Psychiatry and Behavioral Neuroscience, Loyola University Stritch School of Medicine, and Loyola University Medical Center, Maywood, Ill., USA

Abstract

Morbidity and mortality of cardiovascular disease is exceedingly high worldwide. Depressive illness is a serious psychiatric illness that afflicts a significant portion of the population worldwide. Epidemiological studies have confirmed the high co-morbidity between these two entities and the co-morbidity is bidirectional. Systems that are involved in and accountable for this co-morbidity in a major, complex and interactive way include the central and autonomic nervous systems, the neuroendocrine system, the immune system, and the vascular and hematologic systems. Specific pathophysiologic factors across these systems include homeostatic imbalance between the sympathetic and the parasympathetic systems with loss of heart rate variability in depression, sympathoadrenal activation, hypothalamic-pituitary-adrenal axis activation resulting in hypercortisolemia, immune system dysregulation with release of pro-inflammatory cytokines and chemokines, platelet activation and hypercoagulability. All of these abnormalities have been demonstrated in most individuals diagnosed with major depressive disorder. This chapter will focus on inflammatory processes. Inflammation occurs in cardiac and cardiovascular pathology independent of the presence or absence of depression. A chronic pro-inflammatory status has been documented in numerous studies of depression. Inflammation is closely associated with endothelial dysfunction which is a preamble to atherosclerosis and atherothrombosis. Endothelial dysfunction has been detected in depression and may prove to be a trait marker for this illness. Thus, understanding vascular biology in conjunction with psychiatric co-morbidity will be of critical importance. A likely common instigator underlying the co-morbidity between cardiovascular pathology and depression is mental stress. Chronic stress shifts the homeostatic balance in the autonomic nervous system with sustained sympathetic overdrive and diminished vagal tone. Diminished vagal tone contributes to a pro-inflammatory status which affects neurotransmitter regulation, specifically serotonergic transmission. Antidepressant drug therapy is of definite benefit to patients with medical and psychiatric co-morbidity and may reverse the pro-inflammatory status associated with depression.

Cardiovascular disease (CVD) and depressive illness (DI) are two of the world's leading health problems. CVD is the leading cause of mortality worldwide and accounts for approximately 16.7 million deaths every year [1–5]. The most common types of CVD are coronary artery disease (CAD) and cerebrovascular disease (CBVD).

CAD is a leading cause of morbidity and mortality in the industrialized world [6, 7] and it is the leading cause of death and hospital admissions in the United States. It has been estimated that CVD is responsible for close to 1 million deaths and more than six million hospital admissions with an annual cost to the US economy in excess of USD 350 billion [7]. Depressive illness is a very common disorder afflicting an estimated 120 million people worldwide. It is the leading cause of disability worldwide [8]. In the United States depression afflicts approximately 17 million people annually with a direct and indirect cost to the nation exceeding 50 billion in today's dollar estimates. The life-time prevalence of DI is estimated to be as high as 16% and the 12-month prevalence ranges from 3% in Japan to over 9% in the US [9–11]. The incidence of DI has been steadily increasing over the past decades. Using the DALY, Disability Adjusted Life-Years, unipolar major depressive disorder (MDD) was classed in 1990 as the fourth leading burden of disease worldwide for both sexes. By 2004, it advanced to third place, and, according to the World Health Organization's estimate, will rank second to CVD by the year 2020 [12] and will be the leading cause of disease burden by 2030 [11]. The co-morbidity between CVD and DI is remarkably high and is believed to be bidirectional. Numerous studies published over the past two decades have confirmed this association. In a meta-analytic study conducted by Barth et al. [13] clinical depression was identified as a significant risk factor for mortality in patients with coronary heart disease (CHD). The Multiple Risk Factor Intervention Trial (MRFIT) of middle-aged men established an association between depressive symptoms and all-cause mortality with a higher risk of CVD death and more specifically stroke mortality [14]. Several pathophysiological mechanisms are held responsible for the high co-morbidity between CVD and DI, but also other psychopathological disorders and personality traits. In this chapter, I will focus on inflammation-related parameters in attempting to shed some light on the biology underlying this co-morbidity. Inflammation is intricately related to endothelial dysfunction, atheromatosis and atherosclerosis. The autonomic nervous system (ANS) plays a critical role in mediating stress and stress-related disorders, depression being the stress disorder par excellence. Cardiac function is profoundly affected by ANS dysfunction. Cardiac parameters associated with ANS imbalance contribute significantly to the complex factors associated with the co-morbidity. This chapter discusses our current understanding of the co-morbidity between DI and CVD and expands on the reviews we provided previously [15, 16].

Epidemiological Aspects

Depressive Disorder

Epidemiological studies have demonstrated unequivocally that there is a high incidence of depression among patients suffering from ischemic heart disease [17, 18]. Depression is also recognized as an independent risk factor for CVD mortality between six and eighteen months following myocardial infarction (MI) [19, 20] and its presence is associated with increased CVD event rates [21]. In patients with an acute MI, depression is a risk factor for recurrent nonfatal infarction and cardiac mortality independent of cardiac disease severity [22, 23]. Depressed patients have a 2- to 4-fold risk of developing CVD at some point in their lifetime [24–26] and a similarly higher risk of dying after a cardiac event [27, 28]. Depressed patients are at higher risk for developing congestive heart failure (CHF), while patients with CHF and depression have worse outcomes than those with CHF without evidence of depression [29, 30]. Quoting from the Baltimore Epidemiologic Catchment Area Follow-Up Study [31], the diagnosis of depression is associated with an increased odds ratio (OR) for adverse outcomes, for example, an OR of 4.5 for MI and an OR of 2.7 for stroke. The Framingham Heart Study identified a direct relationship between depressive symptomatology and all-cause mortality in a very large cohort of original and offspring cohort participants over a 6-year follow-up period [32]. Several groups of investigators have concluded that the presence of depressive symptomatology predicts future coronary events for initially healthy individuals and a poor prognosis for those who have documented CVD [33–35]. A meta-analytic study sought to establish depression as an etiological or prognostic factor in CHD [36]. In spite of limitations inherent in a meta-analytic study, these authors found significant associations between DI and CHD. However, they were unable to establish depression as an independent risk factor for CHD ostensibly due to 'incomplete and biased availability of adjustments for conventional risk factors and severity of coronary disease'.

Anxiety Disorders

Depression is frequently accompanied by anxiety presenting itself with somatic, emotional and cognitive manifestations. Distinguishing clearly whether depression or anxiety is the primary psychiatric disorder is a challenge for the clinician. To illustrate this point, a recently published study by Celano et al. [37] determined in a cohort of depressed cardiac patients that higher baseline anxiety scores were associated with lesser improvement in depressive symptoms and increased likelihood that the depression would persist at the 6-month follow-up evaluation. The contention that co-occurrence of depression and anxiety may have additive effects on the risk of CHD was addressed in the study by Gomez-Caminero et al. [38]. Patients diagnosed with panic disorder had a 2-fold increased risk for CHD after controlling for confounding factors. By contrast, patients with a co-morbid diagnosis of DI were almost three times more likely to develop CHD again after adjusting for confounds. The Framingham Offspring Study followed 3682 study participants after 10 years for the incidence of CHD, atrial fibrillation (AF) and total mortality. After controlling for confounding factors, they concluded that anxiety was a risk factor for total mortality in both sexes whereas tension was an independent risk factor for incident CHD, AF, and total mortality in men [39]. Evidence in support of a relationship between depression, anxiety, and social isolation and outpatient mortality due to CHF has been presented by Friedman et al. [40].

In an effort to delineate the independent or interdependent contribution of depression versus anxiety to cardiovascular pathology, Scherrer et al. [41] conducted a 7-year retrospective study of a national cohort of Veterans Administration patients who were free of heart disease at baseline. They determined that those patients who suffered from depression, an unspecified anxiety disorder, panic disorder, and post-traumatic stress disorder (PTSD) were at increased risk of developing an incident MI. They additionally concluded that the risk for an incident MI remains high in conjunction with an anxiety disorder even in the absence of concomitant depression.

Posttraumatic Stress Disorder

The bidirectional association between myocardial infarction (MI) or an acute coronary syndrome (ACS) and PTSD is especially intriguing in the context of stress, depression, inflammation and risk for CVD. According to a recent meta-analytic study by Edmondson et al. [42], 1 of every 8 patients with MI or ACS develops PTSD symptoms. These researchers examined data from 24 studies that comprised 2,383 patients with ACS worldwide. They found a 12% prevalence of clinically significant ACS-induced PTSD among the ACS patients and a doubling of risk of mortality and recurrent cardiac events among those ACS patients with PTSD symptomatology. These authors underscored that abundant evidence suggests that psychological disorders are under-recognized and undertreated in cardiac populations [43, 44]. They further commented that 'While awareness of depression has increased in cardiology practice, awareness of the possibility of PTSD due to ACS has lagged, possibly because many still see PTSD as primarily a disorder of combat veterans or sexual assault survivors' [42]. In a related article from the same group of investigators, evidence was presented that PTSD symptoms predict a potentially life threatening pre-hospital delay in patients with ACS requiring emergency care [45]. Reinforcing the findings of these studies, von Känel et al. [46] compared two groups of post-MI patients. One group had developed PTSD symptoms and the other had not. The group with post-MI PTSD had elevated biomarkers of inflammation exemplified in significant elevations of IL-6 and C-reactive protein (CRP). These authors concluded that inflammation may be a link between PTSD and increased risk of CVD events. Similar findings had been presented earlier by Spindler et al. [47]. After reviewing 25 studies that fulfilled their inclusion criteria, these authors established that PTSD had a prevalence rate of 0–38% across studies.

Psychological Distress

The concept of psychological distress subsumes symptoms of depression and anxiety without necessarily attempting to clearly delineate the presence or absence of a distinct nosologic entity. Pertinent studies have used the long or short version of the General Health Questionnaire (GHQ) to profile symptomatology. The Whitehall II Study [48] was a prospective cohort study of London-based civil service employees who were assessed by the GHQ at baseline to predict the incidence of CHD 5 years later. It was determined that psychological distress has an independent effect on CHD risk which is not attributable to psychosocial risk factors. The authors were not able to identify specific aspects of psychological distress that might be associated with CHD risk. A related study was conducted with a Scottish cohort and comparable study design. Rasul et al. [49] reported that psychological distress was associated with a 78% increased risk for CHD in men only. Adjustment for sociodemographic and other CHD risk factors did not attenuate this risk. Interestingly, the authors found no association between psychological distress and CHD incidence in women even after fully adjusted analyses. To further delineate the effect of psychosocial factors on CHD risk, Molloy et al. [50] conducted a similar study by combining data from the Scottish Health Survey from 1995, 1998 and 2003. Their findings reiterated the increased CVD mortality risk for unmarried men and women and for divorced or widowed men. Russ et al. [51] recently published a meta-analysis of 10 large, community-based cohort studies seeking to establish the role of psychological distress as a mortality risk factor for all causes, CBVD, cancer and external causes. They determined that there was a 22% increase in risk of CVD death associated with psychological distress and this association remained robust after adjusting for confounding factors. They also obtained evidence for a 'dose-response effect'.

Personality Traits and Disorders

Psychological distress as a cardiovascular risk marker needs to be expanded to include a general propensity to distress, as has been aptly suggested by several groups of clinician researchers and cogently reviewed by Denollet et al. [52]. Psychological distress is strongly associated with the major mental illnesses, but specific personality traits and certain personality disorders have also been linked to chronic stress and distress and increased cardiovascular morbidity and mortality. Anxiety characteristics were shown by Shen et al. [53] to independently and prospectively predict MI incidence among older men. Anger and, even more significantly, suppressed anger, were shown by Denollet et al. [54] to be associated with major adverse cardiac events. In their cohort, patients with type D personality characteristics had increased vulnerability for CVD events ostensibly due to their exceedingly high rate of suppressed anger. Earlier studies had focused on type A personality traits and behavioral patterns as predictors of incident CVD, as reviewed by Myrtek [55]. Type A traits include not only the customary competitiveness, driveness and perfectionism, but, perhaps more importantly, anger and hostility. Type D personality is defined by 'negative affectivity' and 'social inhibition' with a high level of suppressed anger and pervasive distress. Prospective studies reported by Denollet et al. [52] have shown that type D personality is associated with greater than 3-fold risk of a poor prognosis. While there may be some overlap between DI and type D, they are not identical, nor do they have to co-occur. Type D is more likely a chronic marker for CAD whereas DI is probably a state marker. The association of certain personality disorders, as defined in DSM-IV [56], and incident CVD was investigated by Lee at al. [57] in the context of the Baltimore Epidemiologic Catchment Area (ECA) follow-up study. Their results confirmed previous reports of a cross-sectional association between personality disorders and CVD, but extended them by focusing attention on cluster B personality disorders, such as borderline personality disorder, and incident CVD. In attempting to address the possible pathophysiological determinants of the co-morbidity between personality traits and/or personality disorders and CVD, a 'pro-inflammatory environment' has been postulated to be a key contributory mechanism. Denollet et al. [58] have provided experimental evidence to support this hypothesis.

Factors Accountable for the Co-Morbidity

While there is little doubt that DI and CVD are epidemiologically linked, the definitive mechanisms that interactively link these two complex disease entities elude us. Until we can precisely identify the pathophysiological processes and biobehavioral factors that explain the mechanistic associations between the two entities, specifically targeted pharmacotherapeutic, psychotherapeutic and preventive interventions will not be possible. The reasons accountable for the relative lack of success in identifying the mechanisms these two conditions share in common or the adverse impact they exert upon each other are complex and multifaceted. A host of biological, psychological, socioeconomic, sociodemographic, biobehavioral factors have been unequivocally shown to be significant contributors to the co-morbidity. Stress is a crucial factor and yet there is enormous individual variation in stress susceptibility, responsivity and resilience. Genetic and epigenetic factors exert powerful influences on shaping the individual's coping skills with stressful events, physical and mental illness in particular. Mental illnesses are diverse entities and their psychobiological underpinnings have yet to be fully elucidated. Genetic polymorphisms exert critical effects in conferring resistance or susceptibility to illness. Environmental factors acting via epigenetic mechanisms are only now beginning to be understood.

Cardiac and vascular research avail themselves of procedures and objective and standardized techniques to assess reliably the cardiovascular status of the patient. By contrast, in studies of DI and related psychiatric disorders data are not readily comparable among studies due to differences in choice of rating instruments among test sites and the unavoidable subjectivity of clinician investigators throughout the world. In nonpsychiatric settings there is even greater variability in the choice of criteria used to identify and quantify depressive symptomatology among cardiac patients. Once clinical depression is present, a host of physiological changes occur, including autonomic nervous system activation, neuroendocrine and immunologic changes, rhythm disturbances, and platelet hypercoaguability, all of which exert a negative impact on cardiovascular health. The majority of studies seeking to elucidate the biological underpinnings of this co-morbidity have focused on a single or a handful of parameters making meta-analytic comparisons across studies difficult. Additionally, biobehavioral mechanisms play a critical role in defining the high co-morbidity. For instance, life style factors, such as alcohol consumption and smoking, physical inactivity, and medication nonadherence, partly due to side effects associated with antidepressant medications, all play at least a contributory role to any and all biological factors. Last, but not least, traditional cardiovascular risk factors, such as hypertension, diabetes, dyslipidemia, obesity and substance abuse play variable roles in defining co-morbidity and must always be precisely assessed in such studies.

Taken together, the mechanistic pathways accountable for the co-morbidity between DI and CVD are highly complex and multifaceted. However, in spite of the above barriers, significant findings and promising leads already exist and they can guide future research in this field. It is beyond the scope of this review article to provide a comprehensive discussion of available clinical and research data for each of the putative component mechanisms. However, evidence in support of the likely mechanistic factors leading to and shared by the two conditions will be presented in this chapter.

Mental Stress and Its Consequences

Mental stress can produce profound alterations in the physiology and chemistry of the CNS and ANS, peripheral organs, the endocrine, the vascular and the immune systems. Mental stress can be of many varieties with differences in intensity and duration, and the perception of mental stress is subject to high individual variability and vulnerability. This is based, at least in part, on genetically determined and epigenetically modified stress susceptibility and stress resilience. Any illness, physical or mental, is of itself stressful to the individual. Chronic and especially inescapable stress ultimately leads to pervasive mental status changes and pathological alterations to the function and structure of the cardiovascular system that may lead to irreversible tissue and organ damage. As Selye [59] had postulated, when the demands placed on the organism eventually exceed the available energy and adaptation, the system fails. The complexity of the interactions between stress, and medical and psychiatric illnesses remains a topic of intense scientific endeavor.

As reviewed here, stress (physical, mental or mixed) is a state of short-term or prolonged disturbance of homeostasis leading ultimately to a state of allostasis and allostatic load as suggested by McEwen [60, 61]. In addition to activating the hypothalamic-pituitary-adrenal axis (HPA), stress activates the sympathetic branch of the ANS. This activation leads to a reduction in vagal tone that can have consequences in the body's immune response. The imbalance in ANS function, especially if prolonged, as is invariably the case with psychiatric disorders, has profound effects on cardiovascular physiology and the immune system. These pathophysiological changes contribute significantly to the co-morbidity between CVD and psychiatric disorders. Thus, to quote Thayer and Lane [62], autonomic imbalance and decreased parasympathetic activity in particular, may be the final common pathway to numerous diseases and conditions associated with increased morbidity and mortality.

Psychophysiologic Consequences of Stress

Sympathoadrenal (SA) hyperactivity in depression, with a concomitant decrease in vagal tone, leads to increases in plasma catecholamines, and vasoconstriction, elevated heart rate and platelet activation. The Heart and Soul Study clearly established that in patients with CAD depressive symptoms are associated with elevated levels of norepinephrine [63]. These changes, singly and additively, exert adverse effects on the cardiovascular system. Elevated and sustained catecholaminergic action on the heart, blood vessels and platelets leads to changes in hemodynamic factors with increased shear stress. The decrease in parasympathetic tone may predispose to ventricular arrhythmias and possibly explain the excessive cardiovascular mortality found in patients with

CVD and co-morbid depression. Heart rate variability (HRV) is an index of a healthy heart and good cardiac function with ability to adapt to external and internal stressful demands. Diminished high-frequency HRV reflects decreased parasympathetic tone and has been observed at least in some depressed patients. In a large cohort study (NESDA), Licht et al. [64] demonstrated that depressed patients, when compared to healthy control subjects, have significantly both lower total HRV and lower HRV in the respiratory frequency range, which reflects diminished cardiac vagal control. They concluded that their findings are compatible with theoretical models, notably the Polyvagal Theory coined by Porges [65]. These models have linked the parasympathetic system to the etiopathology of DI based on the premise that low vagal tone is associated with reduced social engagement and an impaired response to environmental stimuli and challenges, as described by Porges [66, 67].

HRV can also be significantly decreased in severe CAD or heart failure. The risk of sudden death after acute MI is significantly higher with decreased HRV. Indeed, HRV is a post-MI prognostic factor (along with age, left-ventricular ejection fraction, frequency of arrhythmia). It has been reported that in patients with CAD, diminished HRV is far more common in the depressed rather than in the matched nondepressed patients.

Neuroendocrine Consequences of Stress

In response to stress, corticotrophin-releasing factor (CRF) containing neurons increase the synthesis and release of adrenocorticotropic hormone (ACTH), β-endorphin, and proopiomelanocortin from the anterior pituitary. Hyperactivity of the HPA axis in DI is well established. Depressed patients show elevated CRF in cerebrospinal fluid, blunting of ACTH response to CRF, nonsuppression of cortisol after dexamethasone, hypercortisolemia, and pituitary and adrenal gland enlargement. In postmortem tissue obtained from depressed patients there are increased numbers of hypothalamic CRF neurons. Resulting hypercortisolemia is toxic to hippocampal cells; it also promotes the development of hypertension and atherosclerosis and accelerates injury to the vascular endothelium [68]. Hypercortisolemia is partially responsible for hypertriglyceridemia, hypertension, injury of vascular endothelium and intima, inhibition of normal healing and induction of atherosclerosis.

The Immune System and Inflammation

This section focuses on the immune system and inflammatory factors that have been consistently identified in patients with cardiac pathology and in patients diagnosed with an affective disorder. Mental stress can be postulated to act as a common instigator initiating the cascade of complex central and peripheral events that lead to sustained activation of the immune system resulting in a pro-inflammatory status with its associated sequelae. However, mental stress alone may not be sufficient to explain the cascade, multiplicity and complexity of events that underlie the co-morbid pathology of CVD and DI. I will provide a brief overview of current thinking linking stress and neuroimmune interactions. It is beyond the scope of this review to elaborate on the complex aspects of the neurobiology of stress. However, a key concept is that stress can be physiological and psychological and in most instances it is mixed. As such, psychological stress poses a threat to the integrity and homeostatic balance of the organism, much like an injury or invasion by pathogenic organisms. The homeostatic challenge is recognized by the innate immune system that surveys the organism for threats by invaders. The inflammatory response aims to restore homeostasis and health, as long as inflammation is commensurate with the threat and limited by anti-inflammatory responses. However, inflammation can persist and expand, both centrally and peripherally, especially in instances of chronic physical or mental illnesses that perpetuate the stressful condition. The relationship between stressful events and the onset of inflammation-related diseases, notably cardiovascular, neurologic and immunologic, has been adequately documented in the literature and is also discussed in greater depth in other chapters in this volume.

The Inflammatory Reflex and the Cholinergic Anti-Inflammatory Pathway

To maintain homeostasis it is vital that a bidirectional communication occurs at all times between brain and peripheral organs. Visceral functions require reciprocal ANS mechanisms, humoral and hormonal factors and neurotransmitter regulation. These complex interactions are prerequisites to the regulation of visceral functions and coordination with 'higher' CNS structures, notably the hypothalamus, amygdala, and forebrain. The concept of the 'inflammatory reflex' was coined by Tracey [69]. This reflex controls innate immunity and inflammation predicated on the bi-directional communication between brain and the immune system and may represent the 'missing link' in neuroimmunomodulation [70]. The 'hard-wired' connection between the nervous and immune systems is represented by the vagus nerve and has been named the 'cholinergic anti-inflammatory pathway' because acetylcholine is the main neurotransmitter of the parasympathetic system [71, 72]. This pathway has been shown experimentally to inhibit localized inflammation; it is represented by efferent vagus nerve fibers that convey the anti-inflammatory signal to the periphery. Acetylcholine released from cholinergic axon terminals, interacts with the α7 subunit of nicotinic receptors on immune cells [73]. Activation of this receptor results in inhibition of pro-inflammatory cytokine release and restoration of immune homeostasis [74]. Tying together the stress response that involves sympathoadrenal activation with diminished vagal tone and the 'cholinergic anti-inflammatory pathway', it becomes plausibly reasonable to hypothesize that stress leads to disinhibition of the body's inflammatory response with loss of ability to re-equilibrate its homeostatic balance. Pervasive disinhibition of inflammatory responses should be manifested in circulating levels of pro-inflammatory cytokines and related biomarkers, as we will discuss below.

Cardiovascular Pathology and Inflammation

Inflammation plays a major role in the development of atherosclerosis and atherothrombosis with the associated clinical sequelae that lead to high morbidity and mortality. Indeed, Ross [75] has shown that atherosclerosis is intrinsically an inflammatory disease. As Lind [76] points out in his review article, immune-competent cells are involved both in the early stages of atherosclerosis and the subsequent events leading to plaque formation and rupture and the ensuing acute coronary syndromes. Over the past two decades, a number of circulating compounds have been identified as markers of inflammation and atherosclerosis. Such compounds include acute-phase proteins, CRP, fibrinogen, immunoglobulins, adhesion molecules, and cytokines. We will focus on inflammation biomarkers in this section, inasmuch as most all of these biomarkers have also been shown to be abnormally regulated in patients with depressive syndromes without any evidence of other inflammatory processes, cardiovascular or immune system pathology. Fibrinogen, CRP and from within the cytokine group tumor necrosis factor-α (TNFα), IL-1 and IL-6 have been discussed extensively in the cardiovascular literature. Indeed, Frasure-Smith et al. [77] have made a case that CRP and depression have prognostic value in predicting adverse cardiac events.

It is widely recognized that inflammatory processes in the atherosclerotic artery may lead to increased blood levels of pro-inflammatory cytokines and other acute-phase reactants. Blood levels of IL-6 and CRP are elevated in patients with an MI or unstable angina, as are other inflammatory biomarkers, such as fibrinogen, IL-7, IL-8 and soluble CD40 ligand. These findings point to inflammation in the coronary arteries. Collectively, these circulating biomarkers indicate the presence of an ACS and may reflect the clinical course of the condition [78].

Heart failure (HF) is a serious syndrome that at best interferes with normal functioning and significantly impairs the capacity of an individual to lead a normal life. Although several models have been proposed to understand the pathophysiology of this syndrome, none have, so far, proven to be critical for the prevention and management of this condition. More recently, however, inflammation has been on center stage. To quote from an earlier publication by Seta et al. [79], 'heart failure progresses, at least in part, as a result of the toxic effects exerted by endogenous cytokine cascades on the heart and the peripheral circulation'. Animal and clinical data support the presence of an inflammatory cascade that is involved at least in the progression of HF [80].

Affective Disorders and Inflammation

Evidence supporting a connection between mood disturbances and the immune system dates back to more than two decades ago [81, 82] and continues to be actively studied [83, 84]. Activation of inflammatory processes as a common link between DI, CVD and CBVD has received increased attention. Pro-inflammatory cytokines have been implicated in the pathogenesis of atherosclerosis and CVD [85–87]. Endothelial damage leads to the release of pro-inflammatory cytokines which induce a sequence of events leading ultimately to thrombus formation and vascular occlusion. Inflammation is a major contributing causal factor to DI and has been postulated to play a major role in endothelial damage of the cerebral vasculature. This has led to the designation of the entity of 'vascular depression' that presents with many of the typical symptoms of depression. Interestingly, elevations in inflammatory markers have been shown in depressed patients with or without CVD [88–90] and in cancer patients with depression [91]. One cytokine, IL-6, secreted in response to stress, is one of the most potent stimulators of the HPA axis, and induces the release of other pro-inflammatory cytokines. These and related observations have led Leonard to postulate that DI is a disease of inflammation in response to chronic psychological stress [92–95]. Since a sustained pro-inflammatory status has been clearly associated with widespread pathology, the assertion can be made that DI is a whole-body disease.

Previous reports have shown that pro-inflammatory cytokines are endogenously overexpressed in depression and other stress-induced disorders [96], but not all reports agree. Some reports have also shown below normal blood levels of pro-inflammatory cytokines. Pro-inflammatory cytokines are causally linked to plaque formation, cardiac irritation, and CAD [97]. As noted above, DI is associated with a high risk for CAD [98–100] and pro-inflammatory cytokine overproduction in DI may be one common link. Although the degree of cytokine elevations in DI are moderate in comparison to those during physical injury and/or infection [101], by being chronically overexpressed in DI, cytokines stand to have long-term adverse consequences on the cardiovascular system [102]. Studies of injected cytokines [103] have also shown that pro-inflammatory cytokines, by themselves, are dysphoric – possessing the capacity to induce a syndrome known as 'sickness behavior' that simulates many of the classical symptoms of depression [104, 105]. The Cytokine Theory of Depression was originally formulated by Smith [106] as 'The Macrophage Theory of Depression' [106] and later expanded by Ur et al. [107]. In essence, this theory postulates that psychological stress, probably in conjunction with genetic factors, increases cytokine production and leads to depressive symptoms when specific neurobiological systems are affected, e.g. serotonin and HPA-axis function. High circulating levels of pro-inflammatory cytokines may thus be central to the pathophysiology of major depressive disorder. Cytokines may firstly participate in the etiology of DI, and later in the high incidence of CAD observed amongst depressed patients [108]. A number of reports indicate beneficial effects of antidepressant drug therapy in normalizing abnormally elevated levels of pro-inflammatory cytokines [109–113]. Such a reduction can only be beneficial as has been clearly described in heart failure patients by Mayer et al. [114]. However, not all reports are consistent in this regard. Differences in study design, mainly the length of treatment and the type of antidepressant agent utilized, appear to be crucial variables that must be controlled in future studies. Additionally, the time course of cytokine normalization may not fully coincide with the time frame of clinical antidepressant response to treatment. The article by Janssen et al. [115] offers a comprehensive review of cytokine involvement in antidepressant treatment response. Clearly more studies are required to clarify the temporal associations between mood improvement and cytokine normalization. Lastly, there is emerging evidence that co-administration of an antidepressant with an anti-inflammatory agent (e.g. a COX-2 inhibitor) may produce a more rapid response in depressed patients and may even allow nonremitters to previous antidepressant trials to show a satisfactory response [116].

Endothelial Dysfunction: Possible Link between CVD and DI

The role of inflammation in the pathogenesis of CVD is now widely accepted. Low-grade chronic inflammation is predictive of MI and ischemic stroke. Endothelial dysfunction is a 'critical intermediate phenotype' in the relationship between low-grade inflammation and CVD [117]. We and others have reported on the consistent presence of a low-grade chronic inflammation in DI exemplified by mild-to-moderate elevations in circulating biomarkers, such as hsCRP, and pro-inflammatory cytokines [118–121]. Nitric oxide (NO) is the principal mediator of smooth muscle cell relaxation and contributes to overall vascular health. Its property as a vasodilator is well established. As levels of CRP increase, basal NO generation decreases significantly [117]. Similarly, TNFα has been shown to reduce the half-life of mRNA encoding NO synthase [122] resulting in decreased basal NO production. This in turn favors increased expression of cell-surface adhesion molecules for leukocytes and platelets inducing a pro-coagulant state. In summary, endothelial dysfunction is a crucial factor in the complex relationship between depression, low-grade chronic inflammation, and CVD. Indeed, some investigators have postulated that endothelial function, viewed as a biomarker of arterial atheromatosis [123, 124] may prove to be a trait marker for depression [125, 126].

Other Factors Contributing to Co-Morbidity

While inflammation plays a key role in CVD, CBVD and DI, there remain missing links requiring further clarification before the co-morbidity can be fully elucidated. Several other mechanisms have been proposed to address the relationship of DI and CVD; they include smooth muscle cell proliferation, extracellular matrix deposition or lysis, and thrombogenesis. These mechanisms appear to be inter-related and interactive but crucial details still elude us. It is beyond the scope of this review to address such factors but platelet factors as they contribute to thrombogenesis and the nitric oxide/arginine pathway will be briefly discussed.

Platelet Factors

Increased β-thromboglobulin (BTG) in patients with CVD was noted many years ago [127] as was platelet factor-4 (PF4) [128]. Laghrissi-Thode et al. [129] reported increased PF4 and BTG in patients with CVD and DI, and these findings were largely confirmed by Kuijpers et al. [130]. Musselman et al. [131] established that platelets from depressed patients showed increased activation before treatment and normalization after treatment with paroxetine [132]. These investigators postulated that platelet activation may be the link between DI and ischemic heart disease [133].

Platelet activation in depression has been demonstrated in numerous studies including our own [134]. We have demonstrated upregulation of platelet imidazoline receptors, which are also present in the brain and are believed to mediate central control of blood pressure and certain aspects of emotionality [135]. Similarly, we have demonstrated sustained elevations in P-selectin in platelets of depressed patients that, unlike the study with paroxetine [132], persisted even after successful antidepressant treatment with bupropion, a nonserotonergic antidepressant, and normalization of mood [134]. Similar findings have been reported by other investigators. It is unclear whether a serotonergic antidepressant is required to downregulate platelet activation.

The platelet/endothelial biomarker substudies by Serebruany et al. [136–138], conducted within the context of the Sertraline Antidepressant Heart Attack Randomized Trial (SADHART) project, provided solid evidence that the SSRI, sertraline, and its primary metabolite, N-desmethylsertraline, suppress platelet function. The ability of SSRIs, if the above results can be extended to other SSRIs, to produce clinically robust and safe antidepressant effects in patient populations with diverse co-morbid conditions while at the same time suppressing pathological platelet activation is of major clinical significance. Thus, successful treatment of depression with SSRIs may have a positive influence on platelet reactivity and possibly also on other dysregulated biomarkers [139].

Nitric Oxide and the Arginine Pathway

The role nitric oxide (NO) may play in the regulation of cardiovascular function is of special interest as it may provide a useful link between DI and stress-induced phenomena. NO stimulates soluble guanylyl cyclase which in turn promotes vasodilation, inhibits platelet adhesion and aggregation and restricts vascular smooth muscle migration to inflamed epithelium [140]. Reduction in NO availability in the endothelium could promote CVD and vascular occlusion. The NO pathway is a side pathway of arginine metabolism, an amino acid that also leads to the formation of agmatine. Agmatine is a confirmed neurotransmitter of relevance to stress, neuroprotection and possibly DI (for review, see Halaris and Piletz [141]). We have demonstrated increases in plasma levels of agmatine in depressed patients that show robust normalization to control levels after successful antidepressant treatment with venlafaxine [unpubl. obs.]. We have hypothesized that arginine may be shunted toward agmatine formation in depression thereby reducing arginine availability for the NO pathway. This hypothesis requires further testing.

Brain-Immune Interaction: Role of Serotonergic Transmission

It is now recognized that depression involves a complex and bidirectional interaction between the brain and the immune system. As the immune system responds to stress with increased production of pro-inflammatory biomarkers, notably cytokines, these inflammatory responses exert a powerful influence on the HPA axis and neurotransmitters critical to mood regulation and cognition. Serotonergic transmission has emerged as a possible 'common denominator' between CVD, DI and the immune system. This is not meant to ignore the role of other neurotransmitters, notably acetylcholine or norepinephrine, but in this section I will focus on serotonin for reasons that will become obvious below.

Abnormalities of serotonergic mechanisms in depression have been described in the literature over the past decades. Leonard [93] and Mendelson [142] have written useful reviews. $Serotonin_{2A}$ receptor upregulation has been described in depression [143, 144]. Studies have reported decreased platelet aggregation in response to serotonin but platelet reactivity to serotonin is significantly increased in depressed patients [145]. Of relevance here is the work of Whyte et al. [146] who demonstrated that depressed patients with the serotonin-transporter-linked promoter region l/l genotype had increased platelet activation compared to depressed patients without this genotype or nondepressed control subjects. These authors hypothesized that persons with this polymorphism are likely to experience higher mortality rates. If this finding is replicated, 'genotyping could be used to help identify the patients with ischemic heart disease at greater risk of depression-associated cardiovascular mortality'.

Another possible mechanism linking serotonergic transmission, DI and CVD is the cytokine connection. In an earlier study, Palazzolo et al. [147] had shown that IL-1 inhibits the release of serotonin. In a more recent study, Zhu et al. [148] showed that two pro-inflammatory cytokines, IL-1β and TNFα, activate serotonin transporters. Both of these mechanisms may act synergistically to reduce serotonin availability at the postsynaptic neuronal membrane. The resulting serotonergic deficiency has been linked pathophysiologically to DI and can be overcome by administration of the widely used serotonin-reuptake-inhibiting antidepressants (SSRIs).

In addition to the effects described above, a more compelling case for the interaction between inflammation and serotonergic transmission can be made on the basis of data involving the tryptophan/kynurenine pathway. Serotonin is synthesized from about 1% of the available tryptophan in the body. Serotonin synthesis occurs predominantly in the gut and only 10–20% occurs in the brain. Under physiological conditions, about 99% of tryptophan is metabolized to kynurenine in the liver by tryptophan 2,3-dioxygenase (TDO). In inflammatory conditions, infections or oxidative stress, the first rate-limiting enzymatic step involving the enzyme, indoleamine 2,3-dioxygenase (IDO), is activated. The IDO enzymatic activity is enhanced by pro-inflammatory cytokines such as IFNγ [89, 90], IL-1, IL-2, IL-6 and TNFα, and inhibited by the anti-inflammatory cytokine IL-4 [91, for review see 149]. Overstimulation of IDO leads to tryptophan depletion and resulting serotonin deficiency in the brain [150]. Additionally, during inflammatory states, peripheral kynurenine formation is increased and is transported through the blood-brain barrier thereby increasing the brain concentration of kynurenine. A metabolic product of the kynurenine pathway is quinolinate that is neurotoxic. An increase in tryptophan breakdown and increased kynurenine formation has been demonstrated in depression [151] and the kynurenine metabolism has been implicated in the pathophysiology of DI, especially the chronicity of the disorder [152]. The upregulation of the initiating step of the kynurenine pathway has been demonstrated in postmortem anterior cingulate cortex from individuals with schizophrenia and affective disorders [153]. This group of investigators [151, 154] has observed a lower plasma tryptophan index and neuroprotective kynurenic acid and higher tryptophan breakdown also in bipolar patients compared to controls.

Utilty of Serotonin Reuptake Inhibitors

Depression is a serious psychiatric illness that must be treated vigorously whether or not it is a concomitant of cardiovascular or other disease entity. Serotonin reuptake inhibitors (SSRIs) are not only effective and safe antidepressants, they may also possess therapeutic potential on the cardiovascular system. Before the introduction of SSRIs into the market, the use of antidepressants in patients with CVD was controversial largely due to their side effect profile. With the advent of the safer SSRIs their use in these patients has grown. Their utility is further supported by the determination that there is reduced risk for MI in patients exposed to SSRIs. The studies by Roose et al. [155], Glassman et al. [156], Muselman et al. [132] and Sauer et al. [157] support the utility of SSRIs and indicate a protective action of these agents against further post-MI morbidity and mortality. The post hoc secondary analyses from the Enhancing Recovery in Coronary Heart Disease (ENRICHD) study strongly support this contention [158, 159].

Conclusion

It is apparent that the relationship between DI and CVD is very complex. Based on the available literature, inflammation emerges as a dominant theme. An active inflammatory process likely precedes the emergence of debilitating symptoms, is definitely present during the active stages of the illness and possibly extends beyond the normalization of mood and symptom remission. To address these complex questions truly multidisciplinary teams of established investigators with diverse expertise in basic laboratory, clinical and cardiodiagnostic settings must come together. Integrated teams of basic and clinical investigators possessing the requisite expertise have only recently begun to work collaboratively. It is only through the cohesive interaction of such multidisciplinary teams that we can succeed in unraveling the complex relationships between mental stress, inflammation, immune responses and DI, CVD and CBVD. The dual efficacy of SSRIs in relieving depression and in reversing at least some of the pathogenetic factors common in both disease entities holds great therapeutic promise. Alternative pharmacotherapeutic agents must also be developed based on discoveries of alternate pathways involving interactively the immune and neurotransmitters systems.

Acknowledgement

Supported by an intramural grant from Loyola University Stritch School of Medicine.

References

1 Dahlof B: Cardiovascular disease risk factors: epidemiology and risk assessment. Am J Cardiol 2010;105(1 suppl):3A–9A.
2 Lloyd-Jones DM, Hong Y, Labarthe D, et al: Defining and setting national goals for cardiovascular health promotion and disease reduction. Circulation 2010;121:586–613.
3 Waldman SA, Terzic A: Cardiovascular health: the global challenge. Clinical Pharmacol Ther 2011;90:483–485.
4 Terzic A, Waldman SA: Chronic diseases: the emerging pandemic. Clin Transl Sci 2011;4:225–226.
5 Kovasic JC, Fuster V: From treating complex coronary disease to promoting cardiovascular health: therapeutic transitions and challenges 2010–2020. Clin Pharmacol Ther 2011;90:509–518.
6 Viles-Gonzalez JF, Fuster V, Badimon JJ: Atherothrombosis: a widespread disease with unpredictable and life-threatening consequences. Eur Heart J 2004;25: 1197–1207.
7 American Heart Association: Heart Disease and Stroke Statistics – 2004 Update. Dallas, American Heart Association, 2004.
8 Murray CJ, Lopez AD: Global mortality, disability and the contribution of risk factors: Global Burden of Disease Study. Lancet 1997;349:1436–1442.
9 Demyttenaere K, Bruffaerts R, Posada-Villa J, et al: Prevalence, severity, and unmet need for treatment of mental disorders in the World Health Organization world Mental Health surveys. JAMA 2004;291:2581–2590.
10 Centers for Disease Control and Prevention (CDC): Current depression among adults – United states, 2006 and 2008. Morb Mortal Wkly Rep 2010;59: 1229–1235.
11 Lepin JP, Briley M: The increasing burden of depression. Neuropsych Dis Treat 2011;7(suppl 1):3–7.
12 Rozanski A, Blumenthal JA, Kaplan J: Impact of psychological factors on the pathogenesis of cardiovascular disease and implications for therapy. Circulation 1999;99:2192–2217.
13 Barth J, Schumacher M, Herrmann-Lingen C: Depression as a risk factor for mortality in patients with coronary heart disease: a meta-analysis. Psychosom. Med 2004;66:802–813.

14 Gump BB, Matthews KA, Eberly LE, Chang YF, MRFIT Research Group: Depressive symptoms and mortality in men: results from Multiple Risk Factor Intervention Trial. Stroke 2005;36: 98–102.
15 Halaris A: Co-morbidity between depression and cardiovascular disease. Int Angiol 2009;28:92–99.
16 Saleptsis VG, Lambropoulos N, Halaris A, Angelopoulos NV, Giannoukas AD: Depression and atherosclerosis. Int Angiol 2011;30:97–104.
17 Fielding R: Depression and acute myocardial infarction: a review and reinterpretation. Soc Sci Med 1991;32:1017–1027.
18 Schleifer SJ, Macari-Hinson MM, Coyle DA: The nature and course of depression following myocardial infarction. Arch Intern Med 1989;149:1785–1789.
19 Frasure-Smith N, Lesperance F, Talajic M: Depression following myocardial infarction: impact of 6-month survival. JAMA 1993;270:1819–1825.
20 Frasure-Smith N, Lesperance F, Talajic M: Depression and 18-month prognosis after myocardial infraction. Circulation 1995;91:999–1005.
21 Lesperance F, Frasure-Smith N, Jueau M, Theroux P: Depression and 1-year prognosis in unstable angina. Arch Intern Med 2000;160:1354–1360.
22 Ahern DK, Gorkin L, Anderson JL, Tierney C, Hallstrom A, Ewart C, Capone RJ, Schron E, Kornfeld D, Herd JA, Richardson DW, Follick MJ: Biobehavioral variables and mortality or cardiac arrest in the Cardiac Arrhythmia Pilot Study (CAPS). Am J Cardiol 1990;66:59–62.
23 Bush DE, Ziegelstein RC, Tayback M, Richter D, Stevens S, Zahalsky H, Fauerbach JA: Even minimal symptoms of depression increase mortality risk after acute myocardial infection. Am J Cardiol 2001;88:337–341.
24 Ford DE, Mead LA, Chang PP: Depression is a risk factor for coronary artery disease in men: the precursors study. Arch Intern Med 1998;158:1422–1426.
25 Ariyo AA, Haan M, Tangen CM, Ruteledge JC, Cushman M, Dobs A: Depressive symptoms and risks of coronary heart disease and mortality in elderly Americans. Cardiovascular Health Study Collaborative Research Group. Circulation 2000;102:1773–1779.
26 Penninx BW, Beekman AT, Honig A, Deeg DJ, Schoevers RA, van Eijk JT: Depression and cardiac mortality: Results from a community-based longitudinal study. Arch Gen Psychiatry 2001; 58:221–227.
27 Jiang W, Alexander J, Christopher E, Kuchibhatia M, Gaulden LH, Cuffe MS: Relationship of depression to increased risk of mortality and rehospitalization in patients with congestive heart failure. Arch Intern Med 2001;161:1849–1856.
28 Lesperance J, Frasure-Smith N, Talajic M, Bourassa M: Five-year risk of cardiac mortality in relation to initial severity and one-year changes in depression symptoms in myocardial infarction. Circulation 2002;105:1049–1053.
29 Junger J, Schellberg D, Muller-Tasch T: Depression increasingly predicts mortality in the course of congestive heart failure. Eur J Heart Fail 2005;7:261–267.
30 Sherwood A, Blumenthal JA, Trivedi R, et al: Relationship of depression to death or hospitalization in patients with heart failure. Arch Intern Med 2007;167: 367–373.
31 Eaton WW, Fogel J, Armenian HK: The Consequences of Psychopathology in the Baltimore Epidemiologic Catchment Area Follow-up. Medical and Psychiatric Co-morbidity over the Lifespan. Washington, American Psychiatric Publishing, 2006, pp 21–38.
32 Wulsin LR, Evans JC, Vasan RS, Murabito JM, Kelly-Hayes M, Benjamin E: Depressive symptoms, coronary heart disease, and overall mortality in the Framingham Heart Study. Psychosomatic Medicine 2005;67:697–702.
33 Musselman DL, Evans DL, Nemeroff CB: The relationship of depression to cardiovascular disease: epidemiology, biology, and treatment. Arch Gen Psychiatry 1998;55:580–592.
34 O'Connor CM, Gurbel PA, Serebruany VL: Depression in ischemic heart disease. Am Heart J 2000;140:63–69.
35 Carney RM, Jaffe AS: Treatment of depression following acute myocardial infarction. JAMA 2002;288:750–751.
36 Nicholson A, Kuper H, Hemingway H: Depression as an aetiologic and prognostic factor in coronary heart disease: a meta-analysis of 6362 events among 146 538 participants in 54 observational studies. Eur Heart J 2006;27:2763–2774.
37 Celano CM, Mastromauro CA, Lenihan EC, Januzzi JL, Rollman BL, Huffman JC: Association of baseline anxiety with depression persistence at 6 months in patients with acute cardiac illness. Psychosom Med 2012;74:93–99.
38 Gomez-Caminero A, Blumentals WA, Russo LJ, Brown RR, Castilla-Puentes R: Does panic disorder increase the risk of coronary heart disease? A cohort study of a national managed care database. Psychosom Med 2005;67:688–691.
39 Eaker ED, Sullivan LM, Kelly-Hayes M, D'Agostino RB, Benjamin EJ: Tension and anxiety and the prediction of the 10-year incidence of coronary heart disease, atrial fibrillation, and total mortality: The Framingham Offspring Study. Psychosom Med 2005;67:692–696.
40 Friedmann E, Thomas SA, Liu F: Relationship of depression, anxiety, and social isolation to chronic heart failure outpatient mortality. Am Heart J 2006;152:940.e1–e8.
41 Sherrer JF, Chrusciel T, Zeringue A, Garfield LD, Hauptman PJ, Lustman PJ, Freedland KE, Carney RM, Bucholz KK, Owen R, True WR: Anxiety disorders increase risk for incident myocardial infarction in depressed and nondepressed Veterans Administration patients. Am Heart J 2010;159:772–779.
42 Edmondson D, Richardson S, Falzon L, Davidson KW, Mills MA, Neria Y: Posttraumatic stress disorder prevalence and risk of recurrence in acute coronary syndrome patients; a meta-analytic review. PLoS ONE 2012;7:e38915.
43 Rutledge T, Reis VA, Linke SE, Greenberg BH, Mills PJ: Depression in heart failure a meta-analytic review of prevalence, intervention effects, and associations with clinical outcomes. J Am Coll Cardiol 2006;48:1527–1537.
44 Berkman LF, Blumenthal J, Burg M, Carney RM, Catellier D, et al: Effects of treating depression and low perceived social support on clinical events after myocardial infarction. The Enhancing Recovery in Coronary Heart Disease Patients (ENRICHD) randomized trial. JAMA 2003;289:3106–3116.
45 Newman JD, Muntner P, Shimbo D, Davidson KW, Shaffer JA, Edmondson D: Post-traumatic stress disorder (PTSD) symptoms predict delay to hospital in patients with acute coronary syndrome. PLoS ONE 2011;6:e27640.

46 von Känel R, Begre S, Abbas CC, Saner H, Gander ML, Schmid JP: Inflammatory biomarkers in patients with posttraumatic stress disorder caused by myocardial infarction and the role of depressive symptoms. Neuroimmunomodulation 2010;17:39–46.
47 Spindler H, Pedersen S: Posttraumatic stress disorder in the wake of heart disease: prevalence, risk factors, and future research directions. Psychosom Medicine 2005;67:715–723.
48 Stansfeld SA, Fuhrer R, Shipley MJ, Marmot MG: Psychological distress as a risk factor for coronary heart disease in the Whitehall II Study. Int J Epidemiol 2002;31:248–255.
49 Rasul F, Stansfeld SA, Hart CL, Smith GD: Psychological distress, physical illness, and risk of coronary heart disease. J Epidemiol Community Health 2005;59:140–145.
50 Molloy GJ, Stamatakis E, Randall G, Hamer M: Marital status, gender and cardiovascular mortality: behavioural, psychological distress and metabolic explanations. Soc Sci Med 2009;69:223–228.
51 Russ TC, Stamatakis E, Hamer M, Starr JM, Kivimaki M, Batty GD: Association between psychological distress and mortality: individual participant pooled analysis of 10 prospective cohort studies. Br Med J 2012;345:e4933.
52 Denollet J, Schiffer AA, Spek V: A general propensity to psychological distress affects cardiovascular outcomes. Circulation 2010;3:546–557.
53 Shen BJ, Avivi YE, Todaro JF, Spiro A, Laurenceau JP, Ward KD, Niaura R: Anxiety characteristics independently and prospectively predict myocardial infarction in men. J Am Coll Cardiol 2008;51:113–119.
54 Denollet J, Gidron Y, Vrints CJ, Conraads VM: Anger, suppressed anger, and risk of adverse events in patients with coronary artery disease. Am J Cardiol 2010;105:1555–1560.
55 Myrtek M: Meta-analyses of prospective studies on coronary heart disease, Type A personality, and hostility. Int J Cardiol 2001;79:245–251.
56 American Psychiatric Association: Diagnostic and Statistical Manual of Mental Disorders, Fourth Edition, Text Revision (DSM-IV-TR), ed 4. Washington, American Psychiatric Association, 2000.
57 Lee HB, Bienvenu JO, Cho SJ, Ramsey CM, Bandeen-Roche K, Eaton WW, Nestadt G: Personality disorders and traits as predictors of incident cardiovascular disease: findings from the 23-year follow-up of the Baltimore ECA Study. Psychosomatics 2010;41:289–296.
58 Denollet J, Vrints CJ, Conraads VM: Comparing type D personality and older age as correlates of tumor necrosis factor-α dysregulation in chronic heart failure. Brain Behav Immun 2008;22:736–743.
59 Selye H: The Stress of Life. McGraw-Hill, New York, 1956.
60 McEwen BS: Protective and damaging effects of stress mediators. N Engl J Med 1998;338:171–179.
61 McEwen BS: Stress, adaptation, and disease allostasis and allostatic load. Ann NY Acad Sci 1998;840:33–44.
62 Thayer JF, Lane RD: The role of vagal function in the risk for cardiovascular disease and mortality. Biol Psychol 2007;74:224–242.
63 Otte C, Neylan TC, Pipkin SS, Browner WS, Whooley MA: Depressive symptoms and 24-hour urinary norepinephrine excretion levels in patients with coronary disease: findings from the Heart and Soul Study. Am J Psychiatry 2005;162:2139–2145.
64 Licht CMM, de Geus EJC, Zitman FG, Hoogendijk WJG, van Dyck R, Penninx BWJH: Association between major depressive disorder and heart rate variability in the Netherlands study of depression and anxiety (NESDA). Arch Gen Psychiat 2008;65912:1358–1367.
65 Porges SW: The polyvagal theory: phylogenetic substrates of a social nervous system. Int J Psychophysiol 2001;42:123–146.
66 Porges SW: Cardiac vagal tone; a physiological index of stress. Neurosci Biobehav Rev 1995;19:225–233.
67 Porges SW: Emotion: an evolutionary by-product of the neural regulation of the autonomic nervous system. Ann NY Acad Sci 1997;807:62–77.
68 Colao A, Pivonello R, Spiezia S, Faggiano A, Ferone D, Filippella M: Persistence of increased cardiovascular risk in patients with Cushing's disease after 5 years of successful cure. J Clin Endocrinol Metab 1999;84:2264–2272.
69 Tracey KJ: The inflammatory reflex. Nature 2002;420:853–859.
70 Blalock JE: Harnessing a neural-immune circuit to control inflammation and shock. J Exp Med 2002;195:F25–F28.
71 Borovikova LV, Ivanova S, Zhang M, Yang H, Botchkina GI, Watkins LR, Wang H, Abumrad N, Eaton JW, Tracey KJ: Vagus nerve stimulation attenuates the systemic inflammatory response to endotoxin. Nature 2000;405:458–461.
72 Borovikova LV, Ivanova S, Nardi D, Zhang M, Yang H, Ombrellino M, Tracey KJ: Auton Neurosci 2000;85:141–147.
73 Wang H, Yu M, Ochani M, Amella CA, Tanovic M, Susarla S, Li JH, Wang H, Yang H, Ulloa L, Al-Abed Y, Czura CJ, Tracey KJ: Nicotinic acetylcholine receptor α7 subunit is an essential regulation of inflammation. Nature 2003;421:384–388.
74 Pavlov VA, Tracey KJ: The cholinergic anti-inflammatory pathway. Brain Behav Immun 2005;19:493–499.
75 Ross R: Atherosclerosis-an inflammatory disease. N Engl J Med 1999;340:115–126.
76 Lind L: Circulating markers of inflammation and atherosclerosis. Atherosclerosis 2003;169:203–214.
77 Frasure-Smith N, Lespérance F, Irwin MR, Sauvé C, Lespérance J, Théroux P: Depression, C-reactive protein and two-year major adverse cardiac events in men after acute coronary syndromes. Biol Psychiatry 2007;62:302–308.
78 Hansson G: Inflammation, atherosclerosis, and coronary artery disease. N Engl J Med 2005;352:1685–1695.
79 Seta Y, Shan K, Bozkurt B, et al: Basic mechanisms in heart failure: the cytokine hypothesis. J Card Fail 1996;2:243–249.
80 Petersen JW, Felker M: Inflammatory biomarkers in heart failure. Congest Heart Fail 2006;12:324–328.
81 Maes M, Bosmans E, Suy E, Vandervorst C, DeJonckheere C, Raus J: Depression-related disturbances in mitogen-induced lymphocyte responses and interleukin-1 beta and soluble interleukin-2 receptor production. Acta Psychiatr Scand 1991;84:379–386.
82 Maes M: Evidence for an immune response in major depression: a review and hypothesis. Prog Neuropsychopharmacol Biol Psychiatry 1995;19:11–38.
83 Wichers M, Maes M: The psychoneuroimmuno-pathophysiology of cytokine-induced depression in humans. Int J Neuropsychopharmacol 2002;5:375–388.

84 Lee KM, Kim YK: The role of IL-12 and TGF-beta1 in the pathophysiology of major depressive disorder. Int Immunopharmacol 2006;6:1298–1304.
85 Koenig, W: Inflammation and coronary heart disease: an overview. Cardiol Rev 2001;9:31–35.
86 Mulvihill NT, Foley JB: Inflammation in acute coronary syndromes. Heart 2002;87:201–204.
87 Robbins M, Topol EJ: Inflammation in acute coronary syndromes. Cleve Clin J Med 2002;69(suppl 2):SII130–SII142.
88 Maes M, Bosmans E, Meltzer HY, Scharpe S, Suy E: Interleukin-1beta: a putative mediator of HPA axis hyperactivity in major depression? Am J Psychiatry 1993;150:1189–1193.
89 Maes M, Van der Planken, Van Gastel A, Desnyder R: Blood coagulation and platelet aggression in major depression. J Affect Disord 1996;40:35–40.
90 Kop WJ, Gottdiener JS, Tangen CM, Fried LP, McBurnie MA, Walston J: Inflammation and coagulation factors on persons >65 years of age with symptoms of depression but without evidence of myocardial ischemia. Am J Cardiol 2002;89:419–424.
91 Musselman DL, Miller AH, Porter MR, Manatunga A, Gao F, Penna S: Higher than normal plasma interleukin-6 concentrations in cancer patients with depression: preliminary findings. Am J Psychiatry 2001;158:1252–1257.
92 Connor TJ, Leonard BE: Depression, stress and immunological activation: the role of cytokines in depressive disorders. Life Sci 1998;62:583–606.
93 Leonard BE: Brain cytokines and the psychopathology of depression; in Leonard BE (ed.): Antidepressants. Basel, Birkhauser, 2001, pp 109–120.
94 Castanon N, Leonard BE, Neveu PJ, Yirmiya R: Effects of antidepressants on cytokine production and actions. Brain Behav Immun 2002;16:569–574.
95 Leonard BE, Myint A: Inflammation and depression: is there a causal connection with dementia? Neutox Res 2006;10: 149–160.
96 von Känel R, Hepp U, Kraemer B, Traber R, Keel M, Mica L, Schnyder U: Evidence for low-grade systemic proinflammatory activity in patients with posttraumatic stress disorder. J Psychiatr Res 2006;41: 744–752.
97 Empana JP, Sykes DH, Luc G, Juhan-Vague I, Arveiler D, Ferrieres J, Amouyel P, Bingham A, Montaye M, Ruidavets JB, Haas B, Evans A, Jouven X, Ducimetiere P, Group PS: Contributions of depressive mood and circulating inflammatory markers to coronary heart disease in healthy European men: the Prospective Epidemiological Study of Myocardial Infarction (PRIME). Circulation 2005; 111:2299–2305.
98 Anda R, Williamson D, Jones D, Macera C, Eaker E, Glassman A, Marks J: Depressed affect, hopelessness, and the risk of ischemic heart disease in a cohort of US adults. Epidemiology 1993;4:285–294.
99 Kamphuis MH, Kalmijn S, Tijhuis MA, Geerlings MI, Giampaoli S, Nissinen A, Grobbee DE, Kromhout D: Depressive symptoms as risk factor of cardiovascular mortality in older European men: the Finland, Italy and Netherlands Elderly (FINE) study. Eur J Cardiovasc Prev Rehabil 2006;13:199–206.
100 Pratt LA, Ford DE, Crum RM, Armenian HK, Gallo JJ, Eaton WW: Depression, psychotropic medication, and risk of myocardial infarction. Prospective data from the Baltimore ECA follow-up. Circulation 1996;94:3123–3129.
101 Di Iorio A, Ferrucci L, Sparvieri E, Cherubini A, Volpato S, Corsi A, Bonafè M, Franceschi C, Abate, Paganelli R: Serum IL-1beta levels in health and disease: a population-based study. 'The InCHIANTI study'. Cytokine 2003;22: 198–205.
102 Biasucci LM: CDC/AHA Workshop on Markers of Inflammation and Cardiovascular Disease: Application to Clinical and Public Health Practice: clinical use of inflammatory markers in patients with cardiovascular diseases: a background paper. Circulation 2004;110:e560–e567.
103 Maes M: Introduction to the special section. Int J Neuropsychopharmacol 2002;5:329–331.
104 Dantzer R: Cytokine-induced sickness behavior: mechanisms and implications. Ann N Y Acad Sci 2001;933:222–234.
105 Vollmer-Conna U, Fazou C, Cameron B, Li H, Brennan C, Luck L, Davenport T, Wakefield D, Hickie I, Lloyd A: Production of pro-inflammatory cytokines correlates with the symptoms of acute sickness behaviour in humans. Psychol Med 2004;34:1289–1297.
106 Smith RS: The macrophage theory of depression. Med Hypotheses 1991;35: 298–306.
107 Ur E, White PD, Grossman A: Hypothesis: cytokines may be activated to cause depressive illness and chronic fatigue syndrome. Eur Arch Psychiatry Clin Neurosci 1992;241:317–322.
108 Lett HS, Blumenthal JA, Babyak MA, Sherwood A, Strauman T, Robins C, Newman MF: Depression as a risk factor for coronary artery disease: evidence, mechanisms, and treatment. Psychosom Med 2004;66:305–315.
109 Lanquillon S, Krieg JC, Bening-Abu-Shach U, Vedder H: Cytokine production and treatment response in major depressive disorder. Neuropsychopharmacology 2000;22:370–379.
110 Kubera M, Lin AH, Kenis G, Bosmans E, van Bockstaele D, Maes M: Anti-inflammatory effects of antidepressants through suppression of the interferon-gamma/interleukin-10 production ratio. J Clin Psychopharmacol 2001;21:199–206.
111 Kubera M, Kenis G, Bosmans E, Kajta M, Basta-Kaim A, Scharpe S, Budziszewska B, Maes M: Stimulatory effect of antidepressants on the production of IL-6. Int Immunopharmacol 2004;4:185–192.
112 Kenis G, Maes M: Effects of antidepressants on the production of cytokines. Int J Neuropsychopharmacol 2002;5:401–412.
113 Himmerich H, Binder EB, Künzel HE, Schuld A, Lucae S, Uhr M, Pollmächer T, Holsboer F, Ising M: Successful antidepressant therapy restores the disturbed interplay between TNF-alpha system and HPA axis. Biol Psychiatry 2006;60: 882–888.
114 Mayer B, Holmer SR, Hengstenberg C, Lieb W, Pfeifer M, Schunkert H: Functional improvement in heart failure patients treated with beta-blockers is associated with a decline of cytokine levels. Int J Cardiol 2005;103:182–186.
115 Janssen DG, Caniato RN, Verster JC, Baune BT: A psychoneuroimmunological review on cytokines involved in antidepressant treatment response. Hum Psychopharmacol 2010;25:201–215.

116 Müller N, Schwarz MJ, Dehning S, Douhe A, Cerovecki A, Goldstein-Müller B, Spellmann I, Hetzel G, Maino K, Kleindienst N, Möller HJ, Arolt V, Riedel M: The cyclooxygenase-2 inhibitor celecoxib has therapeutic effects in major depression: results of a double-blind, randomized, placebo controlled, add-on pilot study to reboxetine. Mol Psychiatry 2006;11:680–684.
117 Cleland SJ, Sattar N, Petrie JR, Forouhi NG, Elliott HL, Connell JMC: Endothelial dysfunction as a possible link between C-reactive protein levels and cardiovascular disease. Clin Sci 2000;98: 531–535.
118 Piletz JE, Halaris A, Iqbal O, et al: Proinflammatory biomarkers in depression: treatment with venlafaxine. World J Biol Psychiatry 2009;10:313–323.
119 Dinan TG: Inflammatory markers in depression. Curr Opin Psychiatry 2008; 22:32–36.
120 Dantzer R, O'Connor JC, Freund GG, Johnson RW, Kelley KW: From inflammation to sickness and depression: when the immune system subjugates the brain. Nat rev Neurosci 2008;9:46–56.
121 Leonard BE: The concept of depression as a dysfunction of the immune system. Curr Immunol Rev 2010;6:205–212.
122 Yoshizumi M, Perella MA, Burnett JC, Lee ME: Circ Res 1991;73:205–209.
123 Tucci M, Quatraro C, Frassanito MA, Silvestris F: Deregulated expression of monocyte chemoattractant protein-1 (MCP-1) in arterial hypertension: role in endothelial inflammation and atheromasia. J Hypertens 2006;24:1307–1318.
124 Verma S, Buchanan MR, Anderson TJ: Endothelial function testing as a biomarker of vascular disease. Circulation 2003;108:2054–2059.
125 Rybakowski JK, Wykretowicz A, Heymann-Szlachcinska A, Wysocki H: Impairment of endothelial function in unipolar and bipolar depression. Biol Psychiatry 2006;60:889–891.
126 Broadley AJ, Korszun A, Jones CJ, Frenneaux MP: Arterial endothelial function is impaired in treated depression. Heart 2002;88:521–523.
127 Smitherman TC, Milam M, Woo J: Elevated beta thromboglobulin in peripheral venous blood of patients with acute myocardial ischemia: direct evidence for enhanced platelet reactivity in vivo. Am J Cardiol 1981;48:395–402.
128 Levine SP, Lindenfield J, Ellis JB: Increased plasma concentrations of platelet factor-4 in coronary artery disease. Circulation 1981;64:626–632.
129 Laghrissi-Thode F, Wagner WR, Pollock BG: Elevated platelet factor 4 and betathromboglobulin plasma levels in depressed patients with ischemic heart disease. Biol Psychiatry 1997;42:290–295.
130 Kuijpers PM, Haumulyak K, Strik JJ: Beta-thromboglobulin and platelet factor 4 levels in post-myocardial infarction patients with major depression. Psychiatry Res 2002;109:207–210.
131 Musselman DL, Tomer A, Manatunga AK, Knight BT, Porter MR, Kasey S, Marzec U, Harker LA, Nemeroff CB: Exaggerated platelet reactivity in major depression. Am J Psychiatry 1996;153: 1313–1317.
132 Musselman DL, Marzec UM, Manatunga A, Penna S, Reemsnyder A, Knight BT, Baron A, Hanson SR, Nemeroff CB: Platelet reactivity in depressed patients treated with paroxetine: preliminary findings. Arch Gen Psychiatry 2000;57: 875–882.
133 Nemeroff CB, Musselman DL: Are platelets the link between depression and ischemic heart disease? Am Heart J 2000;140(4 suppl):57–62.
134 Piletz JE, Zhu H, Madakasira S, Pazzaglia P, Lindsay Devane C, Goldman N, Halaris A: Elevated P-selectin on platelets in depression: response to bupropion. J Psychiatr Res 2000;34:397–404.
135 Halaris A, Piletz JE: Relevance of imidazoline receptors and agmatine to psychiatry. Ann NY Acad Sci 2003;1009:1–19.
136 Serebruany VL, Glassman AH, Malinin AI, Sane DC, Finkel MS, Krishnan R, Atar D, Lekht V, O'Connor CM: Enhanced platelet/endothelial activation in depressed patients with acute coronary syndromes: evidence from recent clinical trials. Blood Coagul Fibrinolysis 2003;14:563–567.
137 Serebruany VL, Glassman AH, Malinin AI, et al: Platelet/endothelial biomarkers in depressed patients treated with the selective serotonin reuptake inhibitor sertraline after acute coronary events: the Sertraline Antidepressant Heart Attack Randomized Trial (SADHART) Platelet Substudy. Circulation 2003;108: 939–944.
138 Serebruany VL, Suckow RF, Cooper TB, et al: Relationship between release of platelet/endothelial biomarkers and plasma levels of sertraline and N-desmethylsertraline in acute coronary syndrome patients receiving SSRI treatment for depression. Am J Psychiatry 2005;162:1165–1170.
139 Basterzi AD, Aydemir C, Kisa C, Aksaray S, Tuzer V, Yazici K, Göka E: IL-6 levels decrease with SSRI treatment in patients with major depression. Hum Psychopharmacol 2005;20:473–476.
140 Ignarro LJ: Nitric oxide as a unique signaling molecule in the vascular system: a historical overview. J Physiol Pharmacol 2002;53:503–551.
141 Halaris A, Piletz J: Agmatine: metabolic pathway and spectrum of activity in brain. CNS Drugs 2007;21:885–900.
142 Mendelson SD: The current status of the platelet 5-HT (2A) receptor in depression. J Affect Disord 2000;57:13–24.
143 Hrdina PD, Bakish D, Chudzik J, Ravindran A, Lapierre YD: Serotonergic markers in platelets of patients with major depression: upregulation of 5-HT2 receptors. J Psychiatr Neurosci 1995;20: 11–19.
144 Neuger J, El Khoury A, Kjellman BF, Wahlund B, Aberg-Wistedt A, Stain-Malgren R: Platelet serotonin functions in untreated major depression. Psychiatry Res 1999;85:189–198.
145 Shimbo D, Child J, Davidson K, Geer E, Osende JI, Reddy S, Dronge A, Fuster V, Badimon JJ: Exaggerated serotonin-mediated platelet reactivity as a possible link in depression and acute coronary syndromes. Am J Cardiol 2002;89: 331–333.
146 Whyte EM, Pollock BG, Wagner WR, Mulsant BH, Ferrell RE, Mazumdar S, Reynolds CF 3rd: Influence of serotonin-transporter-linked promoter region polymorphism on platelet activation in geriatric depression. Am J Psychiatry 2001;158:2074–2076.
147 Palazzolo DL, Quadri SK: Interleukin-1 inhibits serotonin release from the hypothalamus in vitro. Life Sci 1992;51: 1797–1802.
148 Zhu CB, Blakely RD, Hewlett WA: The proinflammatory cytokines interleukin-1beta and tumor necrosis factor-alpha activate serotonin transporters. Neuropsychopharmacology 2006;31:2121–2131.

149 Christmas DM, Potokar JP, Davies SJC: A biological pathway linking inflammation and depression: activation of indoleamine 2,3-dioxygenase. Neurpsychiatr Dis Treat 2011;7:431–439.
150 Wichers M, Maes M: The psychoneuroimmuno-pathophysiology of cytokine-induced depression in humans. Int J Neuropsychopharmacol 2002;5:375–388.
151 Myint AM, Kim YK, Verkerk R, Scharpé S, Steinbusch HWM, Leonard BE: Kynurenine pathway in major depression: evidence of impaired neuroprotection. J Affect Disord 2007;89:143–151.
152 Myint AH, Kim YK: Cytokine-serotonin interaction through IDO; a neurodegeneration hypothesis of depression. Med Hypoth 2003;61:519–525.
153 Miller CL, Llenos IC, Dulay JR, Weis S: Upregulation of the initiating step of the kynurenine pathway in postmortem anterior cingulate cortex from individuals with schizophrenia and bipolar disorder. Brain Res 2006;1073–1074:25–37.
154 Myint AM, Kim YK, Verkerk R, Park SH, Scharpe S, Steinbusch HW, Leonard BE: Tryptophan breakdown pathway in bipolar mania. J Affect Disord 2007;102:65–72.
155 Roose SP, Laghrissi-Thode F, Kennedy JS, Nelson C, Bigger JT, Pollock BG, Gaffney A, Narayan M, Finkel MS, McCafferty J, Gergel I: Comparison of paroxetine and nortriptyline in depressed patients with ischemic heart disease. JAMA 1998;279:287–291.
156 Glassman AH, O'Connor CM, Califf RM, Swedberg K, Schwartz P, Bigger JT, Krishnan KR, van Zyl LT, Swenson JR, Finkel MS, Landau C, Shapiro PA, Pepine CJ, Markedian J, Harrison WM, Barton D, Mclvor M: Sertraline Antidepressant Heart Attack Randomized Trial (SAD-HART) group. JAMA 2002;288:701–709.
157 Sauer WH, Berlin JA, Kimmel SE: Selective serotonin reuptake inhibitors and myocardial infarction. Circulation 2001;104:1894–1898.
158 Taylor CB, Youngblood ME, Catellier D, et al: for the ENRICHD Investigators. Arch Gen Psychiatry 2005;62:792–798.
159 Berkman LF, Blumenthal J, Burg M, Carney RM, et al: Effects of treating depression and low perceived social support on clinical events after myocardial infarction: the Enhancing Recovery in Coronary Heart Disease Patients (ENRICHD) Randomized Trial. JAMA 2003;289: 3106–3116.

Angelos Halaris, MD
Department of Psychiatry and Behavioral Neuroscience
Loyola University Stritch School of Medicine, and Loyola University Medical Center
2160 South First Avenue, Maywood, IL 60153 (USA)
E-Mail ahalaris@lumc.edu

Halaris A, Leonard BE (eds): Inflammation in Psychiatry.
Mod Trends Pharmacopsychiatry. Basel, Karger, 2013, vol 28, pp 162–174 (DOI: 10.1159/000343982)

Inflammatory and Oxidative and Nitrosative Stress Cascades as New Drug Targets in Myalgic Encephalomyelitis and Chronic Fatigue Syndrome

Michael Maes

Maes Clinics @ TRIA, Piyavate Hospital, Bangkok, Thailand

Abstract

Myalgic encephalomyelitis (ME), chronic fatigue syndrome (CFS) and chronic fatigue (CF) are distinct diagnostic categories with regard to clinical symptoms, severity of illness and biomarkers. Patients with ME and CFS show higher scores on fatigue, neurocognitive disorders, hyperalgesia, autonomic symptoms, postexertional malaise and a subjective feeling of infection than patients with CF. ME is characterized by increased postexertional malaise, a subjective feeling of infection and neurocognitive disorders and is a more severe variant than CFS. Fukuda's 1994 CDC criteria are adequate to make a distinction between patients with ME/CFS and CF, while ME/CFS patients should be subdivided into those with and without postexertional malaise into ME and CFS, respectively. Different interrelated pathophysiological mechanisms play a role in ME/CFS, i.e. (1) inflammation and immune activation, (2) oxidative and nitrosative stress and lowered antioxidant defenses, (3) activation of cell signaling networks, e.g. nuclear factor κβ, the 2′,5′-oligoadenylate/RNase-L and/or protein kinase R pathway, (4) a transition towards autoimmune reactions, and (5) bacterial translocation. The inflammatory biomarkers are higher in ME/CFS than in CF and higher in ME than in CFS. The above-mentioned pathways may explain the onset of characteristic ME/CFS symptoms, such as fatigue, malaise, autonomic symptoms, hyperalgesia, and neurocognitive symptoms. Different etiological factors may trigger ME/CFS/CF, e.g. viral and bacterial infections, and (auto)immune and inflammatory disorders, while psychosocial and physical stressors act as modulating factors. New pathophysiologically driven drug candidates for ME and CFS are discussed which target the pathways that play a role in ME/CFS.

Myalgic encephalomyelitis/chronic fatigue syndrome (ME/CFS) is a disabling disorder which can have a greater impact on functional status and well-being than other chronic diseases, including cancer or chronic lung disease [1, 2]. ME/CFS may be associated with a significant lowering in physical functioning and quality of life [3]. ME/CFS patients score significantly lower than patients with hypertension, congestive heart failure, acute myocardial infarction, multiple sclerosis, on most subscales of the 36-item MOS (Medical Outcomes Study) health survey [4]. Full recovery from untreated ME/CFS is rare, i.e. only 5–12%

[5]. The direct and indirect costs of ME/CFS to the society are considerable and can be estimated in the USA at USD 8.5–11 billion annually [6].

Diagnostic Classifications: Evidence for Subgroups, ME, CFS and CF

Recently, we have summarized the different diagnostic classifications and diagnostic criteria for ME/CFS [7]. Although the diagnoses myalgic encephalomyelitis (ME), chronic fatigue syndrome (CFS) and chronic fatigue (CF) are often used interchangeable, ME, CFS and CF are qualitatively distinct diagnostic categories [7]. ME has been described in the medical literature since the 1930s under various names in many cases on account of epidemic outbreaks, e.g. epidemic neuromyasthenia. Characteristic symptoms of ME are postexertional malaise, chronic fatigue, neurocognitive dysfunctions, and autonomic symptoms, with a chronic and relapsing course. The World Health Organization acknowledges ME as a disease of the nervous system [8].

Much later, in the 1980s and 1990s, the diagnosis of CFS was introduced. The definition of CFS is primarily based upon the presence of chronic fatigue lasting more than 6 months. According to a commonly used diagnostic classification of CFS [1], chronic fatigue must be accompanied by four or more of the following symptoms: substantial impairment in short-term memory or concentration; sore throat; tender lymph nodes; muscle pain; multi-joint pain without swelling or redness; headache of a new type, pattern or severity; unrefreshing sleep, and postexertional malaise lasting more than 24 h. ME and CFS are therefore different diagnostic concepts. Postexertional malaise, for example, obligatory for the diagnosis ME, is not mandatory for the diagnosis CFS. This suggests that the diagnosis of CFS according to the criteria of Fukuda et al. [1] defines a heterogeneous population of individuals with chronic fatigue, of which individuals with ME are a subset [7, 9]. Other authors promote even less strict criteria for CFS [9]. For example, CFS is sometimes interpreted as invalidating chronic fatigue or by defining CFS as an increased score on self-rating scales that measure fatigue.

In an attempt to define a more homogenous patient population, a panel of ME/CFS experts have published consensus criteria for ME [6]. They propose to abandon the label CFS and the chronic fatigue criterion and to use the name ME for a patient group characterized by postexertional malaise and neurocognitive symptoms as obligatory features. These new diagnostic criteria [6] are based on a consensus and are not the result of pattern recognition methods (supervised or unsupervised learning techniques) that should have been used to statistically delineate new diagnostic categories or confirm the existence of previous diagnostic classifications [7]. These new consensus diagnostic criteria for ME [6] implicate that many patients with CFS, i.e. those not fulfilling the new ME criteria, are no longer diagnosed as suffering from chronic fatigue or even from a medical condition. In any case, the diagnosis of CFS may refer to different concepts, ranging from increased self-reported fatigue, to chronic fatigue together with somatic symptoms [1], to a symptom complex where chronic fatigue is not even an obligatory criterion [6]. Needless to say, those differences in diagnostic classifications and criteria blurred and will blur the results of scientific research.

Recently, we found – using pattern recognition methods – that ME, CFS and CF are three qualitatively distinct diagnostic classes in a continuum of illness severity progressing from CF to CFS to ME [7]. Thus, ME is characterized by higher scores on postexertional malaise, a subjective feeling of infection and neurocognitive disorders. Patients with ME and CFS show higher scores on fatigue, neurocognitive disorders, hyperalgesia, autonomic symptoms, postexertional malaise and a subjective feeling of infection than patients with CF. We constructed an easily applicable algorithm to subdivide patients with chronic fatigue into ME, CFS and CF. Thus, Fukuda's CDC criteria [1] may be used to distinguish between ME/CFS, on the one hand, and CF on the other. ME/CFS patients should further be divided into those with and without postexertional malaise into ME and CFS, respectively [7]. The clinical diagnoses of ME, CFS and CF are externally validated by biomarkers [7]: patients with ME show more inflammatory disorders that those with CFS, while patients with CFS show more inflammation and immune activation than those with CF. It should be added that individuals with CF sometimes show minor alterations in those pathways when compared with normal controls [7]. Therefore, it is suggested that future research on chronic fatigue should take into account three qualitatively distinct subclasses, i.e. ME, CFS and CF. This review will refer to ME/CFS when reporting results on biomarkers obtained in patients with ME or CFS.

Symptoms of ME/CFS

Chronic fatigue is common to ME, CFS and CF [7]. The majority of ME/CFS patients additionally suffer from other characteristic symptoms [10]:

Chronic Fatigue
The physical and mental fatigue experienced by ME/CFS/CF patients is quite different from commonplace fatigue. It is similar to the fatigue experienced during flu-like infections and it is not substantially alleviated by rest.

Aches and Pains
Many patients with ME/CFS experience widespread muscle and or joint pain, muscle tension, and headaches. The pain symptoms bear similarities with those observed in fibromyalgia.

Muscle Weakness
Many ME/CFS patients report muscle weakness, reduced muscle power and muscular endurance.

Neurocognitive Impairment
Concentration disorders and memory impairments (brain fog) are frequently reported by individuals with ME/CFS. The neurocognitive symptoms are aggravated by exercise [9].

Autonomic Symptoms
Autonomic dysfunction and orthostatic abnormalities are other characteristics of ME/CFS. Autonomic symptoms are associated with the severity of fatigue, and are often accompanied by neurocognitive impairment and cardiovascular abnormalities.

Sleep Disturbances
Many patients report sleep disturbances, e.g. middle insomnia, nightmares, and nonrestorative sleep. Sleep disturbances are accompanied by changes in sleep EEG patterns.

Gastrointestinal Symptoms
Around 50–60% of ME/CFS patients comply with the diagnosis of irritable bowel syndrome (IBS) according to the Rome II criteria. The typical IBS symptoms in individuals with ME/CFS are abdominal discomfort/pain relieved with defecation; more than 3 bowel movements per day; abnormal stool form; abnormal urgency; feeling of incomplete bowel movement, and abdominal pain/cramps.

Depression
Depressive symptoms and major depressive episodes frequently occur during the course of ME/CFS [11]. There is a significant comorbidity between major depression and ME/CFS. Depressive symptoms appear together with fatigue, a flu-like malaise, and autonomic symptoms.

A Subjective Feeling of Infection or a Flu-Like Malaise
ME/CFS patients and in particular patients with ME, experience a subjective feeling of infection. This symptom refers to general symptoms

of infection, a flu-like malaise, or self-reported infections, such as respiratory and urinary tract infection.

Postexertional Fatigue or Malaise

As discussed above, postexertional malaise is a key component of ME. Exertion aggravates many symptoms of ME/CFS, e.g. fatigue, malaise, neurocognitive impairment, hyperalgesia, and autonomic symptoms. Likewise, patients with ME/CFS show a significantly reduced exercise capacity as indicated by reduced anaerobic threshold, oxygen uptake (VO_2), and maximum oxygen uptake (VO_{2max}).

The severity of ME/CFS can be measured by means of validated rating scales, such as the Fibromyalgia and Chronic Fatigue Syndrome (FF) Rating Scale [12].This scale scores 12 characteristic symptoms of ME/CFS and fibromyalgia on a scale between 0 and 6. The manual offers a precise description of the defined steps.

Pathophysiology of ME and CFS

There is some evidence that ME/CFS is frequently accompanied by activation of inflammatory, immune and oxidative and nitrosative stress (IO&NS) pathways. This review paper is based on information gathered from more than 450 papers. Due to space limitations, this paper refers mainly to published reviews and all references mentioned in these papers [13–23].

Inflammation and Immune Activation

Inflammation in ME/CFS is demonstrated by increased levels of pro-inflammatory cytokines (PICs), e.g. TNFα, IL-1, and IL-6 [7, 19, 23]; acute phase reactants, e.g. C-reactive protein (CRP) and the α2 protein fraction obtained during protein electrophoresis, and β_2-microglobulin, PMN-elastase and lysozyme [16]. Increased plasma IL-1 and TNFα are significantly and positively correlated with characteristic symptoms of ME/CFS as measured with the FF scale, e.g. fatigue, sadness, autonomic symptoms, and a flu-like malaise [16]. Increased levels of IL-6, IL-1β, IL-12, IL-8, and IL-13 are correlated with increased pain after exercise [24]. Increased plasma concentrations of PMN-elastase are significantly associated with concentration difficulties, failing memory and a subjective experience of infection [16].

Although there is evidence for immune activation in ME/CFS, conclusions on cytokine profiles of type 1 and type 2 are variable. Some studies observed increased IFNγ, IL-2, IL-12, and IL-2 receptor levels, although the data on IFNγ and IL-2 were contradicted by other studies [16, 19, 23, 24]. Increased serum neopterin, a sensitive marker of cell-mediated immunity, has consistently been observed in ME/CFS [16]. Increased serum neopterin in ME/CFS is significantly and positively correlated with fatigue, autonomic symptoms, and a flu-like malaise [16]. Immune activation is also corroborated by findings on increased expression of cell surface markers, e.g. CD38+ and HLA-DR+ on CD8+ cells [25], an elevated percentage of suppressor/cytotoxic T lymphocytes, and higher numbers of CD8+ cells expressing class II activation markers [26].

Relatively more studies have reported a Th2-like pattern, with increased IL-10, IL-4 and IL-5 [27]. These studies, however, did not report on the IFNγ/IL-10 (or IL-4, IL-5) ratio and, therefore, it remains elusive which type of cytokine pattern prevails in ME/CFS. To complicate matters, many studies in ME/CFS did not report on the drug state of the patients. For example, ME/CFS patients are sometimes using antidepressants, which are known to inhibit Th1-like responses, while enhancing Th2-like responses [28]. In a study performed on a small study group of drug-free ME/CFS patients, increased levels of CD4+ and CD8+ T cells spontaneously producing IFNγ and IL-4, and increased IL-4 levels after stimulation were found [29]. This shows ongoing immune activation with a mixed response pattern in baseline conditions and a Tc2-like response upon stimulation [29].

As in major depression and some (auto)immune disorders, in vivo immune activation is accompanied by ex vivo signs of immunosuppression: decreased natural killer cell activity (NKCA) [30, 31], reduced responses of T cells to mitogens [31], and a significantly lowered expression of the early activation marker CD69 by T and NKCs following ex vivo stimulation [32]. These signs of immunosuppression may play a role in the frequently occurring infections in ME/CFS [17].

Oxidative and Nitrosative Stress Pathways
There is abundant evidence that ME/CFS is accompanied by activated IO&NS pathways in the peripheral blood, e.g. increased reactive oxygen and nitrogen species and O&N damage to lipids, proteins or DNA [11, 13, 14]. These findings are corroborated by studies showing elevated expression of IO&NS genes. ME/CFS is also accompanied by reduced levels of antioxidants, such as serum zinc, dehydroepiandrosterone sulfate, and coenzyme Q10 [11]. Activation of the IO&NS pathways is likely amplified or sustained by the low levels of these antioxidants and the chronic, low-grade inflammation, thereby perpetuating a vicious circle between reduced antioxidant defenses, low-grade inflammation and activated IO&NS pathways [14]. In patients with ME/CFS, markers of lipid peroxidation and protein oxidation/nitrosylation are significantly associated with the severity of fatigue, postexertional malaise, aches and pains and muscular tension [11, 14, 22]. The activated IO&NS processes may play a role in the mitochondrial dysfunction in ME/CFS [14]. In muscle biopsies of patients with postviral CFS, mild-to-severe atrophy of type II fibers, branching and fusion of mitochondrial cristae, mitochondrial degeneration with swelling, vacuolation, myelin figures and secondary lysosomes were found [33]. Highly sensitive and specific alterations in the availability of ATP, oxidative phosphorylation, and the transfer efficiency of ATP into the cytosol were detected in neutrophil mitochondria [34]. Virtually all individuals with combined somatic and depressive complaints showed a significantly lower ATP production in muscle biopsies [35].

Activation of Cell Signaling Networks
Patients with ME/CFS show a significantly higher production of unstimulated TNFα- and PMA-stimulated production of NFκβ) as compared with normal controls [36]. NFκβ production was significantly and positively correlated to the severity of ME/CFS and to aches and pains, muscular tension, fatigue, irritability, sadness, and subjective feeling of infection as measured with the FF scale. The same ME/CFS patients also showed a significantly increased production of cyclo-oxygenase (COX-2) and inducible NO synthase (iNOS) [37]. Both COX-2 and iNOS were significantly related to aches and pains, muscular tension, fatigue, concentration difficulties, failing memory, sadness and a subjective experience of infection [36, 37]. The results show that the IO&NS responses in ME/CFS are at least in part driven by the transcription factor NFκβ and the increased production of iNOS.

There are also findings on activation of the 2′,5′-oligoadenylate (2–5A)/RNase-L and/or the protein kinase R (PKR) pathway in ME/CFS [21]. Both pathways are inducible by interferons and have antiviral and antiproliferative effects. 2′,5′-Oligoadenylate synthetase, the enzyme that synthesizes 2–5A, additionally may be activated by bacterial infections, while PKR may be directly activated by pro-inflammatory stimuli, including IL-1, TNFα, and bacterial lipopolysaccharides (LPS). In ME/CFS, dysregulation of the 2–5A/RNAse pathway manifests itself in RNase-L proteolysis, resulting in dysfunctional low-molecular-weight RNase-L fragments [21].

Bacterial Translocation
Increased IgA and IgM-mediated immune responses against the LPS of Gram-negative bacteria are detected in around 40–60% of patients with ME/CFS in particular those with IBS [38]. This indicates that in patients with ME/CFS and IBS there

is a loss of integrity of the tight-junction barrier. Consequently, commensal bacteria are no longer contained by the mucosal barrier and may be translocated from the gut into the lamina propria and the mesenteric lymph nodes and even into the systemic circulation. In ME/CFS, significant correlations were found between the IgA responses to LPS and increased levels of IL-1, TNFα and neopterin [38]. This indicates that (a) increased bacterial translocation drives, at least in part, the activated inflammatory and immune pathways, and/or (b) IO&NS pathways have caused increased gut permeability and, consequently, bacterial translocation. So, even if the bacterial translocation is secondary to the IO&NS processes in ME/CFS, increased bacterial translocation may amplify the pre-existing IO&NS responses thus aggravating the pathophysiology. Other gastrointestinal aberrations, like gut inflammation and aberrant intestinal immunity have been associated with ME/CFS [38].

Autoimmune Responses

Autoimmune responses in ME/CFS are indicated by increased antibody levels directed against Sa antigen, cell nucleus, serotonin, gangliosides, vasoactive neuropeptides, muscarinic cholinergic receptor, and mitochondrial cardiolipin [10]. Likewise, increased numbers of B cells, i.e. CD19+, CD20+ and CD21+, and B cells co-expressing CD20+ and CD5+ are observed in ME/CFS [39]. Many patients with ME/CFS additionally show increased IgM-mediated (auto)immune responses against a number of self antigens that became immunogenic due to damage by IO&NS [22], e.g. those directed against the anchorage molecules, palmitic and myristic acid, and NO adducts, such as NO-tryptophan and NO-tyrosine. These autoimmune responses may mediate cellular dysfunctions, e.g. by interfering with the binding of hundreds of proteins to the membrane. These mechanisms, called palmitoylation and myristoylation, determine proper cell functioning including intracellular signaling, apoptosis and cellular differentiation.

Sequels and Concomitants of Activated Immune and IO&NS Pathways

Neurological Abnormalities

Brain abnormalities are frequently observed in ME/CFS [40]. These include reduced white [40] and gray matter volume [41], significantly increased numbers of foci of T2-bright signal in the periventricular and subcortical white matter and in the centrum semiovale using MR imaging techniques [42], reduced blood flow and hypoperfusion of specific brain areas [43], and reduced prefrontal cortex oxygenation during incremental exercise [44]. Importantly, a significant association was found between diverse neurological anomalies and the neurocognitive symptoms of ME/CFS [45]. It is tempting to speculate that some of these brain abnormalities reflect a neuroinflammatory pathology as has been described in depression and neurodegenerative disorders [28].

Cardiovascular Abnormalities

Cardiovascular dysfunctions frequently occur in ME/CFS, e.g. circulatory impairment; cardiac hypofunction with reduced cardiac output and stroke volume, low blood volume, and reduced blood flow to brain and muscles [15]. Jason et al. [46] showed that the mean age (57.7 years) of ME/CFS patients who died from heart failure was remarkably lower than that of individuals dying from heart failure in the general population (83.1 years). These findings implicate that ME/CFS confers increased risk towards cardiovascular morbidity and mortality. Obviously, the activated IO&NS pathways may increase the risk to develop cardiovascular disorders in ME/CFS [15].

Abnormal Responses to Stress

Psychological stress may intensify ME/CFS symptoms. Researchers have established hypothalamic-pituitary-adrenal (HPA) axis dysfunction, e.g. decreased production of HPA-axis hormones, alterations in diurnal variation, a blunted response to provocation tests, including administration of corticotropin-releasing hormone and insulin, and altered responses to psychological and physical stressors [47].

Predisposing or Risk Factors

IgG subclass deficiencies are highly prevalent in patients with ME/CFS, e.g. IgG1 and IgG3 subclasses [48]. These deficiencies may increase the vulnerability to infections and recurrent infections; and increase the propensity to develop inflammatory, immune and autoimmune reactions following viral or bacterial infections. A genetic predisposition has frequently been observed in ME/CFS. Examples are: a significant increase in TNF-857 TT and CT genotypes, likely to be associated with increased production of TNFα and IFNγ; human leukocyte antigen class II allele HLA-DQA1*01; C/T genetic polymorphism leading to His161Arg substitution in IL-17F; IFNγ +874T/A and IL-10 –592C/A polymorphisms [14]. Decreased ratios of ω3/ω6 polyunsaturated fatty acids (PUFAs) are detected in patients with ME/CFS [49]. This is important for the inflammatory pathways as ω3 PUFAs, like eicosapentaenoic acid (EPA), have anti-inflammatory effects and inhibit PIC production, whereas ω6 PUFAs, e.g. arachidonic acid, activate inflammatory processes, including PIC production. Thus, the relative depletion of ω3 PUFAs and the relatively higher ω6 contents in ME/CFS may predispose those individuals to develop IO&NS reactions in response to trigger factors.

Trigger, Maintaining and Modulating Factors

Viral and bacterial infections probably play an important role in ME/CFS, either as trigger or maintaining factors [14]. This section on infections is based on information gathered from around 35 different papers on the role of infectious agents in ME/CFS. Due to space limitations, this paper refers only to a few of these [48, 50, 51] and to some reviews [11, 14, 17, 23]. ME/CFS often starts with a flu-like or upper respiratory tract infection [50]. Epidemics of ME/CFS have repeatedly been described in the literature [52]. Several studies have demonstrated the presence of infections, e.g. recurrent and opportunistic infections, during the course of the illness [48]. The immunological anomalies and distinctive gene expression are in accordance with (latent) viral or bacterial infections [14, 23]. Gene expression studies indicate subtype-specific gene expression for EBV, enteroviruses and *Coxiella burnetii* [53]. It is suggested that persistence of herpes viral replication and incomplete herpesvirus multiplications play a role in ME/CFS. Epstein-Barr virus (EBV), *C. burnetii* and Parvo B19 are known to trigger ME/CFS. Other pathogens that are frequently associated with ME/CFS are human herpes virus (HHV) 6 and 7, cytomegalovirus (HCMV, HHV-5), enteroviruses, Borna disease virus, mycoplasma species, *Chlamydia pneumoniae* and *Borrelia burgdorferi*.

There is also a high prevalence of chronic fatigue and malaise, and by inference ME/CFS, in patients with human immunodeficiency virus (HIV) infection and cancers, such as breast carcinoma. Also, in those disorders there is a strong association between activated IO&NS pathways and the onset of ME/CFS symptoms. For example, the fatigue during HIV infection is strongly related to inflammatory biomarkers and a dysfunctional carnitine-dependent energy production, while reduced quality of life is associated with immune activation. Increased IL-6 production may explain the fatigue that occurs in cancer patients [13, 14].

Fatigue has a prevalence of 41–81% in autoimmune disorders, such as systemic lupus erythematosus, rheumatoid arthritis, and Sjogren's syndrome [54]. There is also a high prevalence (48–50%) of fatigue in neuroinflammatory disorders, such as Parkinson's disorder and multiple sclerosis. The prevalence of chronic fatigue is also high (41–82%) in inflammatory disorders, such as post-poliomyelitis, ankylosing spondylitis, psoriatic arthritis, and biliary cirrhosis [54]. There is evidence that increased IL-1 may explain the occurrence of fatigue in autoimmune conditions [14]. In multiple sclerosis, the increased production of IFNγ is significantly related to fatigue. In inflammatory bowel disease, increased PIC production is regarded as a potential cause or contributory factor for fatigue and sleep disturbances [54].

There is also some evidence that psychological stressors may be associated with the onset of ME/CFS acting as modulating factors [55, 56]. Psychological stressors are indeed known to elicit inflammation, immune activation, and immunosuppression and to activate IO&NS pathways [28] and therefore could contribute to the development of ME/CFS. Recently, we have reviewed that physical stressors, such as strenuous exertion, may induce IO&NS pathways and therefore may aggravate ME/CFS, e.g. pain, fatigue, postexertional malaise and neurocognitive impairment [9].

The above-mentioned pathophysiological factors in ME and CFS may be interrelated phenomena that may be induced by the trigger factors through a number of different mechanisms [14, 57]. For example, viral and bacterial infections may be associated with the onset of ME and CFS through activation of NFκβ, the 2–5A/RNase-L and/or PKR pathways and IO&NS cascades in some predisposed individuals, e.g. those with IgG subclass deficiencies or genetic predispositions. As a consequence, the production of COX-2, iNOS and PICs may be induced causing more activation of IO&NS processes. The latter may cause damage to membrane fatty acids, functional proteins, mitochondria and DNA. All these aberrations, in turn, may change self antigens into neoepitopes, which acquire immunogenicity and consequently act as triggers to bypass immunological tolerance. This and other mechanisms [22, 38, 57] may explain the transition from infections to autoimmune reactions, which may aggravate the preexisting inflammation and therefore increase fatigue and malaise. Moreover, IO&NS processes may cause bacterial translocation and suppression of cellular immune functions. Bacterial translocation, in turn, may aggravate the primary IO&NS response and cause autoimmune responses [38, 57]. Immunosuppression may increase the likelihood of viral reactivation or opportunistic infections, which will further amplify the IO&NS responses. In other patients, a primary gut inflammation could induce a loosening of the tight junction barrier followed by bacterial translocation, which in turn may induce NFκβ and thus IO&NS pathways. This implicates that not every patient shows the full spectrum of all these IO&NS alterations. Different combinations of aberrations in the IO&NS pathways, and predisposing, trigger, and maintaining factors determine the biological heterogeneity of ME/CFS.

Mechanistic Explanations How IO&NS Pathways May Cause ME/CFS Symptoms

PICs, including IL-1 and TNFα, Th1-like cytokines, such as IFNγ and IL-2, IO&NS pathways, lowered levels of some antioxidants; mitochondrial dysfunction, and increased bacterial translocation may elicit specific symptoms of ME/CFS [14, 16, 28]. PICs, such as IL-1 and TNFα, play a key role in the development of fatigue [16]. For example, injection to C57BL/6 mice of *Corynebacterium parvum* antigen elicits fatigue in association with increased IL-1 and TNFα mRNA expression. Recovery from muscle damage is characterized by fatigue, macrophage activation and increased IL-1β production in the brain. Systemic administration and microinjections of TNFα in the anterior hypothalamus provokes malaise and autonomic dysfunctions. Inflammatory pathways may cause nociceptor sensitization and an increased responsivity to catecholamines [16].

Administration of LPS may activate the production of PICs by binding to the CD14-Toll-like receptor-4 (TLR4) complex [38]. LPS recognition induces an inflammatory response proceeding through activation of NFκβ and mitogen-activated protein kinases. Systemic LPS administration induces neuroinflammation with rapid increases in brain PICs, such as TNFα. The latter are associated with the onset of malaise, fatigue, and other symptoms of illness. The increased PIC levels may remain elevated for up to 10 months, suggesting that short-term exposure to LPS may have long-lasting neuroinflammatory effects [38].

Decreased levels of specific antioxidants, in particular coenzyme Q10, may cause symptoms reminiscent of ME/CFS [14, 28, 35]. Thus, the low coenzyme Q10 syndrome in ME/CFS is associated with fatigue, myalgia and neurocognitive disorders. Mitochondrial disorders like Prader-Willi syndrome, Friedrich's ataxia and Steinert's myotonic dystrophy, are often accompanied by low coenzyme Q10 values and by fatigue, myalgia, and neurocognitive disorders [28]. Treatment with statins may cause the same symptoms through depletion of serum coenzyme Q10. These symptoms can be reversed following treatment with coenzyme Q10. Moreover, coenzyme Q10 is a strong antioxidant and has anti-inflammatory effects targeting NFκβ; PICs, such as TNFα, and LPS-induced inflammation [28]. Therefore, the low coenzyme Q10 syndrome in ME/CFS may predispose towards inflammatory and IO&NS processes or amplify existing IO&NS processes [13, 14]. Coenzyme Q10 plays a key role in the electron transport chain in the mitochondria and the associated production of ATP [35].

Findings suggest that damage by IO&NS may be a causal factor in the onset of ME/CFS symptoms. For example, in ME/CFS incremental exercise is associated with activation of IO&NS processes and with alterations in muscle membrane excitability [58]. Translational experiments show that experimentally induced activation of IO&NS processes, such as lipid peroxidation and lowered antioxidant levels, associate with ME/CFS symptoms. Moreover, attenuating these activated IO&NS pathways reverses the symptoms [13, 14, 28]. Reactive oxygen and nitrogen species can lead to damage to mitochondria and mitochondrial DNA, and inhibit mitochondrial function [14, 28, 35]. Mitochondrial dysfunctions are accompanied by chronic fatigue (75%), headache (82%), myalgia, gastrointestinal symptoms, neurological symptoms, etc. [35]. The low coenzyme Q10 syndrome in ME/CFS could aggravate the mitochondrial dysfunctions, including inhibition of the mitochondrial respiratory chain and oxidative damage to mitochondrial DNA. NO is known to induce vasodilation. Thus, increased levels of NO may contribute to vascular dysfunction and, in combination with higher COX2 levels, to decreased oxygen uptake [59]. NO is together with increased PIC levels involved in central pain sensitization [60]. Hyperactive RNase L – by degrading RNA – is associated with decreased daily activities and a reduced general health [14, 21].

All in all, the IO&NS pathways that underpin the pathophysiology of ME/CFS offer a plausible explanation for characteristic symptoms of ME/CFS, e.g. fatigue, weakness, pain, neurocognitive impairment, sleep disorders, autonomic symptoms, a flu-like malaise, and postexertional malaise. Figure 1 gives an overview of the trigger and predisposing factors, the pathophysiology and sequels, and the characteristic symptoms of ME/CFS.

New Treatments of ME/CFS

Current treatments for ME and CFS are predominantly psychological, including cognitive behavioral therapy often coupled with gradual exercise therapy. The recent findings on the pathways that are involved in ME and CFS would suggest a range of novel treatments for ME and CFS. Since ME/CFS is biologically a heterogeneous disorder in which different combinations of the above-mentioned pathways, trigger, risk, modulating and maintaining factors are involved, a one-size-fits-all therapy for ME/CFS is unlikely. The therapies most likely to be effective are tailor-made treatments targeting the different underlying pathways, and the etiologic and predisposing factors as well. Most

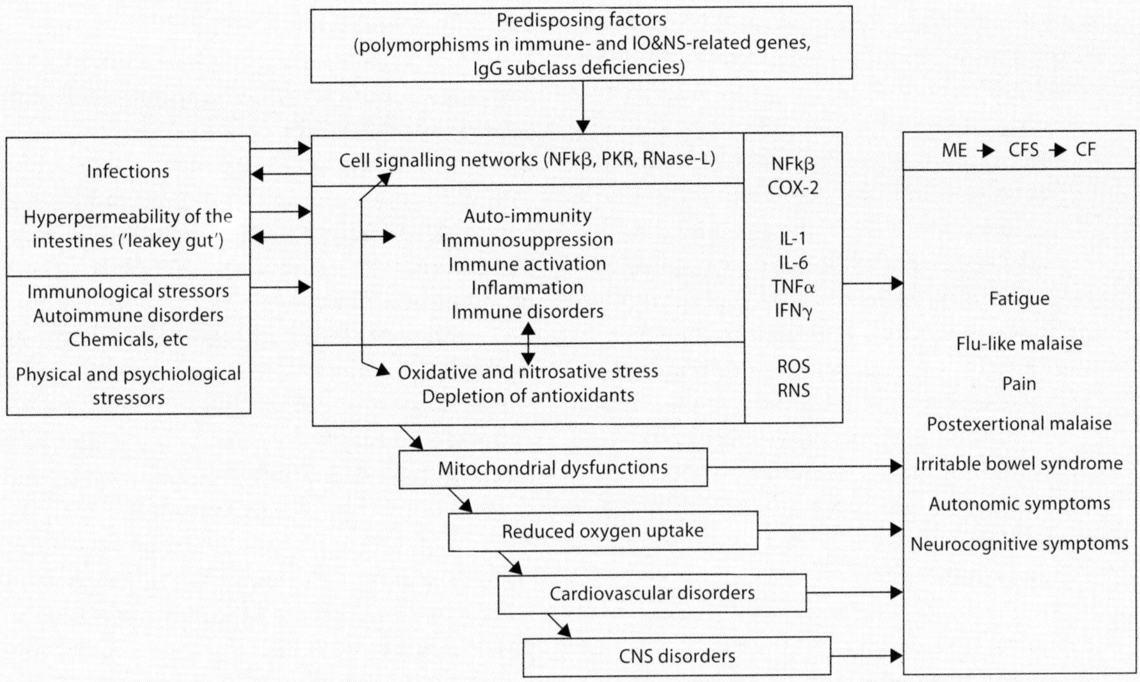

Fig. 1. Interconnections between trigger and predisposing or risk factors, the pathophysiology and the symptoms of ME, CFS and CF. Trigger factors may cause activation of cell signaling pathways, e.g. NFκβ, PKR and 2–5A/RNase-L, and/or inflammatory, IO&NS pathways in some predisposed individuals, e.g. those with IgG subclass deficiencies and immune-gene polymorphisms. The production of COX-2, proinflammatory cytokines, such as IL-1, IL-6, TNFα, and IFNγ and ROS and nitrogen species (RNS) is induced and associated with the onset of characteristic symptoms of ME/CFS/CF. Activation of IO&NS pathways is likely amplified or sustained by the low levels of key antioxidants and the low-grade inflammation, thereby perpetuating a vicious circle between reduced antioxidant defenses, low-grade inflammation and activated IO&NS pathways. There is probably also a transition from infections and oxidatively/nitrosatively damaged epitopes to autoimmune reactions, which may aggravate the preexisting IO&NS processes thereby further amplifying the symptoms. Suppression of cellular immune functions may increase the likelihood of viral reactivation or opportunistic infections, while bacterial translocation (leaky gut) may amplify the primary IO&NS response and cause autoimmune responses. Sequels of these activated networks and pathways are mitochondrial and cardiovascular dysfunction, and metabolic and structural brain defects, which together may further shape the ME/CFS phenotype. Not every patient shows the full spectrum of all these IO&NS alterations. Different combinations of aberrations in the IO&NS pathways, and predisposing, trigger, and maintaining factors determine the biological heterogeneity and symptomatology of ME/CFS.

likely combinatorial treatments of immunomodulating drugs together with antioxidants are essential. The former should damp inflammation, on the one hand, and boost cytotoxic activity of macrophages and NKCs in case of infections, on the other. The latter should target IO&NS processes. We now discuss treatments that have been shown to have some efficacy in ME/CFS or that recently have been proposed. This summary is based on 55 published papers.

1 Hemispherx has developed a drug targeting the RNAse L pathway, i.e. Ampligen [61].

A phase III clinical trial is completed in ME/CFS, but based on the results, the US Food and Drug Administration concluded that there was not sufficient evidence to market Ampligen for treatment of ME/CFS.

2 Based on B cell proliferation and autoimmune reactions some authors have proposed to use monoclonal antibodies, i.e. rituximab, to deplete B cell populations [62].

3 Treatment with immunoglobulins intravenously has been advocated to treat recurrent infections in patients with IgG subclass deficiencies and attenuate bacterial translocation as well as autoimmune and IO&NS responses [58].

4 In order to restore bacterial translocation, a combination of natural anti-inflammatory and anti-oxidative substances is advocated, i.e. NAC + zinc + glutamine [57].

5 Antimicrobial therapies may have some clinical efficacy. Examples of these pathogen-targeted treatments are: valacyclovir for EBV, valganciclovir for HCMV, doxycycline or minocycline for *Coxiella burnetti*, doxycycline for *Mycoplasma* species, azithromycin for *C. pneumoniae*, and foscarnet (foscavir) and cidofovir for HHV-6 and HHV-7 infections.

Combination treatments with natural anti-inflammatory and antioxidative substances (NAIOS) are other potentially effective strategies [57]. Some of these NAIOSs that may have beneficial effects are: glutathione or precursors, e.g. N-acetylcysteine, ω3 fatty acids and in particular EPA, zinc for its antioxidant properties and it role as a copper/zinc superoxide dismutase precursor and immune modulator, L-carnitine, coenzyme Q10, curcumin, quercetin, and epigallocatechin gallate. Other authors proposed new treatments, e.g. IFNβ, a drug used in the treatment for relapsing-remitting MS and rheumatoid arthritis. Treatment with IFNβ has anti-inflammatory effects since it inhibits IL-1 and TNFα and increases IL-10 and the IL-1 receptor antagonist, and increases NKCA and T cell cytotoxicity. Drugs that inhibit the production of TNFα, e.g. adalimumab, etanercept and infliximab, and vitamin D binding protein-macrophage-activating factor are advocated as possible new treatments.

Future research on new treatments for ME and CFS should target the IO&NS pathways, possible pathogens, and risk, modulating and maintaining factors. Therefore, future research should examine subgroups of ME and CFS patients, defined by their biological profiles, e.g. inflammation, immune activation, immunosuppression, IO&NS, bacterial translocation, and viral infections.

References

1 Fukuda K, Straus SE, Hickie I, Sharpe M, Dobbins JG, Komaroff AL: The chronic fatigue syndrome: a comprehensive approach to its definition and study. Ann Intern Med 1994;121:953–959.

2 Nacul LC, Lacerda EM, Campion P, Pheby D, Drachler Mde L, Leite JC, Poland F, Howe A, Fayyaz S, Molokhia M: The functional status and well-being of people with myalgic encephalomyelitis/chronic fatigue syndrome and their carers. BMC Publ Health 2011;11:402.

3 Herrell R, Goldberg J, Hartman S, Belcourt M, Schmaling K, Buchwald D: Chronic fatigue and chronic fatigue syndrome: a co-twin control study of functional status. Qual Life Res 2002; 11:463–471.

4 Komaroff AL, Fagioli LR, Doolittle TH, Gandek B, Gleit MA, Guerriero RT, Kornish RJ 2nd, Ware NC, Ware JE Jr, Bates DW: Health status in patients with chronic fatigue syndrome and in general population and disease comparison groups. Am J Med 1996;101:281–290.

5 Cairns R, Hotopf M: A systematic review describing the prognosis of chronic fatigue syndrome. Occup Med (Lond) 2005;55:20–31.

6 Carruthers BM, van de Sande MI, De Meirleir KL, Klimas NG, Broderick G, Mitchell T, Staines D, Powles AC, Speight N, Vallings R, Bateman L, Baumgarten-Austrheim B, Bell DS, Carlo-Stella N, Chia J, Darragh A, Jo D, Lewis D, Light AR, Marshall-Gradisbik S, Mena I, Mikovits JA, Miwa K, Murovska M, Pall ML, Stevens S: Myalgic encephalomyelitis. International Consensus Criteria. J Intern Med 2011;270: 327–338.

7 Maes M, Twisk FNM, Johnson C: Myalgic encephalomyelitis (ME), chronic fatigue syndrome (CFS), and chronic fatigue (CF) are distinguished accurately: results of supervised learning techniques applied on clinical and inflammatory data. Psychiatr Res 2012, Apr 20, Epub ahead of print.
8 WHO: International Classification of Diseases, revision 10 (ICD-10). 1992; G93.3.
9 Twisk FN, Maes M: A review on cognitive behavorial therapy (CBT) and graded exercise therapy (GET) in myalgic encephalomyelitis (ME)/chronic fatigue syndrome (CFS): CBT/GET is not only ineffective and not evidence-based, but also potentially harmful for many patients with ME/CFS. Neuro Endocrinol Lett 2009;30:284–299.
10 Maes M: Nooit meer moe: CVS ontmaskerd. Brugge, Zorro Uitgevers, 2012.
11 Maes M: An intriguing and hitherto unexplained co-occurrence: depression and chronic fatigue syndrome are manifestations of shared inflammatory, oxidative and nitrosative (IO&NS) pathways. Prog Neuropsychopharmacol Biol Psychiatry 2011;35:784–794.
12 Zachrisson O, Regland B, Jahreskog M, Kron M, Gottfries CG: A rating scale for fibromyalgia and chronic fatigue syndrome (the FibroFatigue scale). J Psychosom Res 2002;52:501–509.
13 Maes M: Inflammatory and oxidative and nitrosative stress pathways underpinning chronic fatigue, somatization and psychosomatic symptoms. Curr Opin Psychiatry 2009;22:75–83.
14 Maes M, Twisk FN: Chronic fatigue syndrome: Harvey and Wessely's (bio)psychosocial model versus a bio(psychosocial) model based on inflammatory and oxidative and nitrosative stress pathways. BMC Med 2010;8:35.
15 Maes M, Twisk FN: Why myalgic encephalomyelitis/chronic fatigue syndrome (ME/CFS) may kill you: disorders in the inflammatory and oxidative and nitrosative stress (IO&NS) pathways may explain cardiovascular disorders in ME/CFS. Neuro Endocrinol Lett 2009; 30:677–693.
16 Maes M, Twisk FN, Kubera M, Ringel K: Evidence for inflammation and activation of cell-mediated immunity in myalgic encephalomyelitis/chronic fatigue syndrome (ME/CFS): increased interleukin-1, tumor necrosis factor-α, PMN-elastase, lysozyme and neopterin. J Affect Disord 2012;136:933–939.
17 Komaroff AL, Cho TA: Role of infection and neurologic dysfunction in chronic fatigue syndrome. Semin Neurol 2011; 31:325–337.
18 Arnett SV, Alleva LM, Korossy-Horwood R, Clark IA: Chronic fatigue syndrome: a neuroimmunological model. Med Hypotheses 2011;77:77–83.
19 Fletcher MA, Zeng XR, Barnes Z, Levis S, Klimas NG: Plasma cytokines in women with chronic fatigue syndrome. J Transl Med 2009;7:96.
20 Lorusso L, Mikhaylova SV, Capelli E, Ferrari D, Ngonga GK, Ricevuti G: Immunological aspects of chronic fatigue syndrome. Autoimmun Rev 2009;8:287–291.
21 Vojdani A, Lapp CW: Interferon-induced proteins are elevated in blood samples of patients with chemically or virally induced chronic fatigue syndrome. Immunopharmacol Immunotoxicol 1999;21:175–202.
22 Maes M, Mihaylova I, Leunis JC: Chronic fatigue syndrome is accompanied by an IgM-related immune response directed against neopitopes formed by oxidative or nitrosative damage to lipids and proteins. Neuro Endocrinol Lett 2006;27:615–621.
23 Broderick G, Fuite J, Kreitz A, Vernon SD, Klimas N, Fletcher MA: A formal analysis of cytokine networks in chronic fatigue syndrome. Brain Behav Immun 2010;24:1209–1217.
24 White AT, Light AR, Hughen RW, Bateman L, Thomas B, Martins TB, et al: Severity of symptom flare after moderate exercise is linked to cytokine activity in chronic fatigue syndrome. Psychophysiology 2010;47:615–624.
25 Landay AL, Jessop C, Lennette ET, Levy JA: Chronic fatigue syndrome: clinical condition associated with immune activation. Lancet 1991;338:707–712.
26 Klimas NG, Salvato FR, Morgan R, Fletcher MA: Immunologic abnormalities in chronic fatigue syndrome. J Clin Microbiol 1990;28:1403–1410.
27 Patarca-Montero R, Antoni M, Fletcher MA, Klimas NG: Cytokine and other immunologic markers in chronic fatigue syndrome and their relation to neuropsychological factors. Appl Neuropsychol 2001;8:51–64.
28 Leonard B, Maes M: Mechanistic explanations how cell-mediated immune activation, inflammation and oxidative and nitrosative stress pathways and their sequels and concomitants play a role in the pathophysiology of unipolar depression. Neurosci Biobeh Rev 2012; 36:764–785.
29 Skowera A, Cleare A, Blair D, Bevis L, Wessely SC, Peakman M: High levels of type 2 cytokine-producing cells in chronic fatigue syndrome. Clin Exp Immunol 2004;135:294–302.
30 Fletcher MA, Zeng XR, Maher K, Levis S, Hurwitz B, Antoni M, Broderick G, Klimas NG: Biomarkers in chronic fatigue syndrome: evaluation of natural killer cell function and dipeptidyl peptidase IV/CD26. PLoS One 2010;5:e10817.
31 Klimas NG, Salvato FR, Morgan R, Fletcher MA: Immunologic abnormalities in chronic fatigue syndrome. J Clin Microbiol 1990;28:1403–1410.
32 Mihaylova I, DeRuyter M, Rummens JL, Bosmans E, Maes M: Decreased expression of CD69 in chronic fatigue syndrome in relation to inflammatory markers: evidence for a severe disorder in the early activation of T lymphocytes and natural killer cells. Neuro Endocrinol Lett 2007;28:477–483.
33 Behan WM, More IA, Behan PO: Mitochondrial abnormalities in the postviral fatigue syndrome. Acta Neuropathol 1991;83:61–65.
34 Myhill S, Booth NE, McLaren-Howard J: Chronic fatigue syndrome and mitochondrial dysfunction. Int J Clin Exp Med 2009;2:1–16.
35 Gardner A, Johansson A, Wibom R, Nennesmo I, von Döbeln U, Hagenfeldt L, Hällström T: Alterations of mitochondrial function and correlations with personality traits in selected major depressive disorder patients. J Affect Disord 2003;76:55–68.
36 Maes M, Mihaylova I, Bosmans E: Not in the mind of neurasthenic lazybones but in the cell nucleus: patients with chronic fatigue syndrome have increased production of nuclear factor kappa beta. Neuro Endocrinol Lett 2007;28:456–462.

37 Maes M, Mihaylova I, Kubera M, Bosmans E: Not in the mind but in the cell: increased production of cyclo-oxygenase-2 and inducible NO synthase in chronic fatigue syndrome. Neuro Endocrinol Lett 2007;28:463–469.
38 Maes M, Twisk FN, Kubera M, Ringel K, Leunis JC, Geffard M: Increased IgA responses to the LPS of commensal bacteria is associated with inflammation and activation of cell-mediated immunity in chronic fatigue syndrome. J Affect Disord 2012;136:909–917.
39 Tirelli U, Marotta G, Improta S, Pinto A: Immunological abnormalities in patients with chronic fatigue syndrome. Scand J Immunol 1994;40:601–608.
40 Barnden LR, Crouch B, Kwiatek R, Burnet R, Mernone A, Chryssidis S, Scroop G, Del Fante P: A brain MRI study of chronic fatigue syndrome: evidence of brainstem dysfunction and altered homeostasis. NMR Biomed 2011;24: 1302–1312.
41 Okada T, Tanaka M, Kuratsune H, Watanabe Y, Sadato N: Mechanisms underlying fatigue: a voxel-based morphometric study of chronic fatigue syndrome. BMC Neurol 2004;4:14.
42 Schwartz RB, Komaroff AL, Garada BM, Gleit M, Doolittle TH, Bates DW, Vasile RG, Holman BL: SPECT imaging of the brain: comparison of findings in patients with chronic fatigue syndrome, AIDS dementia complex, and major unipolar depression. AJR Am J Roentgenol 1994; 162:943–951.
43 Biswal B, Kunwar P, Natelson BH: Cerebral blood flow is reduced in chronic fatigue syndrome as assessed by arterial spin labeling. J Neurol Sci 2011;301:9–11.
44 Patrick Neary J, Roberts AD, Leavins N, Harrison MF, Croll JC, Sexsmith JR: Prefrontal cortex oxygenation during incremental exercise in chronic fatigue syndrome. Clin Physiol Funct Imaging 2008; 28:364–372.
45 Schmaling KB, Lewis DH, Fiedelak JI, Mahurin R, Buchwald DS: Single-photon emission computerized tomography and neurocognitive function in patients with chronic fatigue syndrome. Psychosom Med 2003;65:129–136.
46 Jason LA, Corradi K, Gress S, Williams S, Torres-Harding S: Causes of death among patients with chronic fatigue syndrome. Health Care Women Int 2006; 27:615–626.
47 Scott LV, Medbak S, Dinan TG: Blunted adrenocorticotropin and cortisol responses to corticotropin-releasing hormone stimulation in chronic fatigue syndrome. Acta Psychiatr Scand 1998;97: 450–457.
48 Hilgers A, Frank J: Chronic fatigue syndrome: immune dysfunction, role of pathogens and toxic agents and neurological and cardial changes. Wien Med Wochenschr 1994;144:399–406.
49 Maes M, Mihaylova I, Leunis JC: In chronic fatigue syndrome, the decreased levels of omega-3 poly-unsaturated fatty acids are related to lowered serum zinc and defects in T cell activation. Neuro Endocrinol Lett 2005;26:745–751.
50 Naess H, Sundal E, Myhr KM, Nyland HI: Postinfectious and chronic fatigue syndromes: clinical experience from a tertiary-referral centre in Norway. In Vivo 2010;24:185–188.
51 Nicolson GL, Gan R, Haier J: Multiple co-infections (mycoplasma, chlamydia, human herpes virus-6) in blood of chronic fatigue syndrome patients: association with signs and symptoms. APMIS 2003;111:557–566.
52 Strickland PS, Levine PH, Peterson DL, O'Brien K, Fears T: Neuromyasthenia and chronic fatigue syndrome (CFS) in Northern Nevada/California: a ten-year follow-up of an outbreak. J Chron Fatigue Syndr 2001;9:3–14.
53 Zhang L, Gough J, Christmas D, Mattey DL, Richards SC, Main J, Enlander D, Honeybourne D, Ayres JG, Nutt DJ, Kerr JR: Microbial infections in eight genomic subtypes of chronic fatigue syndrome/myalgic encephalomyelitis. J Clin Pathol 2010;63:156–164.
54 van Langenberg DR, Gibson PR: Systematic review: fatigue in inflammatory bowel disease. Aliment Pharmacol Ther 2010;32:131–143.
55 Kang HK, Natelson BH, Mahan CM, Lee KY, Murphy FM: Post-traumatic stress disorder and chronic fatigue syndrome-like illness among Gulf War veterans: a population-based survey of 30,000 veterans. Am J Epidemiol 2003;57:141–148.
56 Dobbins JG, Natelson BH, Brassloff I, Drastal S, Sisto S-A: Physical, behavioral and psychological risk factors for chronic fatigue syndrome: a central role for stress? J Chron Fatigue Syndr 1995; 1:43–58.
57 Maes M, Coucke F, Leunis JC: Normalization of the increased translocation of endotoxin from Gram-negative enterobacteria (leaky gut) is accompanied by a remission of chronic fatigue syndrome. Neuro Endocrinol Lett 2007;28:739–744.
58 Jammes Y, Steinberg JG, Mambrini O, Bregeon F, Delliaux S: Chronic fatigue syndrome: assessment of increased oxidative stress and altered muscle excitability in response to incremental exercise. J Intern Med 2005;257:299–310.
59 Heinonen I, Saltin B, Kemppainen J, Sipilä HT, Oikonen V, Nuutila P, Knuuti J, Kalliokoski K, Hellsten Y: Skeletal muscle blood flow and oxygen uptake at rest and during exercise in humans: a PET study with nitric oxide and cyclooxygenase inhibition. Am J Physiol Heart Circ Physiol 2011;300:H1510–H1517.
60 Latremoliere A, Woolf CJ: Central sensitization: a generator of pain hypersensitivity by central neural plasticity. J Pain 2009;10:895–926.
61 Suhadolnik RJ, Reichenbach NL, Hitzges P, Adelson ME, Peterson DL, Cheney P, Salvato P, Thompson C, Loveless M, Müller WE, et al: Changes in the 2–5A synthetase/RNase L antiviral pathway in a controlled clinical trial with poly(I)-poly(C12U) in chronic fatigue syndrome. In Vivo 1994;8:599–604.
62 Fluge , Mella O: Clinical impact of B-cell depletion with the anti-CD20 antibody rituximab in chronic fatigue syndrome: a preliminary case series. BMC Neurol 2009;9:28.

Prof. Dr. Michael Maes, MD, PhD
Maes Clinics @ TRIA, Piyavate Hospital
998 Rimklongsamsen Road
Bangkok 10310 (Thailand)
E-Mail dr.michaelmaes@hotmail.com

Halaris A, Leonard BE (eds): Inflammation in Psychiatry.
Mod Trends Pharmacopsychiatry. Basel, Karger, 2013, vol 28, pp 175–187 (DOI: 10.1159/000346362)

Peripheral Inflammation and Cognitive Aging

Alvin Lim · Katarina Krajina · Anna L. Marsland

Department of Psychology, University of Pittsburgh, Pittsburgh, Pa., USA

Abstract

Evidence suggests that inflammation, an innate immune response facilitating recovery from injury and pathogenic invasion, is positively associated with age-related cognitive decline and may play a role in risk for dementia. Physiological pathways linking the peripheral immune and central nervous systems are outlined, and studies linking inflammation with neurocognitive function are overviewed. We also present recent studies from our laboratory showing that midlife inflammation is related to cognitive function and brain morphology. Finally, potential implications for treatment, future directions, and limitations are discussed.

Age-related cognitive decline is characterized by the gradual and progressive deterioration of several domains of cognitive ability, including executive function, working and episodic memory, processing speed and attention [1]. Compared to younger individuals, elderly adults perform significantly worse on tasks involving delayed recall [2], recognition [3] and mental flexibility [3, 4]. These declines typically begin in middle adulthood and progress at a consistent rate across the rest of life [5]. In addition to negatively impacting the quality of life of older adults, deteriorating cognitive function imposes significant risk for dementia, injuries, hospitalization and death [5, 6]. Accordingly, recent research has focused on the identification of factors that predict risk for age-related cognitive decline and may be targeted by preventative intervention. In this regard, it is known that traditional cardiovascular risk factors contribute to the prediction of risk for accelerated neurocognitive aging. Recent evidence suggests that systemic inflammation may also play a role [7–9]. In this chapter, evidence linking inflammation with neurocognitive function will be overviewed, with a particular focus on recent work highlighting morphometric brain changes that accompany the elevations in peripheral inflammatory activity that occur in later life.

Acute Inflammatory Response and Chronic Systemic Inflammation

The acute inflammatory response is initiated when tissue-residing macrophages are activated by tissue damage or pathogen invasion, resulting in the release of several pro-inflammatory cytokines including interleukin-1β (IL-1β), IL-6, and tumor necrosis factor-α (TNF-α). Locally, these cytokines serve several functions, including stimulating the expression of epithelial adhesion molecules, increasing vascular permeability, and supporting leukocyte trafficking through chemotactic signaling. Proinflammatory cytokines also enter the peripheral circulatory system and are involved in signaling a systemic inflammatory response characterized by hepatic synthesis and release of acute-phase proteins including C-reactive protein (CRP) [10, 11]. Although TNF-α and IL-1β decay rapidly, IL-6 and CRP have longer half-lives and are reliably detected in human plasma/serum. In the absence of acute infection, it is widely accepted that circulating concentrations of IL-6 and CRP reflect an index of chronic system-wide inflammation at the time of the blood draw. Although generally stable over extended periods, circulating levels of IL-6 and CRP increase with age and predict risk for accelerated cognitive decline among elderly populations [12].

Pathways Linking the Peripheral Immune and Central Nervous Systems

For several decades, it was believed that the immune system and the brain functioned in isolation due to the anatomical separation of leukocytes from the central nervous system (CNS) by the blood-brain barrier. More recent evidence, however, challenges this notion and shows that peripheral inflammatory mediators exert an influence on brain function and play a critical role in the induction of sickness symptoms that accompany inflammatory disease. For example, human research shows that systemic infusion of IL-1 and TNF-α leads to suppression of food intake, decreased social exploration, and poorer memory consolidation [13–15]. Impairments in cognitive function have also been observed in murine models following the peripheral administration of lipopolysaccharide (LPS) – an endotoxin that stimulates the release of proinflammatory cytokines [16]. Together, results from these and other studies have led to current understanding that peripheral cytokines elicit an array of affective, behavioral, and cognitive adaptations that facilitate recovery during the course of acute infection. Three distinct molecular pathways link peripheral immune stimuli to changes in the CNS. First, peripheral cytokines can activate vagal afferent nerves that stimulate the production of proinflammatory cytokines by central tissues [17]. Second, peripheral cytokines can stimulate brain vascular endothelial cells to release secondary messengers in the CNS that promote the central release of proinflammatory cytokines [18]. Finally, peripheral cytokines can be actively transported in the paraventricular regions of the blood-brain barrier [19]. These pathways provide a link between the peripheral expression of proinflammatory cytokines in response to infection or injury and the central immune responses that result in sickness behaviors that include neurocognitive symptoms.

Inflammation in the Central Nervous System: Animal Studies

Along with the transient affective and behavioral symptoms that accompany acute illness, neuroinflammatory processes in the brain play a role in the long-term modulation of neurocognitive function and age-related cognitive decline [20]. Despite distinct morphological differences, microglial cells and macrophages are derived from common progenitor cells and serve analogous roles in their respective systems [21]. When activated by proinflammatory cytokines microglial cells adopt an inflammatory phenotype that resembles activated macrophages, including the expression of surface antigens and the production of central proinflammatory cytokines [22–24]. Of particular relevance for neurocognitive processes, receptors for proinflammatory cytokines are highly expressed on microglia found in the hippocampus and prefrontal brain regions [25–27].

Elevations in central proinflammatory cytokines interfere with a variety of neuromolecular processes, including hippocampal neurogenesis, long-term potentiation (LTP), synaptic plasticity, and synaptic scaling [28], and eventually result in dendritic atrophy with negative consequences for learning and memory [29, 30]. For example, central administration of IL-1 impairs performance on some [31–33], but not all [34, 35] hippocampal-dependent spatial learning tasks. Similarly, transgenic mice that overexpress central levels of IL-6 show deficits in synaptic plasticity and impaired avoidance learning [36], whereas the administration of IL-6 receptor antagonists to intact mice prevents the decreases in hippocampal LTP and neurogenesis, and the subsequent cognitive sequelae that accompany peripheral and central inflammation [37]. Taken together, these findings suggest that proinflammatory cytokines play a critical role in modulating the neuromolecular processes that underlie cognitive processes, such as learning and memory.

Much of the early animal literature focused on the impact of acute inflammatory processes on cognitive function; however, recent attention has turned to the consequences of chronic elevations in inflammation that more closely resemble the changes that accompany aging. Here, studies show that chronic stress results in an increase in peripheral inflammation that sensitizes or 'primes' a proinflammatory shift in microglial phenotype, resulting in an increase in central proinflammatory cytokines and concomitant deficits in learning and memory [38, 39]. For example, mice subjected to social isolation over a 4-week period show elevated levels of hippocampal IL-1β, decreased hippocampal neurogenesis, and specific impairment of hippocampal-dependent memory [40]. It is also suggested that age-related elevations in systemic inflammation affect cognitive function [41]. Aging has been associated with proinflammatory cytokine expression in the periphery [42], IL-6 levels in hippocampal and prefrontal brain regions [43], and activation of microglia [44]. Furthermore, when compared with younger mice, aged mice show greater elevations in central cytokine expression and larger deficits in cognitive performance following peripheral immune challenge [45]. In sum, animal evidence suggests that age-related microglial activation results in exaggerated central responses to peripheral inflammation, which may play a role in the neurocognitive decline that accompanies aging [46].

Inflammation in the Central Nervous System: Human Studies

In humans, several independent lines of experimental and epidemiologic research support a role of inflammation in the modulation of cognitive function. Utilizing a similar approach to the animal research, a number of experimental studies have examined the cognitive sequelae of acute immune activation. This technique typically involves examining changes in cognitive performance that accompany the peripheral administration of an inflammatory stimulus. For example, in an early study, Reichenberg et al. [29] administered low-dose *Salmonella abortus equi* endotoxin and saline placebo to 20 male volunteers in a double-blind balanced crossover design. When compared with response to saline, endotoxin was associated with significant impairments in cognitive performance. Furthermore, this effect was dependent on the magnitude of the IL-6 response, with higher responses predicting the greatest impairments in performance. Similar patterns of results were reported in subsequent studies by Krabbe et al. [42] and Brydon et al. [47] that examined the impact of low-dose *Escherichia coli* endotoxin and typhoid vaccination, respectively. Both studies found inverse associations of IL-6 production and performance on declarative memory and executive function tasks. Finally, more direct evidence for a role of IL-6 comes from a study conducted by Spath-Schwalbe et al. [48], which showed that peripheral administration of recombinant IL-6 decreased self-reported attentional capacity when compared to placebo. Findings from these human experimental studies parallel results from animal models, and show that acute elevations in peripheral inflammation are accompanied by declines in several cognitive domains including memory, attention, and executive function.

Corroborating experimental evidence, epidemiological studies also support an inverse association of peripheral inflammation with cognitive function. Insight can be derived from studies of neurodegenerative diseases that typically involve deficits in memory, including Alzheimer's disease and vascular dementia. These syndromes are generally associated with higher than normal circulating levels of CRP, IL-6, and IL-1β [49–51]. For example, Zuliani et al. [52] showed that compared to healthy individuals, those diagnosed with dementia had higher levels of circulating TNF-α, IL-1β, and IL-6. Cross-sectional studies of this nature do not shed light on the direction of the observed associations. Indeed, it is plausible that elevations in peripheral cytokines may reflect a consequence, rather than a cause, of neurodegenerative processes [42, 53].

To disentangle these possibilities, prospective longitudinal designs have been utilized. In general, these studies show that peripheral inflammation predicts future cognitive declines and subsequent risk for Alzheimer's disease and other dementias [54, 55]. For example, a recent study showed that compared to healthy participants, individuals with midlife rheumatoid arthritis, a chronic inflammatory condition associated with elevated levels of IL-6, have a 1.96-fold greater risk for developing mild cognitive impairment, and a 2.43-fold increased risk for developing Alzheimer's disease over a 20-year follow-up period [56]. Likewise, midlife levels of circulating CRP and production of IL-1β and TNF-α by stimulated peripheral blood cells have been positively linked with risk for Alzheimer's disease and vascular dementia in late life [57, 58]. Although most studies suggest that inflammation predicts future risk for dementia, not all findings are consistent. For example, plasma CRP levels were unrelated to dementia risk among older adults followed over a 13-year period in the Framingham Heart Study [59].

In addition to clinical studies of dementia, a growing literature suggests an association of proinflammatory cytokines with normative variation in cognitive function. While the specific assessments of cognitive ability vary, several studies show inverse associations of proinflammatory markers and cognitive functioning [60–63], but again, not all results are consistent [64, 65]. Reasons for these inconsistencies are unclear, but it is speculated that age may play a moderating role [66–68], with the progressive increases in peripheral inflammation that occur across adult life accelerating neurocognitive aging. However, to date, most studies have examined these associations in the elderly when issues relating to differential survivorship, immunoscenescence, and existing health conditions complicate the picture. Findings from longitudinal studies that examine inflammation in midlife are more consistent, and show that accelerated patterns of cognitive decline and increased risk for dementia accompany higher levels of systemic inflammation [57, 69].

Interleukin-6 and Cognitive Performance among Midlife Community Volunteers

We have reported some of the first findings to show an inverse association of plasma IL-6 with working memory/attention and executive function among healthy middle-aged adults [70]. In this study, we recruited 500 community volunteers (age 30–54; 51% male) and drew blood samples for the assessment of plasma IL-6 levels. Subjects also completed a battery of neuropsychological tests to assess memory and executive function. Based on animal findings showing that peripheral proinflammatory cytokines interfere with hippocampal processes and evidence that receptors for IL-6 are concentrated in prefrontal and temporal brain regions, we hypothesized that IL-6 would covary inversely with performance on tests that involve these brain regions, including verbal and nonverbal learning and memory, executive function, and attention. Participants were administered the Wechsler Memory Scale, ed 3 [71] to assess memory and attention, and the Trail Making Test [72] and Stroop Color-word Test [73] to assess executive function.

As expected, circulating levels of IL-6 were inversely associated with performance on tests of executive function, attention/working memory, and auditory recognition memory before and after controlling for demographics including age, race, gender, and years of education. Results are shown in figures 1 and 2. For these figures, plasma IL-6 levels were separated into tertiles (low, medium, and high IL-6 means = 0.74, 1.32, and 3.35 pg/ml, respectively). These results remained significant after controlling for lifestyle factors including smoking, blood pressure, physical activity, sleep duration, alcohol use, and BMI. Our findings are consistent with other studies conducted among older adults, and suggest that inverse associations of IL-6 with cognitive function are independent of other established risk factors of cognitive decline, including hypertension.

These findings contribute to a growing literature that links chronic inflammation to poorer cognitive functioning in domains that show the largest age-related declines, and extend these findings to a midlife sample. This raises the possibility that circulating levels of IL-6 are a biomarker of future risk for accelerated cognitive aging. Although scores on neuropsychological tests fell within the normal range in our sample, it is plausible that inflammation-related cognitive declines begin well before clinically verifiable deficits are observed. Our findings support the hypothesis that inflammation precedes, and may play an active role in, the pathogenesis of early cognitive decline, possibly contributing to increased risk for dementia in later life.

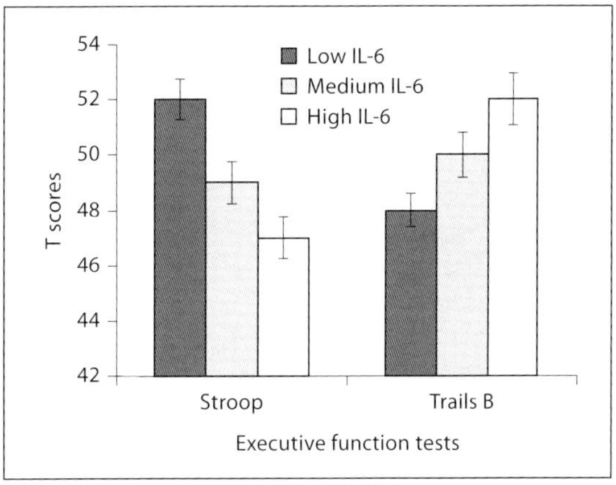

Fig. 1. Mean performance on tests of executive function (Trail Making Test and Stroop Color-Word Test) among individuals with low, medium, and high plasma IL-6. Higher scores reflect better performance on the Stroop task and worse performance on Trail Making. Reprinted from Marsland et al. [70], with permission from Wolters Kluwer Health.

Although our findings are consistent with animal evidence that suggests that inflammation modulates cognitive function, reverse causality is also possible, with poor cognitive function predicting greater levels of inflammation. For example, cognitive function may influence lifestyle choices and health behaviors. Alternatively, cognitive function and systemic inflammation could be related to a third independent factor, such as genetic vulnerability or subclinical neuroinflammatory processes. For this reason, prospective longitudinal studies are required, following individuals over time to elucidate the direction of observed effects.

Inflammation and Specific Brain Regions: Prefrontal Cortex and Hippocampus

Whether systemic inflammation reflects a marker of poor function in discrete brain regions or of a global neurodegenerative process also remains

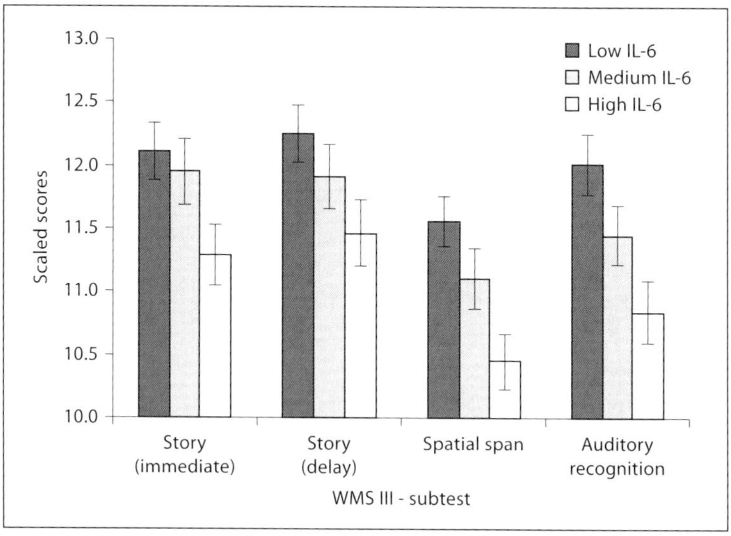

Fig. 2. Mean performance on the Wechsler Memory III scales assessing auditory and visual attention/working memory among individuals with low, medium, and high plasma IL-6. Reprinted from Marsland et al. [70], with permission from Wolters Kluwer Health.

to be determined. Our findings provide support for both possibilities. We observed an inverse association of IL-6 with performance on executive function tasks that involve the prefrontal cortex (e.g. set shifting and cognitive flexibility), on working memory tasks that involve the prefrontal cortex and hippocampus, and on attentional capacity tasks that typically involve global neurocognitive function. It is also possible that inflammation may induce progressive global brain alterations, with declines in executive function and attention/working memory being early effects.

Evidence from another of our studies contributes to a growing body of evidence that systemic inflammation is associated with structural and functional alterations of specific brain regions where receptors for proinflammatory cytokines are most abundantly expressed. These regions include the hippocampus [74] and prefrontal cortex [75], with animal evidence suggesting that hippocampal-dependent functions are particularly susceptible to disruption by inflammatory mechanisms. For this reason, we hypothesized that midlife adults with higher levels of IL-6 would show lower grey matter volume of the hippocampus, a morphometric change that has been linked to memory impairments [72].

Interleukin-6 and Hippocampal Grey Matter Volume among Middle-Aged Adults

To explore this possibility, we utilized a structural neuroimaging method paired with optimized voxel-based morphometry [76] to determine hippocampal grey matter volume among 76 midlife volunteers on whom we had previously determined plasma IL-6 levels and cognitive function [70]. Results from this study are shown in figure 3. Consistent with our hypothesis, IL-6 was inversely related to hippocampal grey matter volume, and this relationship remained significant after adjustment for demographic and lifestyle factors including age, race, sex, years of education, percent body fat, blood pressure, physical activity, smoking, alcohol use, sleep duration, and total grey matter volume.

Our findings are consistent with animal evidence that peripheral inflammation and

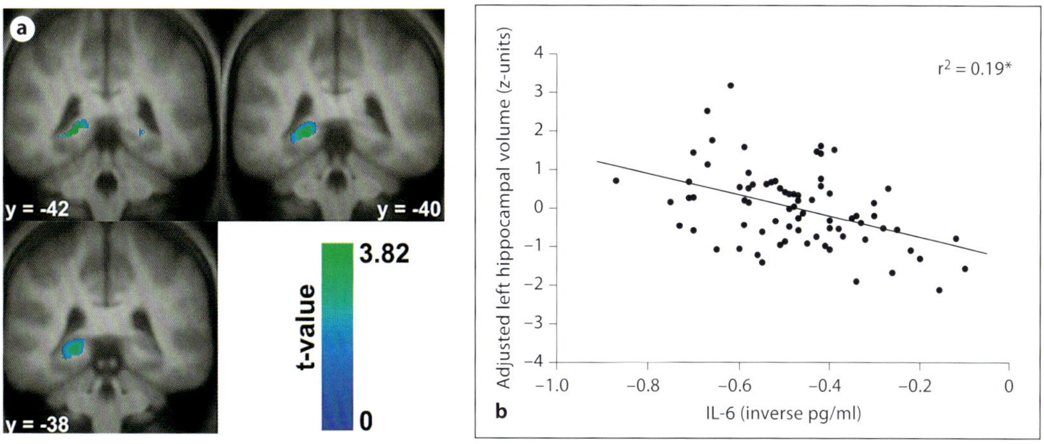

Fig. 3. Higher levels of plasma IL-6 are associated with decreased grey matter volume in left, but not right, hippocampus at a family-wise error rate corrected level of significance. **a** Statistical parametric maps profiling clusters of the left and right hippocampus where higher IL-6 was associated with decreased grey matter volume after controlling for age, sex, race and total grey matter volume. **b** Plotted along the y-axis are extracted left hippocampal grey matter volumes in standardized units adjusted for age, sex, race and total grey matter volume. Plotted along the x-axis are transformed plasma IL-6 values (less negative values indicate higher IL-6 levels). * $p < 0.01$. Reprinted from Marsland et al. [86], with permission from Elsevier.

subsequent activation of central inflammatory processes results in hippocampal remodeling, particularly neurodegeneration and atrophy [77]. To the extent that hippocampal atrophy plays a role in memory-related deficits, these results further support the possibility that systemic inflammation may play a role in age-related declines in learning and memory.

In addition to these findings, exploratory whole brain analyses revealed an inverse association of peripheral IL-6 with grey matter volume in the prefrontal cortex. This association is consistent with our previous results showing associations of IL-6 with cognitive performance on tasks involving working memory and executive function that typically recruit prefrontal regions [70]. Indeed, functional neuroimaging studies suggest that a network of brain regions are activated while encoding stimuli that are subsequently remembered, and these regions include the prefrontal cortex along with hippocampal areas [78]. These findings warrant future investigation to further elucidate the structure and function of brain regions involved in inflammatory-related cognitive function.

Peripheral Markers of Inflammation and White Matter Hyperintensities

Other studies have extended the assessment of inflammation-related brain structure to an examination of white matter hyperintensities (WMH), providing a marker of white matter damage and lesions [79]. WMH increase with age [80] and have been shown to predict future incidence of stroke, dementia, cognitive decline, and death [81]. The possibility that inflammatory processes play a role in age-related increases in WMH has been the focus of a number of recent investigations [82–84]. For example, Satizabal et al. [85] assessed plasma IL-6 and CRP levels in 1,841 older adults who also completed a structural MRI protocol. Consistent with our findings [86], IL-6 levels were inversely associated with total and hippocampal grey matter volumes. Additionally, IL-6, and to a lesser extent CRP, were positively associated with WMH volume. A similar inverse association of CRP with WMH was also found in the Rotterdam Scan Study [87]. Furthermore, a recent study by Furney et al. [88] examined the contribution of structural MRI results and inflammation to the prediction of risk for dementia. Although cytokine levels and MRI findings contributed independently to the prediction of risk, a model that combined them both accounted for greater predictive variance. To the extent that inflammation plays a role in the future progression of dementia, these findings raise the possibility that traditional structural MRI techniques may not adequately detect the entirety of inflammatory-related effects.

Peripheral Markers of Inflammation and Microstructural White Matter Integrity: Diffusion Tensor Imaging

Complementing studies that examine WMH, diffusion tensor imaging (DTI) offers a promising technique for the elucidation of specific neural networks that may be vulnerable to cytokine-related disruptions. This approach is typically used to infer white matter microstructural integrity, and is based on differential patterns of water molecule diffusion occurring along the length of axons [89]. Water molecule movement that is perpendicular to the length of the axon (radial diffusivity) is compared to movement that is parallel (axial diffusivity). The asymmetry in molecule movement is termed fractional anisotropy (FA), and is believed to reflect global characteristics of the tract, including fiber densities, axonal diameters, and myelination [90]. In DTI, FA is utilized as a standard measure of white matter integrity, with higher levels of FA reflecting greater integrity. Similar to WMH, aging is associated with reductions in global FA and increases in radial diffusivity throughout the brain [91], indicating increased diffusion of water molecules across myelin boundaries among older adults. These findings likely reflect a decrease in myelin integrity.

We recently investigated the possibility that microstructural white matter integrity, as measured by DTI, may covary with plasma IL-6 and CRP among a sample of 155 cognitively normal, midlife adults participating in the Pittsburgh Imaging Project (age 30–50 years, 50% female) [Marsland et al., unpubl. findings]. Our findings show an inverse association of IL-6 and CRP with global FA scores. These associations were retained in analyses that controlled for age, gender, and BMI. IL-6 and CRP levels were also associated with reductions in total white matter volume and with impaired performance on the Stroop word-color task, a test of executive function, suggesting that cytokine-related microstructural changes manifest at both the anatomical and behavioral levels. An examination of the white matter tract skeleton suggests that IL-6 and CRP are associated with reductions in white matter integrity in multiple brain regions (fig. 4, 5). Our results are consistent with a recent DTI study conducted by Wersching et al. [92], which reported a similar inverse association of CRP with global FA and regional FA scores in the frontal lobes. Taken together, these findings provide further human support for a neurobiological pathway linking inflammation to brain networks implicated in learning, memory and executive function.

Summary, Treatment Implications and Future Directions

The studies overviewed in this chapter suggest that circulating markers of inflammation predict future cognitive risk and possibly contribute to the pathophysiology of preclinical neurocognitive decline. Recent human studies combining

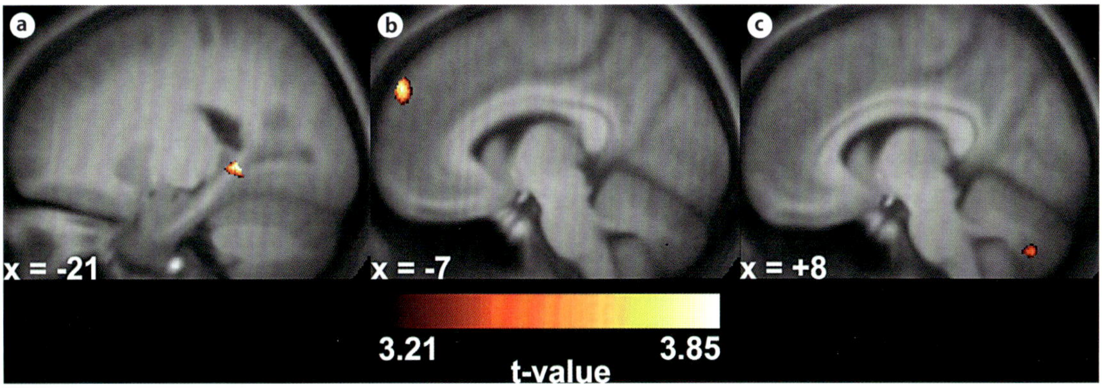

Fig. 4. Diffusion tensor-based images of axial (**a**) and sagittal (**b**) brain slices showing IL-6-related changes in fractional anisotropy (FA). The white matter skeleton is shown in turquoise, while red shows significantly decreased FA related to IL-6 ($p < 0.01$ and $k > 50$ continuous voxels).

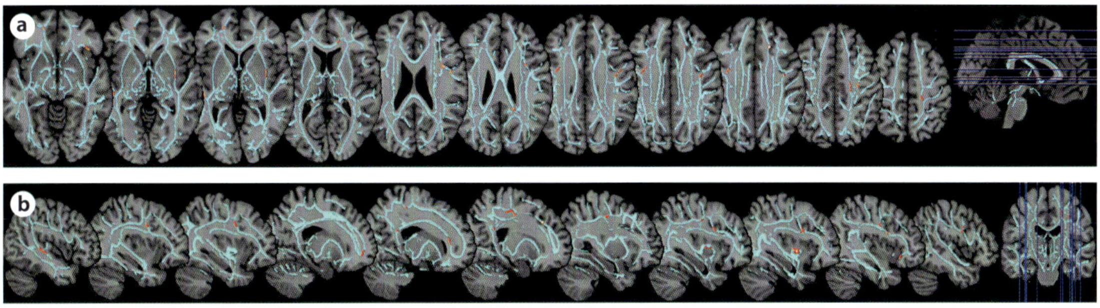

Fig. 5. Diffusion tensor-based images of axial (**a**) and sagittal (**b**) brain slices showing CRP-related changes in fractional anisotropy (FA). The white matter skeleton is shown in turquoise, while red shows significantly decreased FA related to CRP ($p < 0.01$ and $k > 50$ continuous voxels).

neuroimaging methods, cognitive performance testing, and markers of inflammation are consistent with the animal literature and support the notion that peripheral inflammatory processes can adversely affect CNS processes involved in working and episodic memory and executive function. In general, studies show that inflammation contributes to lower cognitive performance over and above more traditional cardiovascular risk factors, including education, hypertension, metabolic factors, smoking, subclinical atherosclerosis and symptoms of cardiovascular disease [93]. However, inflammation covaries significantly with other cardiovascular risk factors in the prediction of cognitive aging and it is difficult to isolate their unique contributions. Indeed, recent evidence suggests that a synergistic combination of traditional and inflammatory risk factors predicts greater risk than single risk factors alone [94]. Thus, it is likely that profiles will be of greater use in the future prediction of risk.

The current findings have potential implications for the early detection and treatment of inflammation. Findings have shown that anti-inflammatory drugs can restore hippocampal neurogenesis in rats [77], and some human epidemiological evidence suggests that nonsteroidal anti-inflammatory drugs (NSAIDs) may slow the progression of memory loss in patients with dementia [95], and decrease the future risk for Alzheimer's disease [51]. However, not all findings are consistent and clinical trials show no benefit of NSAIDs in preventing cognitive decline in patients with dementia [96, 97]. To date, the majority of these clinical trials have focused on secondary prevention in elderly populations who are already diagnosed with dementia. It remains to be determined whether NSAIDs are beneficial if provided earlier in the disease course. Initial evidence suggests that this may be the case, with results from a recent study showing that anti-inflammatory drug use protects against age-related atrophy of grey and white matter among older women [98].

In addition to immune cells, adipose tissue is a key contributing source of proinflammatory mediators, including IL-6. High BMI is also associated with age-related cognitive declines [99] and risk for future dementia [100], and, thus, it is plausible that interventions aimed at reducing adiposity may protect against cognitive impairment. While weight loss by dieting and bariatric surgery has been associated with reductions in peripheral inflammation [101], it is unknown whether the magnitude of these effects conveys protection against accelerated cognitive aging.

In addition to biobehavioral and lifestyle choices, psychosocial characteristics including chronic stress and clinical depression have been shown to influence peripheral inflammatory activity. Evidence shows that chronic psychological stressors, including low socioeconomic status, loneliness, and caregiving for a terminally ill family member, are associated with increased systemic inflammation [102]. Psychological stress has also been linked to poorer cognitive function, and with stress hormones including cortisol thought to play a role in modulating hippocampal processes [103]. The possibility that inflammation mediates associations of chronic stress and cognitive function remains to be determined.

There are several limitations of the current literature that warrant attention in future work. While some prospective longitudinal designs have been utilized, studies investigating associations of inflammation and cognitive function are largely cross-sectional, precluding causal interpretations. Animal studies support a role of inflammation in modulating cognitive performance; however, the reverse is also a possibility with elevations in peripheral inflammation reflecting 'spillover' from neurodegenerative processes in the CNS [104]. It is also possible that associations may stem from a third factor, possibly relating to individual differences in genetics. Another limitation of the current literature is the use of single assessments of inflammatory markers. Evidence suggests that these markers are relatively stable over extended periods [105]; however, multiple factors are known to impact circulating levels of inflammatory mediators, including acute inflammatory disease, psychological stress, and physical activity. Thus, a more reliable assessment of stable individual difference would be derived from assessing means across multiple testing occasions. In the future, prospective investigations are needed beginning in middle adulthood and examining serial assessments of inflammation, brain structure and cognitive function. This work will enable an assessment of whether variation in systemic inflammation among mid-life adults contributes to risk for accelerated cognitive decline and dementia, shedding light on possible human neural pathways of cognitive aging.

References

1 Buckner RL: Memory and executive function in aging and AD: multiple factors that cause decline and reserve factors that compensate. Neuron 2004;44:195–208.

2 Davis HP, et al: Acquisition, recall, and forgetting of verbal information in long-term memory by young, middle-aged, and elderly individuals. Cortex 2003;39:1063–1091.

3 Huh TJ, et al: Response bias and aging on a recognition memory task. J Int Neuropsychol Soc 2006;12:1–7.

4 Wecker NS, et al: Mental flexibility: age effects on switching. Neuropsychology 2005;19:345–352.

5 Salthouse TA: When does age-related cognitive decline begin? Neurobiol Aging 2009;30:507–514.

6 Barberger-Gateau P, et al: Functional impairment in instrumental activities of daily living: an early clinical sign of dementia? J Am Geriatr Soc 1999;47:456–462.

7 Yaffe K: Metabolic syndrome and cognitive disorders: is the sum greater than its parts? Alzheimer Dis Assoc Disord 2007;21:167–171.

8 Yaffe K, et al: The metabolic syndrome, inflammation, and risk of cognitive decline. JAMA 2004;292:2237–2242.

9 Brown GC, Neher JJ: Inflammatory neurodegeneration and mechanisms of microglial killing of neurons. Mol Neurobiol 2010;41:242–247.

10 Heinrich PC, Castell JV, Andus T: Interleukin-6 and the acute phase response. Biochem J 1990;265:621–636.

11 Bataille R, Klein B: C-reactive protein levels as a direct indicator of interleukin-6 levels in humans in vivo. Arthritis Rheum 1992;35:982–983.

12 Maggio M, et al: Interleukin-6 in aging and chronic disease: a magnificent pathway. J Gerontol A Biol Sci Med Sci 2006;61: 575–584.

13 Bluthe RM, et al: Synergy between tumor necrosis factor alpha and interleukin-1 in the induction of sickness behavior in mice. Psychoneuroendocrinology 1994;19:197–207.

14 Rachal Pugh C, et al: The immune system and memory consolidation: a role for the cytokine IL-1beta. Neurosci Biobehav Rev 2001;25:29–41.

15 Dantzer R: Cytokine-induced sickness behaviour: a neuroimmune response to activation of innate immunity. Eur J Pharmacol 2004;500:399–411.

16 Lee JW, et al: Neuro-inflammation induced by lipopolysaccharide causes cognitive impairment through enhancement of beta-amyloid generation. J Neuroinflammation 2008;5:37.

17 Tracey KJ: The inflammatory reflex. Nature 2002;420:853–859.

18 Ek M, et al: Inflammatory response: pathway across the blood-brain barrier. Nature 2001;410:430–431.

19 Laflamme N, Rivest S: Effects of systemic immunogenic insults and circulating proinflammatory cytokines on the transcription of the inhibitory factor kappaB alpha within specific cellular populations of the rat brain. J Neurochem 1999;73:309–321.

20 Streit WJ, Mrak RE, Griffin WS: Microglia and neuroinflammation: a pathological perspective. J Neuroinflammation 2004;1:14.

21 Prinz M, Mildner A: Microglia in the CNS: immigrants from another world. Glia 2011;59:177–187.

22 Aloisi F: Immune function of microglia. Glia 2001;36:165–179.

23 Dickson DW, et al: Microglia and cytokines in neurological disease, with special reference to AIDS and Alzheimer's disease. Glia 1993;7:75–83.

24 Hanisch U-K: Microglia as a source and target of cytokines. Glia 2002;40:140–155.

25 Gadient RA, Otten U: Expression of interleukin-6 (IL-6) and interleukin-6 receptor (IL-6R) mRNAs in rat brain during postnatal development. Brain Res 1994;637:10–14.

26 Vallieres L, et al: Reduced hippocampal neurogenesis in adult transgenic mice with chronic astrocytic production of interleukin-6. J Neurosci 2002;22:486–492.

27 Vitkovic L, et al: Cytokine signals propagate through the brain. Mol Psychiatry 2000;5:604–615.

28 McAfoose J, Baune BT: Evidence for a cytokine model of cognitive function. Neurosci Biobehav Rev 2009;33:355–366.

29 Reichenberg A, et al: Cytokine-associated emotional and cognitive disturbances in humans. Arch Gen Psychiatry 2001;58:445–452.

30 Yirmiya R, Goshen I: Immune modulation of learning, memory, neural plasticity and neurogenesis. Brain Behav Immun 2011;25:181–213.

31 Oitzl MS, et al: Interleukin-1 beta, but not interleukin-6, impairs spatial navigation learning. Brain Res 1993;613:160–163.

32 Gibertini M, et al: Spatial learning impairment in mice infected with *Legionella pneumophila* or administered exogenous interleukin-1-beta. Brain, Behavior, and Immunity 1995;9:113–128.

33 Song C, Horrobin D: Omega-3 fatty acid ethyl-eicosapentaenoate, but not soybean oil, attenuates memory impairment induced by central IL-1beta administration. J Lipid Res 2004;45:1112–1121.

34 Lacosta S, Anisman H, Merali Z: Influence of acute and repeated interleukin-2 administration on spatial learning, locomotor activity, exploratory behaviors, and anxiety. Behav Neurosci 1999;113:1030–1041.

35 Gibertini M: Cytokines and cognitive behavior. NeuroImmunoModulation 1998;5:160–165.

36 Heyser CJ, et al: Progressive decline in avoidance learning paralleled by inflammatory neurodegeneration in transgenic mice expressing interleukin 6 in the brain. Proc Natl Acad Sci USA 1997;94:1500–1505.

37 Balschun D, et al: Interleukin-6: a cytokine to forget. FASEB J 2004;18:1788–1790.

38 Frank MG, et al: Microglia serve as a neuroimmune substrate for stress-induced potentiation of CNS pro-inflammatory cytokine responses. Brain Behav Immun 2007;21:47–59.

39 Perry VH, Cunningham C, Holmes C: Systemic infections and inflammation affect chronic neurodegeneration. Nat Rev Immunol 2007;7:161–167.

40 Ben Menachem-Zidon O, et al: Intrahippocampal transplantation of transgenic neural precursor cells overexpressing interleukin-1 receptor antagonist blocks chronic isolation-induced impairment in memory and neurogenesis. Neuropsychopharmacology 2008;33:2251–2262.

41 Simen AA, et al: Cognitive dysfunction with aging and the role of inflammation. Ther Adv Chron Dis 2011;2:175–195.
42 Krabbe KS, Pedersen M, Bruunsgaard H: Inflammatory mediators in the elderly. Exp Gerontol 2004;39:687–699.
43 Ye SM, Johnson RW: Increased interleukin-6 expression by microglia from brain of aged mice. J Neuroimmunol 1999;93:139–148.
44 Perry VH, Matyszak MK, Fearn S: Altered antigen expression of microglia in the aged rodent CNS. Glia 1993;7:60–67.
45 Chen J, et al: Neuroinflammation and disruption in working memory in aged mice after acute stimulation of the peripheral innate immune system. Brain Behav Immun 2008;22:301–311.
46 Sparkman NL, Johnson RW: Neuroinflammation associated with aging sensitizes the brain to the effects of infection or stress. NeuroImmunoModulation 2008;15:323–330.
47 Brydon L, et al: Synergistic effects of psychological and immune stressors on inflammatory cytokine and sickness responses in humans. Brain Behav Immun 2009;23:217–224.
48 Spath-Schwalbe E, et al: Acute effects of recombinant human interleukin-6 on endocrine and central nervous sleep functions in healthy men. J Clin Endocrinol Metab 1998;83:1573–1579.
49 Swardfager W, et al: A meta-analysis of cytokines in Alzheimer's disease. Biol Psychiatry 2010;68:930–941.
50 Guerreiro RJ, et al: Peripheral inflammatory cytokines as biomarkers in Alzheimer's disease and mild cognitive impairment. Neurodegener Dis 2007;4:406–412.
51 McGeer PL, McGeer EG: The inflammatory response system of brain: implications for therapy of Alzheimer and other neurodegenerative diseases. Brain Res Brain Res Rev 1995;21:195–218.
52 Zuliani G, et al: Plasma cytokines profile in older subjects with late onset Alzheimer's disease or vascular dementia. J Psychiatr Res 2007;41:686–693.
53 Licastro F, et al: Increased plasma levels of interleukin-1, interleukin-6 and alpha-1-antichymotrypsin in patients with Alzheimer's disease: peripheral inflammation or signals from the brain? J Neuroimmunol 2000;103:97–102.
54 Engelhart MJ, et al: Inflammatory proteins in plasma and the risk of dementia: the Rotterdam study. Arch Neurol 2004;61:668–672.
55 Holmes C, et al: Systemic inflammation and disease progression in Alzheimer disease. Neurology 2009;73:768–774.
56 Wallin K, et al: Midlife rheumatoid arthritis increases the risk of cognitive impairment two decades later: a population-based study. J Alzheimers Dis 2012;31:669–676.
57 Schmidt R, et al: Early inflammation and dementia: a 25-year follow-up of the Honolulu-Asia Aging Study. Ann Neurol 2002;52:168–174.
58 Tan ZS, et al: Inflammatory markers and the risk of Alzheimer disease: the Framingham Study. Neurology 2007;68:1902–1908.
59 van Himbergen TM, et al: Biomarkers for insulin resistance and inflammation and the risk for all-cause dementia and Alzheimer disease: results from the Framingham Heart Study. Arch Neurol 2012;69:594–600.
60 Teunissen CE, et al: Inflammation markers in relation to cognition in a healthy aging population. J Neuroimmunol 2003;134:142–150.
61 Tilvis RS, et al: Predictors of cognitive decline and mortality of aged people over a 10-year period. J Gerontol A Biol Sci Med Sci 2004;59:268–274.
62 Weaver JD, et al: Interleukin-6 and risk of cognitive decline: MacArthur studies of successful aging. Neurology 2002;59:371–378.
63 Yaffe K, et al: Inflammatory markers and cognition in well-functioning African-American and white elders. Neurology 2003;61:76–80.
64 Alley DE, et al: Inflammation and rate of cognitive change in high-functioning older adults. J Gerontol A Biol Sci Med Sci 2008;63:50–55.
65 Dik MG, et al: Contribution of metabolic syndrome components to cognition in older individuals. Diabetes Care 2007;30:2655–2660.
66 Gustafson DR, et al: 37 years of body mass index and dementia: observations from the prospective population study of women in Gothenburg, Sweden. J Alzheimer's Dis 2012;28:163–171.
67 Hoozemans JJ, et al: Neuroinflammation in Alzheimer's disease wanes with age. J Neuroinflammation 2011;8:171.
68 Krstic D, Knuesel I: Deciphering the mechanism underlying late-onset Alzheimer disease. Nat Rev Neurol 2012.
69 Laurin D, et al: Midlife C-reactive protein and risk of cognitive decline: a 31-year follow-up. Neurobiol Aging 2009;30:1724–1727.
70 Marsland AL, et al: Interleukin-6 covaries inversely with cognitive performance among middle-aged community volunteers. Psychosom Med 2006;68:895–903.
71 Wechsler D: The WMS-III Administration and Scoring Manual. Antonio, The Psychological Corporation, 1997.
72 Reitan RM, Wolfson D: The Halstead-Reitan Neuropsychological Test Battery. Tuscon, Neuropsychology Press, 1985.
73 Golden JC: Stroop Color and Word Test. Chicago, Stoelting, 1978.
74 Jack CR Jr, et al: Rates of hippocampal atrophy correlate with change in clinical status in aging and AD. Neurology 2000;55:484–489.
75 Mattay VS, et al: Neurophysiological correlates of age-related changes in working memory capacity. Neurosci Lett 2006;392:32–37.
76 Ashburner J, Friston KJ: Voxel-based morphometry – the methods. Neuroimage 2000;11:805–821.
77 Monje ML, Toda H, Palmer TD: Inflammatory blockade restores adult hippocampal neurogenesis. Science 2003;302:1760–1765.
78 Wagner AD, et al: Building memories: remembering and forgetting of verbal experiences as predicted by brain activity. Science 1998;281:1188–1191.
79 Fazekas F, et al: Pathologic correlates of incidental MRI white matter signal hyperintensities. Neurology 1993;43:1683–1689.
80 Gunning-Dixon FM, et al: Aging of cerebral white matter: a review of MRI findings. Int J Geriatr Psychiatry 2009;24:109–117.
81 Debette S, Markus HS: The clinical importance of white matter hyperintensities on brain magnetic resonance imaging: systematic review and meta-analysis. BMJ 2010;341:c3666.
82 Baune BT, et al: Association between cytokines and cerebral MRI changes in the aging brain. J Geriatr Psychiatry Neurol 2009;22:23–34.

83 Furney SJ, et al: Combinatorial markers of mild cognitive impairment conversion to Alzheimer's disease – cytokines and MRI measures together predict disease progression. J Alzheimer's Dis 2011;26:395–405.
84 Miralbell J, et al: Structural brain changes and cognition in relation to markers of vascular dysfunction. Neurobiol Aging 2012;33:20.
85 Satizabal CL, et al: Circulating IL-6 and CRP are associated with MRI findings in the elderly: the 3C-Dijon Study. Neurology 2012;78:720–727.
86 Marsland AL, et al: Interleukin-6 covaries inversely with hippocampal grey matter volume in middle-aged adults. Biol Psychiatry 2008;64:484–490.
87 van Dijk EJ, et al: C-reactive protein and cerebral small-vessel disease: the Rotterdam Scan Study. Circulation 2005;112:900–905.
88 Furney SJ, et al: Combinatorial markers of mild cognitive impairment conversion to Alzheimer's disease – cytokines and MRI measures together predict disease progression. J Alzheimers Dis 2011;3:395–405.
89 Hagmann P, et al: Understanding diffusion MR imaging techniques: from scalar diffusion-weighted imaging to diffusion tensor imaging and beyond. Radiographics 2006;26(suppl 1): S205–S223.
90 Le Bihan D, et al: Diffusion tensor imaging: concepts and applications. J Magn Reson Imaging 2001;13:534–546.
91 Giorgio A, et al: Age-related changes in grey and white matter structure throughout adulthood. Neuroimage 2010;51:943–951.
92 Wersching H, et al: Serum C-reactive protein is linked to cerebral microstructural integrity and cognitive function. Neurology 2010;74:1022–1029.
93 Schram MT, et al: Systemic markers of inflammation and cognitive decline in old age. J Am Geriatr Soc 2007;55:708–716.
94 Roberts RO, et al: Metabolic syndrome, inflammation, and nonamnestic mild cognitive impairment in older persons: a population-based study. Alzheimer Dis Assoc Disord 2010;24:11–18.
95 in t' Veld BA, et al: Nonsteroidal antiinflammatory drugs and the risk of Alzheimer's disease. N Engl J Med 2001;345:1515–1521.
96 Jaturapatporn D, et al: Aspirin, steroidal and non-steroidal anti-inflammatory drugs for the treatment of Alzheimer's disease. Cochrane Database Syst Rev 2012;2:CD006378.
97 de Craen AJ, et al: Meta-analysis of nonsteroidal anti-inflammatory drug use and risk of dementia. Am J Epidemiol 2005;161:114–120.
98 Walther K, et al: Anti-inflammatory drugs reduce age-related decreases in brain volume in cognitively normal older adults. Neurobiol Aging 2011;32:497–505.
99 Fitzpatrick AL, et al: Midlife and late-life obesity and the risk of dementia: cardiovascular health study. Arch Neurol 2009;66:336–342.
100 Ward MA, et al: The effect of body mass index on global brain volume in middle-aged adults: a cross sectional study. BMC Neurol 2005;5:23.
101 Forsythe LK, Wallace JM, Livingstone MB: Obesity and inflammation: the effects of weight loss. Nutr Res Rev 2008;21:117–133.
102 Kiecolt-Glaser JK, et al: Chronic stress and age-related increases in the proinflammatory cytokine IL-6. Proc Natl Acad Sci USA 2003;100:9090–9095.
103 McEwen BS, Sapolsky RM: Stress and cognitive function. Curr Opin Neurobiol 1995;5:205–216.
104 Meyer JS, et al: Risk factors for cerebral hypoperfusion, mild cognitive impairment, and dementia. Neurobiol Aging 2000;21:161–169.
105 Rao KM, et al: Variability of plasma IL-6 and crosslinked fibrin dimers over time in community dwelling elderly subjects. Am J Clin Pathol 1994;102:802–805.

Anna L. Marsland, PhD, RN
University of Pittsburgh
603 Old Engineering Hall
Pittsburgh, PA 15260 (USA)
E-Mail marsland@pitt.edu

Halaris A, Leonard BE (eds): Inflammation in Psychiatry.
Mod Trends Pharmacopsychiatry. Basel, Karger, 2013, vol 28, pp 188–202 (DOI: 10.1159/000346085)

Inflammation in Suicidality: Implications for Novel Treatment Options

Shorena Janelidze[a] · Lena Brundin[a,b]

[a]Psychoimmunology Unit, Section for Psychiatry, Department of Clinical Sciences, Lund University, Lund, Sweden;
[b]Department of Translational Science and Molecular Medicine, Michigan State University, Grand Rapids, Mich., USA

Abstract

Suicide is an increasing health problem worldwide and the most severe complication of multiple psychiatric disorders. It is also prevalent in several somatic conditions. In recent years, significant advances have been made in understanding the neurobiology of suicidal behaviors. Several lines of evidence have emerged suggesting that inflammation may contribute to the pathophysiology of suicide. Basic and clinical data indicate that the effects of inflammation on mood and behavior are likely to be mediated by the kynurenine pathway metabolites and glutamatergic neurotransmission. At the same time, the triggers of inflammatory changes observed in suicidal patients are largely unknown but may include stress, infectious agents and autoimmune diseases. As available treatment options against suicidality are only moderately effective, targeting the inflammatory system may provide novel therapeutic opportunities. For this goal to be achieved, however, we need to gain better insight into the origin, mechanisms and outcomes of inflammation in suicidal behavior.

Suicide, the termination of ones own life on purpose, is a global health problem. At least 1 million deaths by suicide take place around the world each year [1]. The number of suicide attempts is generally 10–20 times higher than the number of completed suicides. Suicide is the tenth leading cause of death worldwide with a higher prevalence in men than in women [2], although suicide statistics vary considerably from country to country [3]. The phenomenon is sometimes the subject of cultural taboos that could cause an underestimation of true suicide rates. Both the deaths by suicide and the attempts are signs of extreme psychological suffering that unaffected persons find difficult to comprehend. Each completed suicide carries with it deep emotional pain and grief that stand to impact family members for years to come.

A previous suicide attempt [4] and mental illness are the strongest risk factors for suicidal behavior. Suicide rates are high among patients with major depressive disorders (MDD) [5], bipolar disorders [6], anxiety disorders [7], alcohol abuse [8], schizophrenia [9] and several other psychiatric conditions [2]. The percentage of patients with a diagnosis of MDD who will attempt suicide is about 14–17% [10–12], and the corresponding number for completed suicide is 4–6% [13,

14]. Certain personality traits, including aggressiveness, impulsivity [15, 16] and hopelessness/helplessness [17, 18], are also associated with increased risk of suicide. However, frequent psychiatric comorbidity complicates the identification of true associations between specific mental disorders and suicidality [19].

Despite increased pharmacological treatment of suicidal individuals over the past decade, the incidence rates of suicide and suicide attempts have been increasing. It is estimated that 1.5 million people will die from suicide yearly in 2020, if current trends remain unchanged [20]. Moreover, the health care system is often unable to accurately detect suicidality, in spite of the fact that almost half of these patients contact primary care providers in the month prior to their attempt [21]. Consequently, there is a great need for both improved methods for detection of suicide risk and for specific treatments of suicidal patients [22].

Treatment of Suicidality

Currently there is no treatment of choice for suicidal patients. Treatment depends on the primary psychiatric diagnosis, if any, and often includes serotonin reuptake inhibitors and anxiolytic medication [23]. Severe cases can be treated with electroconvulsive therapy in many countries, in addition to the pharmacological treatment. Interestingly, there is a known increase in the risk for suicidal behavior during the first few weeks of treatment with antidepressants, especially in children and adolescents [24]. This has been attributed to reduced psychomotor retardation and elevated energy levels, which may occur prior to the improvement of mood [25]. At the neurobiological level, there is no explanation so far as to why there is approximately a 1-month time lag from the initiation of treatment with an antidepressant to the improvement of mood given that neurotransmitter levels normalize almost immediately.

A number of recent studies have found a strikingly rapid anti-suicidal effect of the N-methyl-D-aspartate (NMDA)-receptor antagonist, ketamine, in MDD and bipolar depression. In contrast to conventional antidepressants, ketamine reduces suicidal ideation and depressive symptoms within minutes or hours after intravenous administration [26–29]. In the study by Price et al. [26], treatment-resistant medication-free depressed patients were given a single dose or five repeated iv infusions of ketamine. In the single-dose regimen, suicidality, rated using the suicidality item on the Montgomery-Asberg Depression Rating Scale (MADRS-SI) and Implicit Association Test, was reduced in 8 of 13 patients 24 h postinfusion. The reduction in suicidal ideation was also observed in 9 of 10 patients who had received repeated ketamine infusions. In a similar group of 33 patients, DiazGranados et al. [27] showed that the anti-suicidal effects of ketamine, assessed with the Scale for Suicide Ideation, were already detectable 40 min after the initiation of treatment and lasted for at least 4 h. Moreover, it has been later demonstrated by Larkin and Beautrais [28] that reduction in the MADRS-SI was maintained for 10 days following a singe intravenous ketamine bolus injection in 13 of 14 MDD patients. As reported by Zarate et al. [29], ketamine also resolved suicidal ideation in patients with treatment-resistant bipolar depression. These findings suggest enhanced NMDA-receptor signaling as part of the pathophysiology of suicidal behavior, especially when considering the rapid onset of antisuicidal effects of ketamine administration.

Neurobiology of Suicidality

It is well accepted that suicidal behavior is driven by highly complex interactions between genetic and environmental factors. Nevertheless, the neurobiological substrates of suicide are poorly understood. To date, the most commonly reported changes in the neurobiology of suicidal behavior are abnormalities in serotonergic neurotransmission [30] and hypothalamic-pituitary-adrenal (HPA) axis activity [31]. Accumulating evidence also suggests that inflammation contributes to the pathophysiology of suicidality. The mediators of inflammation are known to interact closely and reciprocally with the HPA axis, and can also induce substantial changes in the serotonin system, as will be described below.

Serotonin System and Suicidality

The idea of deficient serotonergic neurotransmission in suicidality emerged in the 1970s when Asberg et al. [32] described reduced levels of the serotonin metabolite, 5-hydroxy-indole-acetic acid in violent suicide attempters. Numerous subsequent studies focusing on suicidal patients further uncovered impairments in various aspects of serotonergic function. Low serotonin synthesis was demonstrated in the ventromedial and orbital prefrontal cortex (PFC) of suicide attempters [33]. Hypofunction of the serotonergic system in suicidality was further supported by findings of a blunted prolactin response following fenfluramine challenge in suicide attempters [34–36]. Prolactin levels are used as an index of serotonergic function as its release increases dose-dependently in response to activation by serotonin agonists, such as fenfluramine. In addition, binding of the serotonin transporter (SERT) is reduced in suicide victims in the ventral PFC [37–39] indicating lower serotonergic innervation in this area. Previous studies also demonstrated upregulation of serotonin receptors, 5-HT_{1A} and 5-HT_{2A}, in the PFC cortex of individuals who died by suicide [40, 41]. Upregulation of 5-HT receptors in suicidal behavior is thought to represent a feedback mechanism compensating for deficient serotonergic function [31]. Finally, genetic evidence has implicated polymorphisms in SERT and in tryptophan hydroxylase, the rate-limiting enzyme catalyzing serotonin synthesis from tryptophan, in suicidal behavior [42, 43]. However, postmortem microarray studies revealed no change in the expression of genes related to serotonergic neurotransmission in suicide victims compared to non-suicidal patients with matching psychiatric diagnosis [44, 45], suggesting regulation of serotonin-related genes at the posttranslational level [46].

It should be noted that findings concerning serotonergic dysfunction in suicide and in depression are sometimes conflicting. Several investigations have failed to detect differences in 5-HT receptor binding sites between victims of suicide and control subjects in the relevant brain regions [47–49]. Reduced as well as unaltered 5-HT_{2A} binding has been reported in the hippocampus of suicide victims [50, 51]. These discrepancies could be partly due to the variations in experimental approaches and heterogeneity of patient sample.

Stress Axis and Suicidality

HPA axis hyperactivity, as manifested by nonsuppression in the dexamethasone (Dex) suppression test (DST), elevated basal plasma and salivary cortisol levels, increased urinary free cortisol excretion, and elevated cortisol response in the combined Dex/CRH test, is commonly seen in patients with MDD [52]. Results have been less consistent, however, with regard to suicidal patients. Several studies have reported an association between suicidal behavior and non-suppression of endogenous cortisol production after a DST [53–57]. In contrast, two large-scale prospective studies did not support these findings [53, 58]. Coryell et al. [53] did, however, demonstrate a link between future suicide risk and DST nonsuppression in a subgroup of inpatients with manifest suicidality at the time of the test. Secunda et al. [59] found significantly lower pre-DST plasma cortisol levels in suicide attempters compared to nonsuicidal depressed patients, while no significant group differences were observed regarding post-DST cortisol, CSF or urine cortisol levels. Consistent with these findings, a large-scale retrospective study reported significantly lower plasma levels of adrenocorticotropic hormone and cortisol after Dex/CRH in suicidal compared to nonsuicidal patients [60]. Three recent studies have all shown low plasma cortisol levels in patients with a history of suicidal behavior [61–63].

Interactions between the Inflammatory System, Serotonin System and the HPA Axis

Interestingly, inflammation induces changes in monoamine metabolism and transmitter systems in animal models [64], which could provide a mechanistic framework for the serotonergic dysfunction described above for suicidal patients. For example, the proinflammatory cytokines, IL-1β and IL-6, stimulate release of serotonin in several brain regions in rodents [64–68]. A chronic elevation in cytokine levels may result in sustained serotonin release and ultimately depletion of the neurotransmitter stores in a similar way as it has been proposed for various stressors [64].

Extracellular serotonin concentration could also be influenced by altered SERT expression or activity. In fact, both IL-1β and tumor necrosis factor-α (TNF-α) were found to increase SERT activity and stimulate serotonin uptake in the rat embryonic raphe cell line as well as in mouse midbrain and striatal synaptosomes [69]. In addition, cytokines have been shown to differentially modulate 5-HT receptor expression in specific brain regions. Acute intracerebroventricular administration of interferon-α (IFN-α) in mice, which was followed by enhanced production of IL-1β, IL-6 and TNF-α, downregulated 5-HT_{1B} and 5-HT_{2A} mRNA in the PFC and 5-HT_{1A} mRNA in the hippocampus [70].

It is well established that inflammation is reciprocally connected with the HPA axis. There is, for example, ample evidence that stress triggers the release of proinflammatory cytokines. In human subjects, social stress in particular has been associated with high levels of soluble TNF-α receptors, IL1β and IL-6 [71–73]. In laboratory animals, cytokine elevations are found in different experimental stress paradigms [74, 75]. At the same time, lipopolysaccharide (a potent inducer of proinflammatory cytokines), IL-1β and IL-6 increase peripheral adrenocorticotropic hormone and corticosterone/cortisol levels in rodents and in humans [76–79], illustrating the reciprocity of the system.

How Can Inflammation Affect Suicidal Ideation and Behavior?

Activation of the immune system, whether due to stress, infections or autoimmune disorders, can exert profound effects on mood and behavior. A number of models have been proposed to explain how peripherally produced cytokines reach and convey signals to the CNS [80]. At the same time, it has been increasingly recognized that cytokines are also secreted locally in the brain and play an important role in the maintenance of brain function under physiological conditions. Existing experimental data suggest several different mechanisms by which cytokines can contribute to behavioral symptoms of suicidality. First, cytokine receptors are present on neurons in specific brain regions in both animals and man. For example, IL-6 receptors are expressed on serotonergic neurons in the medulla oblongata, as well as in the hypothalamus, hippocampus, cerebellum and selective cortical areas [81–84]. IL-1β receptors are located on hippocampal, amygdaloidal and thalamic neurons, on 5-HT2 receptor-expressing neurons in the hypothalamus and on cerebellar Purkinje cells [85, 86]. The presence of cytokine receptors on neurons might well mean that they have specific and direct effects on neuronal function. In fact, IL-6, IL-1β and TNF-α have all been implicated in the regulation of synaptic transmission and plasticity [87–90]. As mentioned above, there is also ample evidence that cytokines distinctly modulate the concentration of monoaminergic neurotransmitters and their metabolites in various regions of the CSN [64, 67]. It is thus possible that by affecting neuronal function, and, in particular, monoaminergic neurotransmission, cytokines are involved in the development of pathological states associated with suicidal behavior.

Inflammation, especially the cytokine interferon-γ, activates the kynurenine pathway of tryptophan metabolism. When this pathway is stimulated, the essential amino acid tryptophan is shunted into generation of neuroactive inflammatory metabolites instead of serotonin production. Thus, inflammation could cause a decrease in serotonin levels by this shift to production of kynurenine metabolites [91]. Patients with cancer who underwent cytokine therapy and developed depression were found to have decreased tryptophan levels in the serum, providing evidence to this concept [92]. This mechanism could potentially contribute to the lowered levels of monoamine metabolites found in CSF of suicide attempters [30, 32, 93]. However, it has not yet been firmly established that inflammation does indeed cause depletion of brain tryptophan and serotonin levels in patients with IFN-α-induced, primary depression or suicidality [94].

The first and rate-limiting enzymatic step of the kynurenine pathway is indoleamine 2,3-dioxygenase (IDO), expressed in microglia cells in the brain. Its activation initiates a cascade of enzymatic reactions generating neuroactive compounds, including quinolinic acid (QUIN) and kynurenic acid (KYNA). QUIN is an NMDA-receptor agonist, acting through receptors containing the NR1 + NR2A and the NR1 + NR2B subunits [95, 96]. KYNA, on the other hand, antagonizes the glycine site of the NMDA receptor and blocks the cholinergic α7 nicotinic receptor [95, 97]. KYNA has neuroprotective and anticonvulsive properties [98, 99], whereas QUIN is a potent excitotoxin with pro-inflammatory properties [98–100]. Increased levels of QUIN could therefore contribute to the cell loss and structural changes in cortical and subcortical regions that has been reported in patients with suicidal behavior [100–106]. In addition to being a direct NMDA receptor agonist, QUIN increases neuronal glutamate release and decreases glutamate uptake and recycling by astrocytes [100]. Increased extracellular levels of QUIN thus facilitate glutamate neurotransmission prior to induction of apoptosis.

Signs of Inflammation in Suicidality

To our knowledge, the first publication connecting inflammation and suicidal behavior appeared in 1992, showing increased levels of the soluble IL-2 receptor (IL-2R) in the plasma of suicide attempters [107]. It took more than a decade for new reports to follow. In 2008, a pioneering study by Steiner et al. [108] suggested that inflammatory changes occur in the central nervous system of individuals who commit suicide. At postmortem examination, the authors found microgliosis in brains of suicide victims with depression and schizophrenia. Remarkably, the micogliosis was specific for suicide victims irrespective of primary psychiatric diagnosis; patients from the same diagnostic groups who died from other causes did not display any alterations in microglia density. The same year, another group demonstrated increased mRNA transcription of IL-4 and IL-13 in the orbitofrontal cortical area of female and male suicide victims, respectively [109]. In support of these observations, Lindqvist et al. [110] found elevated CSF levels of IL-6 in suicide attempters compared to healthy controls and a correlation between IL-6 levels and severity of depression, as evaluated using the Montgomery-Asberg Depressiom Rating Scale (MADRS). Recently, increased IL-1β, IL-6 and TNF-α at both the mRNA and protein levels was shown in the anterior PFC (Brodmann area 10) of teenage suicide victims [111]. Moreover, in a novel postmortem study Steiner et al. [112] found increased counts of quinolinic acid reactive microglia cells in the subgenual anterior cingulate cortex and anterior midcingulate cortex of suicide victims with depression.

Suicidal behavior is accompanied by changes in peripheral cytokine levels as well. Higher IL-6 and TNF-α levels, and lower IL-2 levels, were observed in suicidal patients compared to both nonsuicidal patients and healthy controls [113]. Kim et al. [114] used in vitro-stimulated whole blood from nonsuicidal and suicidal MDD patients, and found a decreased production of both IL-2 and IL-6 in the suicidal individuals. In line with the original postmortem data from Steiner et al. [115], the studies on peripheral markers thus suggest that suicidal patients may be distinguished from non-suicidal depressive patients. However, additional studies on peripheral cytokines in suicidal patients are clearly needed to establish potential useful biomarkers, especially since cytokine levels are known to display a large biological spread and a high rate of fluctuation. The protein S100B is another marker of inflammation, with peripheral levels increasing during injury and inflammatory conditions of the central nervous system (CNS) with damage to the blood-brain barrier (BBB). Another group analyzed serum S100B in adolescents with depression and psychosis, and found the levels to be related to the intensity of suicidal ideation irrespective of diagnosis.

Sublette et al. [116] have recently demonstrated that blood levels of kynurenine, one of the first metabolites produced in the kynurenine pathway, is significantly elevated in suicide attempters with depression compared to nonsuicidal patients with depression. This provides evidence of a peripherally activated kynurenine pathway in suicide attempters compared to nonattempters. On the other hand, the neuroactive metabolites, QUIN and KYNA, do not cross the BBB, and therefore central measures are necessary to establish if any changes are present with the potential to affect glutamatergic neurotransmission. In a study which measured KYNA and QUIN in the CSF of suicide attempters, QUIN-levels were found to be more than twice as high than in healthy controls, while KYNA was unaltered [117]. This gave rise to an increased QUIN/KYNA quotient, indicative of a net positive effect on NMDA-receptor transmission in suicidality. Taken together, these findings could provide a neurobiological rationale for the rapid remedial effects of the NMDA receptor antagonist, ketamine, on suicidal ideation [26–29]. Moreover, QUIN correlated with CSF IL-6 levels, suggesting that this metabolite was indeed generated by an ongoing inflammatory process. The study provided further evidence that QUIN was associated with suicidality by establishing that the levels correlated with the degree of suicidal intent [117].

The regulation of the enzymes in the kynurenine pathway still has to be described in detail, as well as their potential role in different psychiatric conditions. It is currently not known whether specific cytokines activate some of the enzymes in the kynurenine pathway in preference over others, or why some metabolites (such as QUIN) would accumulate in favor of another. In this respect, it has been suggested that genetic variants could pay a role [118].

Symptoms and Traits of Importance for Suicidality

For obvious reasons, there are no animal models directly aimed at reflecting suicidal behavior. However, since suicidality critically depends on several important personality trait/symptoms, notably aggression and helplessness [119], animal studies focusing on these traits have indicated that cytokines might indeed influence behavior linked to suicidality. Specifically, IL-1β and IL-2 have been shown to affect feline aggression albeit through different mechanisms. When injected into the medial hypothalamus or midbrain periaqueductal gray (PAG), IL-1β acted on 5-HT_2 receptors and potentiated aggressive behavior [120, 121]. The outcomes of IL-2 administration varied between the regions studied. It promoted aggression in the PAG through neurokinin NK(1) receptors, but the opposite effect was observed in the medial hypothalamus and this aggression-suppressing effect was mediated through $GABA_A$ receptors [122]. This example shows that the effects of cytokines on behavior are complex and vary across different brain regions. TNF-α is also likely to be involved in the regulation of aggression, as TNF-α receptor-deficient mice do not exhibit aggressive behavior in the resident-intruder test [123]. Deletion of IL-6 in knockout mice, on the other hand, led to a less helpless phenotype compared to wild-type controls in the learned helplessness paradigm [124].

In line with the animal studies, clinical data have also linked peripheral cytokines to aggressive behavior and hopelessness. IL-6 was found to positively correlate with anger [125] and hopelessness [126], whereas IL-1β was associated with hostility [127]. Increased anger/hostility has repeatedly been shown in hepatitis C patients undergoing IFN-α treatment [128, 129]. Interestingly, an increase in labile anger in these patients was linked to genetic variability in the TNF-α gene [130]. Future studies should determine if associations between inflammation and suicide-related personality traits and symptoms also exist in individuals with suicidal behavior.

Suicidality in Somatic Inflammatory Conditions

The above-described studies provide evidence of inflammation in both suicidal behavior per se, and in specific symptoms and traits with a relevance for suicidality. On the basis of this evidence, we propose that inflammation in patients with primary psychiatric conditions is part of a pathogenetic mechanism underlying suicidal symptoms. Peripheral and central inflammation is associated with a variety of somatic conditions. If our hypothesis is true, then suicidality should also be increased in at least some of these conditions. Moreover, it is not known whether specific inflammatory factors are more important for suicidality than others. This knowledge is of vital importance for future generation of optimal pharmacological tools. To better understand how inflammation of different kinds and origin may contribute, it is therefore of interest to investigate whether suicidal ideation and behavior are increased in somatic conditions associated with inflammation.

Interferons for the Treatment of Somatic Conditions and Suicidality

Chronic administration of IFN-α in patients suffering from melanoma and hepatitis C induces or aggravates depressive symptoms approximately 1 month after the beginning of medication [131–133]. The immunotherapy trials have shown a causal relationship between induced inflammation and subsequent onset of depression and suicidality [94, 134]. These studies have also provided neurobiological proof of principle that peripheral inflammation induces changes in central levels of kynurenine metabolites. After the injection of IFN-α, both quinolinic and kynurenic acid, as well as CSF cytokines, were significantly elevated in the patients, and these elevations correlated with depressive symptoms [94, 134]. Several case reports also describe the development of suicidal ideation [135, 136] and even suicide attempts in this patient population [137]. Similarly, suicide ideation and attempts have been documented in previously psychiatrically healthy individuals with multiple sclerosis during and after treatment with IFN-β [138].

Suicide, Asthma and Allergies

Asthma or allergies are accompanied by an array of innate and adaptive immune responses locally but also systemically, in the peripheral blood [139–141]. To our knowledge, there are neither preclinical nor clinical studies to date on the inflammatory changes in CSF and brain in these conditions. However, this topic is of great interest, considering that several epidemiological studies have shown a link between suicide and asthma/allergies. There is a confirmed increase in the risk for suicidal ideation, completed suicides and suicide attempts in asthmatic patients [142]. A large epidemiological study from Taiwan estimated suicide rates in more than 160,000 high-school students, with or without asthma, over 12 years [143]. The authors reported that the suicide rate in students with asthma was more than double that of the control group, with no difference in the number of natural deaths. In another large sample of over 6,500 individuals from the United States both suicidal ideation and suicide attempts were found to be associated with current but not former asthma [144]. The risk increase for suicidal ideation and attempts seems to be approximately 2- to 3-fold, with numbers varying between different studies.

It is difficult to determine whether the increased risk for suicidal behavior and ideation in asthmatics is due to the disease itself, or to disease-related factors. There are indeed reports and studies suggesting a link between at least two types of asthma medications, the leukotriene antagonist Monteleukast and theophylline, and suicidality. In several countries, warnings have been issued regarding a possible association between suicidal ideation and Monteleukast [145]. Recently, Favreau at al. [146] examined the relationship between suicidal ideation and anti-asthma medication, e.g. theophylline and beta-2 adrenergic agonists. The study, which controlled for asthma severity and comorbid depression, found that theophylline, but not beta-2 adrenergic agonists, increased the risk of suicidal ideation.

Several reports indicate a possibly increased risk for suicidality in allergy. For example, there is an association between tree pollen peak seasons and nonviolent suicides in females, although this association was not controlled for the confounding effects of medication [147]. Another study investigated the effects of anti-allergy medications on suicidality, involving 120,000 suicides in the USA over 3 years. The prescription volume of intranasal corticosteroids was associated with a lower suicide risk while that of antihistamines was associated with a modestly greater suicide risk, independently of antidepressant medications [148]. Thus, both asthma and allergies in general seem to be associated with an increased risk for suicidality, although it is currently not established whether the relationship is mediated through the disease itself, by the medications used, or possibly a combination of both.

Autoimmunity and Suicidality

Two autoimmune disorders where increased rates of suicidality have been described are systemic lupus erythematosus (SLE) and multiple sclerosis (MS). SLE is a chronic autoimmune disease which may involve any organ and system, including peripheral, autonomic, and central nervous systems [149]. The SLE neurological and psychiatric manifestations are usually reported together, and specific data on SLE psychopathology are limited. A recent review concluded that depression may be present in 39% of SLE patients and cognitive dysfunction in up to 80% [150]. Interestingly, in neuropsychiatric SLE, antibodies against the NMDA receptor are often found in patients, in combination with microglial activation and BBB dysfunction [151]. However, the exact pathology of SLE and neuropsychiatric SLE is not known. Karassa et al. [152] examined the prevalence of suicide attempts in SLE and found that all attempts were confined to patients with neuropsychiatric SLE. Furthermore, in a sample of 285 Chinese SLE patients, the prevalence of suicidal ideation appeared to be as high as 34% [153].

MS in a chronic demyelinating disease of the peripheral- and central nervous systems [154]. Depressive symptoms are common in as many as 40% of MS patients [155]. Fredrikson et al. [156] found an increased suicide risk (the standardized mortality ratio = 2.3) in a group of over 12,000 hospitalized patients in Sweden, diagnosed with primary or secondary MS, when compared to the general population. Similarly, a twofold higher risk of dying by suicide was reported in a smaller samples of 115 MS patient in Denmark [157].

Epilepsy and Suicidality

Epilepsy is another condition which is associated with an increased risk of suicide. An increasing body of evidence suggests that neuroinflammation has an important role in the generation and recurrence of epileptic seizures [158]. A recent, large epidemiological study compared the incidence rates and prevalence of psychiatric conditions including suicide attempts in 3,700 patients with epilepsy and in 14,000 healthy controls [159]. Interestingly, the prevalence of suicide attempts was significantly increased in the patient group both before and after the diagnosis of epilepsy was made. Although the exact mechanism responsible for epileptic seizures is still not defined, several available medications are thought to control epileptic activity through inhibition of glutamatergic neurotransmission [160]. It is also clear that inflammation and fever reduces the seizure threshold, both in healthy individuals and those with a diagnosis of epilepsy. One might therefore speculate that also in epilepsy, excess stimulation of glutamatergic receptors, perhaps on the basis of inflammation, could represent a pathogenetic mechanism of suicidality.

Infections and Suicidality

Infectious agents might greatly increase CSF levels of cytokines, kynurenine and its metabolites [161]. Therefore, important insights into the role of inflammation in suicide could be gained by establishing whether suicidality is increased in connection with CNS infections. One of the most studied areas in this context is HIV and AIDS. The later disease stages are often associated with brain infection and neuropsychiatric manifestations. A systematic review came to the conclusion that suicide and suicidal ideation are prevalent in HIV patients, and that there should be an increased clinical awareness and evaluation of such symptoms [162]. Other neuro-inflammatory infections, such as neuroborreliosis and neurosyphilis, lead to neuropsychiatric symptoms, although the prevalence and incidence rates of suicidality in these conditions still remain to be established. Moreover, depression and suicide rates after other serious illnesses affecting the CNS, such as bacterial and viral meningitis, would be of great interest to characterize, as such patients might be of heightened risk and would benefit from increased surveillance.

Five recent studies have proven a significantly increased risk for suicidal behavior in individuals who are positive for *Toxoplasma gondi* infection [163–167]. *T. gondii* is a neurotropic protozoan parasite [168], which is common all over the world and affects approximately one-third of the human population. In immunologically healthy individuals, the infection resolves with minimal or no symptoms. After infection the parasite enters the brain where it resides in intracellular cysts in neurons and glial cells. Animal studies have shown that the *T. gondii* cysts are localized, for example, in the PFC and amygdala [169], areas that are of importance for emotions and behavior.

At the neurobiological level, it has been shown that this protozoan parasite contains two genes encoding tyrosine hydroxylase, the enzyme responsible for producing L-dopa [170], which in turn may lead to an increase in dopamine. Alterations in dopamine levels have been hypothesized to affect the ability to control suicidal impulses and overcoming fear of death [168]. Changes in impulsivity and regulation of fear might increase the risk for suicidal behavior, especially in combination with inflammatory changes in the CNS and concomitant depressive emotions.

Concluding Remarks: The Path Forward

In summary, there is a growing body of evidence that inflammation, as manifested by increased levels of pro-inflammatory cytokines and inflammatory metabolites, is present in patients with suicidal behavior and ideation. The inflammatory changes are present in the periphery, in the cerebrospinal fluid and brain parenchyma of suicidal patients and suicide victims. Together with clinical trials showing a rapid antisuicidal effect of the NMDA receptor antagonist, ketamine, this suggests that inflammation activates the kynurenine pathway with a net result of kynurenine metabolites affecting glutamate neurotransmission as important downstream effectors, responsible for symptom generation. In addition, cytokines have distinct effects on behavior and emotion in specific brain areas. Our knowledge in this field is just beginning to evolve. The triggers of inflammation may vary, and could involve stress as well as infectious agents and autoimmunity. In order to find the optimal treatment regimen for suicidal symptoms, it is of critical importance to identify both the initial trigger of inflammation, as well as the important neurobiological effectors. To this end, future clinical trials could, for example, evaluate the effects of eradicating *T. gondii* infection in patients with suicidal behavior. Other novel therapeutic strategies could include inhibition of the enzyme IDO-1 or microglial activation. The latter could be tested using the antibiotic minocycline, a treatment that is currently tested for efficacy in bipolar depression in several ongoing clinical trials. As knowledge increases about the role of specific cytokines for symptom generation, specific cytokine blockers, potentially with access to the CNS, could also be tested.

References

1 WHO: World Report on Violence and Health. Geneva, World Health Organization, 2002.
2 Hawton K, van Heeringen K: Suicide. Lancet 2009;373:1372–1381.
3 Varnik P: Suicide in the world. Int J Environ Res Public Health 2012;9:760–771.
4 Tidemalm D, Langstrom N, Lichtenstein P, Runeson B: Risk of suicide after suicide attempt according to coexisting psychiatric disorder: Swedish cohort study with long term follow-up. BMJ 2008;337:a2205.
5 Cavanagh JT, Carson AJ, Sharpe M, Lawrie SM: Psychological autopsy studies of suicide: a systematic review. Psychol Med 2003;33:395–405.
6 Finseth PI, Morken G, Andreassen OA, Malt UF, Vaaler AE: Risk factors related to lifetime suicide attempts in acutely admitted bipolar disorder inpatients. Bipolar Disord 2012.
7 Boden JM, Fergusson DM, Horwood LJ: Anxiety disorders and suicidal behaviours in adolescence and young adulthood: findings from a longitudinal study. Psychol Med 2007;37:431–440.
8 Flensborg-Madsen T, Knop J, Mortensen EL, Becker U, Sher L, Gronbaek M: Alcohol use disorders increase the risk of completed suicide – irrespective of other psychiatric disorders. A longitudinal cohort study. Psychiatry Res 2009;167:123–130.
9 Hor K, Taylor M: Suicide and schizophrenia: a systematic review of rates and risk factors. J Psychopharmacol 2010;24:81–90.
10 Holma KM, Melartin TK, Haukka J, Holma IA, Sokero TP, Isometsa ET: Incidence and predictors of suicide attempts in DSM-IV major depressive disorder: a five-year prospective study. Am J Psychiatry 2010;167:801–808.
11 Claassen CA, Trivedi MH, Rush AJ, Husain MM, Zisook S, Young E, Leuchter A, Wisniewski SR, Balasubramani GK, Alpert J: Clinical differences among depressed patients with and without a history of suicide attempts: findings from the STAR*D trial. J Affect Disord 2007;97:77–84.
12 Oquendo MA, Bongiovi-Garcia ME, Galfalvy H, Goldberg PH, Grunebaum MF, Burke AK, Mann JJ: Sex differences in clinical predictors of suicidal acts after major depression: a prospective study. Am J Psychiatry 2007;164:134–141.
13 Coryell W, Young EA: Clinical predictors of suicide in primary major depressive disorder. J Clin Psychiatry 2005;66:412–417.
14 Schneider B, Philipp M, Muller MJ: Psychopathological predictors of suicide in patients with major depression during a 5-year follow-up. Eur Psychiatry 2001;16:283–288.
15 Dumais A, Lesage AD, Alda M, Rouleau G, Dumont M, Chawky N, Roy M, Mann JJ, Benkelfat C, Turecki G: Risk factors for suicide completion in major depression: a case-control study of impulsive and aggressive behaviors in men. Am J Psychiatry 2005;162:2116–2124.
16 Perroud N, Baud P, Mouthon D, Courtet P, Malafosse A: Impulsivity, aggression and suicidal behavior in unipolar and bipolar disorders. J Affect Disord 2011;134:112–118.
17 McMillan D, Gilbody S, Beresford E, Neilly L: Can we predict suicide and non-fatal self-harm with the Beck Hopelessness Scale? A meta-analysis. Psychol Med 2007;37:769–778.
18 David Klonsky E, Kotov R, Bakst S, Rabinowitz J, Bromet EJ: Hopelessness as a predictor of attempted suicide among first admission patients with psychosis: a 10-year cohort study. Suicide Life Threat Behav 2012;42:1–10.
19 Nock MK, Hwang I, Sampson NA, Kessler RC: Mental disorders, comorbidity and suicidal behavior: results from the National Comorbidity Survey Replication. Mol Psychiatry 2010;15:868–876.
20 Bertolote JM, Fleischmann A: Suicide and psychiatric diagnosis: a worldwide perspective. World Psychiatry 2002;1:181–185.

21 Luoma JB, Martin CE, Pearson JL: Contact with mental health and primary care providers before suicide: a review of the evidence. Am J Psychiatry 2002;159:909–916.
22 Da Cruz D, Pearson A, Saini P, Miles C, While D, Swinson N, Williams A, Shaw J, Appleby L, Kapur N: Emergency department contact prior to suicide in mental health patients. Emerg Med J 2011;28:467–471.
23 Wasserman D, Rihmer Z, Rujescu D, Sarchiapone M, Sokolowski M, Titelman D, Zalsman G, Zemishlany Z, Carli V: The European Psychiatric Association (EPA) guidance on suicide treatment and prevention. Eur Psychiatry 2012;27:129–141.
24 Hammad TA, Laughren T, Racoosin J: Suicidality in pediatric patients treated with antidepressant drugs. Arch Gen Psychiatry 2006;63:332–339.
25 Licinio J, Wong ML: Depression, antidepressants and suicidality: a critical appraisal. Nat Rev Drug Discov 2005;4:165–171.
26 Price RB, Nock MK, Charney DS, Mathew SJ: Effects of intravenous ketamine on explicit and implicit measures of suicidality in treatment-resistant depression. Biol Psychiatry 2009;66:522–526.
27 DiazGranados N, Ibrahim LA, Brutsche NE, Ameli R, Henter ID, Luckenbaugh DA, Machado-Vieira R, Zarate CA Jr: Rapid resolution of suicidal ideation after a single infusion of an N-methyl-D-aspartate antagonist in patients with treatment-resistant major depressive disorder. J Clin Psychiatry 2010;71:1605–1611.
28 Larkin GL, Beautrais AL: A preliminary naturalistic study of low-dose ketamine for depression and suicide ideation in the emergency department. Int J Neuropsychopharmacol 2011;14:1127–1131.
29 Zarate CA, Jr., Brutsche NE, Ibrahim L, Franco-Chaves J, Diazgranados N, Cravchik A, Selter J, Marquardt CA, Liberty V, Luckenbaugh DA: Replication of ketamine's antidepressant efficacy in bipolar depression: a randomized controlled add-on trial. Biol Psychiatry 2012;71:939–946.
30 Bach H, Arango V: Neuroanatomy of serotonergic abnormalities in suicide; in Dwivedi Y (ed): The Neurobiological Basis of Suicide. Boca Raton, CRC Press, 2012.
31 Mann JJ: Neurobiology of suicidal behaviour. Nat Rev Neurosci 2003;4:819–828.
32 Asberg M, Traskman L, Thoren P: 5-HIAA in the cerebrospinal fluid. A biochemical suicide predictor? Arch Gen Psychiatry 1976;33:1193–1197.
33 Leyton M, Paquette V, Gravel P, Rosa-Neto P, Weston F, Diksic M, Benkelfat C: Alpha-[^{11}C]methyl-L-tryptophan trapping in the orbital and ventral medial prefrontal cortex of suicide attempters. Eur Neuropsychopharmacol 2006;16:220–223.
34 Coccaro EF, Siever LJ, Klar HM, Maurer G, Cochrane K, Cooper TB, Mohs RC, Davis KL: Serotonergic studies in patients with affective and personality disorders. Correlates with suicidal and impulsive aggressive behavior. Arch Gen Psychiatry 1989;46:587–599.
35 Mann JJ, McBride PA, Malone KM, DeMeo M, Keilp J: Blunted serotonergic responsivity in depressed inpatients. Neuropsychopharmacology 1995;13:53–64.
36 Keilp JG, Oquendo MA, Stanley BH, Burke AK, Cooper TB, Malone KM, Mann JJ: Future suicide attempt and responses to serotonergic challenge. Neuropsychopharmacology 2010;35:1063–1072.
37 Mann JJ, Huang YY, Underwood MD, Kassir SA, Oppenheim S, Kelly TM, Dwork AJ, Arango V: A serotonin transporter gene promoter polymorphism (5-HTTLPR) and prefrontal cortical binding in major depression and suicide. Arch Gen Psychiatry 2000;57:729–738.
38 Austin MC, Whitehead RE, Edgar CL, Janosky JE, Lewis DA: Localized decrease in serotonin transporter-immunoreactive axons in the prefrontal cortex of depressed subjects committing suicide. Neuroscience 2002;114:807–815.
39 Underwood MD, Kassir SA, Bakalian MJ, Galfalvy H, Mann JJ, Arango V: Neuron density and serotonin receptor binding in prefrontal cortex in suicide. Int J Neuropsychopharmacol 2012;15:435–447.
40 Arango V, Underwood MD, Gubbi AV, Mann JJ: Localized alterations in pre- and postsynaptic serotonin binding sites in the ventrolateral prefrontal cortex of suicide victims. Brain Res 1995;688:121–133.
41 Pandey GN, Dwivedi Y, Rizavi HS, Ren X, Pandey SC, Pesold C, Roberts RC, Conley RR, Tamminga CA: Higher expression of serotonin 5-HT(2A) receptors in the postmortem brains of teenage suicide victims. Am J Psychiatry 2002;159:419–429.
42 Bellivier F, Chaste P, Malafosse A: Association between the TPH gene A218C polymorphism and suicidal behavior: a meta-analysis. Am J Med Genet B Neuropsychiatr Genet 2004;124B:87–91.
43 Li D, He L: Meta-analysis supports association between serotonin transporter (5-HTT) and suicidal behavior. Mol Psychiatry 2007;12:47–54.
44 Klempan TA, Sequeira A, Canetti L, Lalovic A, Ernst C, ffrench-Mullen J, Turecki G: Altered expression of genes involved in ATP biosynthesis and GABAergic neurotransmission in the ventral prefrontal cortex of suicides with and without major depression. Mol Psychiatry 2009;14:175–189.
45 Kim S, Choi KH, Baykiz AF, Gershenfeld HK: Suicide candidate genes associated with bipolar disorder and schizophrenia: an exploratory gene expression profiling analysis of post-mortem prefrontal cortex. BMC Genomics 2007;8:413.
46 Ernst C, Mechawar N, Turecki G: Suicide neurobiology. Prog Neurobiol 2009;89:315–333.
47 Lowther S, De Paermentier F, Cheetham SC, Crompton MR, Katona CL, Horton RW: 5-HT1A receptor binding sites in post-mortem brain samples from depressed suicides and controls. J Affect Disord 1997;42:199–207.
48 Stockmeier CA, Dilley GE, Shapiro LA, Overholser JC, Thompson PA, Meltzer HY: Serotonin receptors in suicide victims with major depression. Neuropsychopharmacology 1997;16:162–173.
49 Rosel P, Arranz B, Vallejo J, Oros M, Menchon JM, Alvarez P, Navarro MA: High affinity [^{3}H]imipramine and [^{3}H]paroxetine binding sites in suicide brains. J Neural Transm 1997;104:921–929.
50 Rosel P, Arranz B, San L, Vallejo J, Crespo JM, Urretavizcaya M, Navarro MA: Altered 5-HT(2A) binding sites and second messenger inositol trisphosphate (IP(3)) levels in hippocampus but not in frontal cortex from depressed suicide victims. Psychiatry Res 2000;99:173–181.
51 Anisman H, Du L, Palkovits M, Faludi G, Kovacs GG, Szontagh-Kishazi P, Merali Z, Poulter MO: Serotonin receptor subtype and p11 mRNA expression in stress-relevant brain regions of suicide and control subjects. J Psychiatry Neurosci 2008;33:131–141.
52 Pariante CM, Lightman SL: The HPA axis in major depression: classical theories and new developments. Trends Neurosci 2008;31:464–468.

53 Coryell W, Schlesser MA: Suicide and the dexamethasone suppression test in unipolar depression. Am J Psychiatry 1981;138:1120–1121.
54 Lopez-Ibor JJ Jr, Saiz-Ruiz J, Perez de los Cobos JC: Biological correlations of suicide and aggressivity in major depressions (with melancholia): 5-hydroxyindoleacetic acid and cortisol in cerebral spinal fluid, dexamethasone suppression test and therapeutic response to 5-hydroxytryptophan. Neuropsychobiology 1985;14:67–74.
55 Yerevanian BI, Feusner JD, Koek RJ, Mintz J: The dexamethasone suppression test as a predictor of suicidal behavior in unipolar depression. J Affect Disord 2004;83:103–108.
56 Mann JJ, Currier D, Stanley B, Oquendo MA, Amsel LV, Ellis SP: Can biological tests assist prediction of suicide in mood disorders? Int J Neuropsychopharmacol 2006;9:465–474.
57 Jokinen J, Carlborg A, Martensson B, Forslund K, Nordstrom AL, Nordstrom P: DST non-suppression predicts suicide after attempted suicide. Psychiatry Res 2007;150:297–303.
58 Black DW, Monahan PO, Winokur G: The relationship between DST results and suicidal behavior. Ann Clin Psychiatry 2002;14:83–88.
59 Secunda SK, Cross CK, Koslow S, Katz MM, Kocsis J, Maas JW, Landis H: Biochemistry and suicidal behavior in depressed patients. Biol Psychiatry 1986;21:756–767.
60 Pfennig A, Kunzel HE, Kern N, Ising M, Majer M, Fuchs B, Ernst G, Holsboer F, Binder EB: Hypothalamus-pituitary-adrenal system regulation and suicidal behavior in depression. Biol Psychiatry 2005;57:336–342.
61 Lindqvist D, Isaksson A, Traskman-Bendz L, Brundin L: Salivary cortisol and suicidal behavior – a follow-up study. Psychoneuroendocrinology 2008;33:1061–1068.
62 Jokinen J, Ouda J, Nordstrom P: Noradrenergic function and HPA axis dysregulation in suicidal behaviour. Psychoneuroendocrinology 2010;35:1536–1542.
63 McGirr A, Diaconu G, Berlim MT, Turecki G: Personal and family history of suicidal behaviour is associated with lower peripheral cortisol in depressed outpatients. J Affect Disord 2011;131:368–373.
64 Anisman H, Merali Z, Hayley S: Neurotransmitter, peptide and cytokine processes in relation to depressive disorder: comorbidity between depression and neurodegenerative disorders. Prog Neurobiol 2008;85:1–74.
65 Zalcman S, Green-Johnson JM, Murray L, Nance DM, Dyck D, Anisman H, Greenberg AH: Cytokine-specific central monoamine alterations induced by interleukin-1, -2 and -6. Brain Res 1994;643:40–49.
66 Zhang J, Terreni L, De Simoni MG, Dunn AJ: Peripheral interleukin-6 administration increases extracellular concentrations of serotonin and the evoked release of serotonin in the rat striatum. Neurochem Int 2001;38:303–308.
67 Dunn AJ: Effects of cytokines and infections on brain neurochemistry. Clin Neurosci Res 2006;6:52–68.
68 Anisman H, Gibb J, Hayley S: Influence of continuous infusion of interleukin-1beta on depression-related processes in mice: corticosterone, circulating cytokines, brain monoamines, and cytokine mRNA expression. Psychopharmacology (Berl) 2008;199:231–244.
69 Zhu CB, Blakely RD, Hewlett WA: The proinflammatory cytokines interleukin-1beta and tumor necrosis factor-alpha activate serotonin transporters. Neuropsychopharmacology 2006;31:2121–2131.
70 Hayley S, Scharf J, Anisman H: Central administration of murine interferon-alpha induces depressive-like behavioral, brain cytokine and neurochemical alterations in mice: a mini-review and original experiments. Brain Behav Immun 2012.
71 Kiecolt-Glaser JK, Preacher KJ, MacCallum RC, Atkinson C, Malarkey WB, Glaser R: Chronic stress and age-related increases in the proinflammatory cytokine IL-6. Proc Natl Acad Sci USA 2003;100:9090–9095.
72 Steptoe A, Hamer M, Chida Y: The effects of acute psychological stress on circulating inflammatory factors in humans: a review and meta-analysis. Brain Behav Immun 2007;21:901–912.
73 Chiang JJ, Eisenberger NI, Seeman TE, Taylor SE: Negative and competitive social interactions are related to heightened proinflammatory cytokine activity. Proc Natl Acad Sci USA 2012;109:1878–1882.
74 Takaki A, Huang QH, Somogyvari-Vigh A, Arimura A: Immobilization stress may increase plasma interleukin-6 via central and peripheral catecholamines. Neuroimmunomodulation 1994;1:335–342.
75 You Z, Luo C, Zhang W, Chen Y, He J, Zhao Q, Zuo R, Wu Y: Pro- and anti-inflammatory cytokines expression in rat's brain and spleen exposed to chronic mild stress: involvement in depression. Behav Brain Res 2011;225:135–141.
76 Naitoh Y, Fukata J, Tominaga T, Nakai Y, Tamai S, Mori K, Imura H: Interleukin-6 stimulates the secretion of adrenocorticotropic hormone in conscious, freely-moving rats. Biochem Biophys Res Commun 1988;155:1459–1463.
77 Dunn AJ: Systemic interleukin-1 administration stimulates hypothalamic norepinephrine metabolism paralleling the increased plasma corticosterone. Life Sci 1988;43:429–435.
78 Rivier C: Stimulatory effect of interleukin-1 beta on the hypothalamic-pituitary-adrenal axis of the rat: influence of age, gender and circulating sex steroids. J Endocrinol 1994;140:365–372.
79 Vedder H, Schreiber W, Yassouridis A, Gudewill S, Galanos C, Pollmacher T: Dose-dependence of bacterial lipopolysaccharide (LPS) effects on peak response and time course of the immune-endocrine host response in humans. Inflamm Res 1999;48:67–74.
80 Dantzer R, O'Connor JC, Freund GG, Johnson RW, Kelley KW: From inflammation to sickness and depression: when the immune system subjugates the brain. Nature Reviews 2008;9:46–56.
81 Schobitz B, de Kloet ER, Sutanto W, Holsboer F: Cellular localization of interleukin 6 mRNA and interleukin 6 receptor mRNA in rat brain. Eur J Neurosci 1993;5:1426–1435.
82 Vallieres L, Rivest S: Regulation of the genes encoding interleukin-6, its receptor, and gp130 in the rat brain in response to the immune activator lipopolysaccharide and the proinflammatory cytokine interleukin-1beta. J Neurochem 1997;69:1668–1683.
83 Rognum IJ, Haynes RL, Vege A, Yang M, Rognum TO, Kinney HC: Interleukin-6 and the serotonergic system of the medulla oblongata in the sudden infant death syndrome. Acta Neuropathol 2009;118:519–530.
84 Schele E, Fekete C, Egri P, Fuzesi T, Palkovits M, Keller E, Liposits Z, Gereben B, Karlsson-Lindahl L, Shao R, Jansson JO: Interleukin-6 receptor alpha is co-localised with melanin-concentrating hormone in human and mouse hypothalamus. J Neuroendocrinol 2012;24:930–943.

85 Yabuuchi K, Minami M, Katsumata S, Satoh M: Localization of type I interleukin-1 receptor mRNA in the rat brain. Brain Res Mol Brain Res 1994;27:27–36.
86 Ericsson A, Liu C, Hart RP, Sawchenko PE: Type 1 interleukin-1 receptor in the rat brain: distribution, regulation, and relationship to sites of IL-1-induced cellular activation. J Comp Neurol 1995;361:681–698.
87 McAfoose J, Baune BT: Evidence for a cytokine model of cognitive function. Neurosci Biobehav Rev 2009;33:355–366.
88 Kawasaki Y, Zhang L, Cheng JK, Ji RR: Cytokine mechanisms of central sensitization: distinct and overlapping role of interleukin-1beta, interleukin-6, and tumor necrosis factor-alpha in regulating synaptic and neuronal activity in the superficial spinal cord. J Neurosci 2008;28:5189–5194.
89 Garcia-Oscos F, Salgado H, Hall S, Thomas F, Farmer GE, Bermeo J, Galindo LC, Ramirez RD, D'Mello S, Rose-John S, Atzori M: The stress-induced cytokine interleukin-6 decreases the inhibition/excitation ratio in the rat temporal cortex via trans-signaling. Biol Psychiatry 2012;71:574–582.
90 Nelson TE, Olde Engberink A, Hernandez R, Puro A, Huitron-Resendiz S, Hao C, De Graan PN, Gruol DL: Altered synaptic transmission in the hippocampus of transgenic mice with enhanced central nervous systems expression of interleukin-6. Brain Behav Immun 2012;26:959–971.
91 Maes M, Leonard BE, Myint AM, Kubera M, Verkerk R: The new '5-HT' hypothesis of depression: cell-mediated immune activation induces indoleamine 2,3-dioxygenase, which leads to lower plasma tryptophan and an increased synthesis of detrimental tryptophan catabolites (TRYCATs), both of which contribute to the onset of depression. Prog Neuro-Psychopharmacol Biol Psychiatry 2011;35:702–721.
92 Capuron L, Ravaud A, Neveu PJ, Miller AH, Maes M, Dantzer R: Association between decreased serum tryptophan concentrations and depressive symptoms in cancer patients undergoing cytokine therapy. Mol Psychiatry 2002;7:468–473.
93 Placidi GP, Oquendo MA, Malone KM, Huang YY, Ellis SP, Mann JJ: Aggressivity, suicide attempts, and depression: relationship to cerebrospinal fluid monoamine metabolite levels. Biol Psychiatry 2001;50:783–791.
94 Raison CL, Dantzer R, Kelley KW, Lawson MA, Woolwine BJ, Vogt G, Spivey JR, Saito K, Miller AH: CSF concentrations of brain tryptophan and kynurenines during immune stimulation with IFN-alpha: relationship to CNS immune responses and depression. Mol Psychiatry 2010;15:393–403.
95 Stone TW: Neuropharmacology of quinolinic and kynurenic acids. Pharmacol Rev 1993;45:309–379.
96 de Carvalho LP, Bochet P, Rossier J: The endogenous agonist quinolinic acid and the nonendogenous homoquinolinic acid discriminate between NMDAR2 receptor subunits. Neurochem Int 1996;28:445–452.
97 Hilmas C, Pereira EF, Alkondon M, Rassoulpour A, Schwarcz R, Albuquerque EX: The brain metabolite kynurenic acid inhibits alpha7 nicotinic receptor activity and increases non-alpha7 nicotinic receptor expression: physiopathological implications. J Neurosci 2001;21:7463–7473.
98 Erhardt S, Olsson SK, Engberg G: Pharmacological manipulation of kynurenic acid: potential in the treatment of psychiatric disorders. CNS Drugs 2009;23:91–101.
99 Schwarcz R, Bruno JP, Muchowski PJ, Wu HQ: Kynurenines in the mammalian brain: when physiology meets pathology. Nat Rev Neurosci 2012;13:465–477.
100 Guillemin GJ: Quinolinic acid: neurotoxicity. FEBS J 2012;279:1355.
101 Monkul ES, Hatch JP, Nicoletti MA, Spence S, Brambilla P, Lacerda AL, Sassi RB, Mallinger AG, Keshavan MS, Soares JC: Fronto-limbic brain structures in suicidal and non-suicidal female patients with major depressive disorder. Mol Psychiatry 2007;12:360–366.
102 Aguilar EJ, Garcia-Marti G, Marti-Bonmati L, Lull JJ, Moratal D, Escarti MJ, Robles M, Gonzalez JC, Guillamon MI, Sanjuan J: Left orbitofrontal and superior temporal gyrus structural changes associated to suicidal behavior in patients with schizophrenia. Prog Neuro-Psychopharmacol Biol Psychiatry 2008;32:1673–1676.
103 Hwang JP, Lee TW, Tsai SJ, Chen TJ, Yang CH, Lirng JF, Tsai CF: Cortical and subcortical abnormalities in late-onset depression with history of suicide attempts investigated with MRI and voxel-based morphometry. J Geriatr Psychiatry Neurol 2010;23:171–184.
104 Desmyter S, van Heeringen C, Audenaert K: Structural and functional neuroimaging studies of the suicidal brain. Prog Neuro-Psychopharmacol Biol Psychiatry 2011;35:796–808.
105 Wagner G, Koch K, Schachtzabel C, Schultz CC, Sauer H, Schlosser RG: Structural brain alterations in patients with major depressive disorder and high risk for suicide: evidence for a distinct neurobiological entity? Neuroimage 2011;54:1607–1614.
106 Soloff PH, Pruitt P, Sharma M, Radwan J, White R, Diwadkar VA: Structural brain abnormalities and suicidal behavior in borderline personality disorder. J Psychiatr Res 2012;46:516–525.
107 Nassberger L, Traskman-Bendz L: Increased soluble interleukin-2 receptor concentrations in suicide attempters. Acta Psychiatr Scand 1993;88:48–52.
108 Steiner J, Bielau H, Brisch R, Danos P, Ullrich O, Mawrin C, Bernstein HG, Bogerts B: Immunological aspects in the neurobiology of suicide: elevated microglial density in schizophrenia and depression is associated with suicide. J Psychiatr Res 2008;42:151–157.
109 Tonelli LH, Stiller J, Rujescu D, Giegling I, Schneider B, Maurer K, Schnabel A, Moller HJ, Chen HH, Postolache TT: Elevated cytokine expression in the orbitofrontal cortex of victims of suicide. Acta Psychiatr Scand 2008;117:198–206.
110 Lindqvist D, Janelidze S, Hagell P, Erhardt S, Samuelsson M, Minthon L, Hansson O, Bjorkqvist M, Traskman-Bendz L, Brundin L: Interleukin-6 is elevated in the cerebrospinal fluid of suicide attempters and related to symptom severity. Biol Psychiatry 2009;66:287–292.
111 Pandey GN, Rizavi HS, Ren X, Fareed J, Hoppensteadt DA, Roberts RC, Conley RR, Dwivedi Y: Proinflammatory cytokines in the prefrontal cortex of teenage suicide victims. J Psychiatr Res 2012;46:57–63.

112 Steiner J, Walter M, Gos T, Guillemin GJ, Bernstein HG, Sarnyai Z, Mawrin C, Brisch R, Bielau H, Meyer zu Schwabedissen L, Bogerts B, Myint AM: Severe depression is associated with increased microglial quinolinic acid in subregions of the anterior cingulate gyrus: evidence for an immune-modulated glutamatergic neurotransmission? J Neuroinflammation 2011;8:94.
113 Janelidze S, Mattei D, Westrin A, Traskman-Bendz L, Brundin L: Cytokine levels in the blood may distinguish suicide attempters from depressed patients. Brain Behav Immun 2011;25:335–339.
114 Kim YK, Lee SW, Kim SH, Shim SH, Han SW, Choi SH, Lee BH: Differences in cytokines between non-suicidal patients and suicidal patients in major depression. Prog Neuro-Psychopharmacol Biol Psychiatry 2008;32:356–361.
115 Donato R, Sorci G, Riuzzi F, Arcuri C, Bianchi R, Brozzi F, Tubaro C, Giambanco I: S100B's double life: intracellular regulator and extracellular signal. Biochim Biophys Acta 2009;1793:1008–1022.
116 Sublette ME, Galfalvy HC, Fuchs D, Lapidus M, Grunebaum MF, Oquendo MA, Mann JJ, Postolache TT: Plasma kynurenine levels are elevated in suicide attempters with major depressive disorder. Brain Behav Immun 2011;25:1272–1278.
117 Erhardt S, Lim CK, Linderholm KR, Janelidze S, Lindqvist D, Samuelsson M, Lundberg K, Postolache TT, Träskman-Bendz L, Guillemin GJ, Brundin L: Connecting inflammation with glutamate agonism in suicidality. Neuropsychopharmacology; accepted article preview 3 December 2012; doi: 10.1038/npp.2012.248.
118 Claes S, Myint AM, Domschke K, Del-Favero J, Entrich K, Engelborghs S, De Deyn P, Mueller N, Baune B, Rothermundt M: The kynurenine pathway in major depression: haplotype analysis of three related functional candidate genes. Psychiatry Res 2011;188:355–360.
119 Malkesman O, Pine DS, Tragon T, Austin DR, Henter ID, Chen G, Manji HK: Animal models of suicide-trait-related behaviors. Trends Pharmacol Sci 2009;30:165–173.
120 Hassanain M, Zalcman S, Bhatt S, Siegel A: Interleukin-1 beta in the hypothalamus potentiates feline defensive rage: role of serotonin-2 receptors. Neuroscience 2003;120:227–233.
121 Bhatt S, Bhatt R, Zalcman SS, Siegel A: Role of IL-1 beta and 5-HT2 receptors in midbrain periaqueductal gray (PAG) in potentiating defensive rage behavior in cat. Brain Behav Immun 2008;22:224–233.
122 Bhatt S, Siegel A: Potentiating role of interleukin 2 (IL-2) receptors in the midbrain periaqueductal gray (PAG) upon defensive rage behavior in the cat: role of neurokinin NK(1) receptors. Behav Brain Res 2006;167:251–260.
123 Patel A, Siegel A, Zalcman SS: Lack of aggression and anxiolytic-like behavior in TNF receptor (TNF-R1 and TNF-R2) deficient mice. Brain Behav Immun 2010;24:1276–1280.
124 Chourbaji S, Urani A, Inta I, Sanchis-Segura C, Brandwein C, Zink M, Schwaninger M, Gass P: IL-6 knockout mice exhibit resistance to stress-induced development of depression-like behaviors. Neurobiol Dis 2006;23:587–594.
125 Carroll JE, Low CA, Prather AA, Cohen S, Fury JM, Ross DC, Marsland AL: Negative affective responses to a speech task predict changes in interleukin (IL)-6. Brain Behav Immun 2011;25:232–238.
126 Sjogren E, Leanderson P, Kristenson M, Ernerudh J: Interleukin-6 levels in relation to psychosocial factors: studies on serum, saliva, and in vitro production by blood mononuclear cells. Brain Behav Immun 2006;20:270–278.
127 Westling S, Ahren B, Traskman-Bendz L, Brundin L: Increased IL-1beta reactivity upon a glucose challenge in patients with deliberate self-harm. Acta Psychiatr Scand 2011;124:301–306.
128 Kraus MR, Schafer A, Faller H, Csef H, Scheurlen M: Psychiatric symptoms in patients with chronic hepatitis C receiving interferon alfa-2b therapy. J Clin Psychiatry 2003;64:708–714.
129 Bezemer G, Van Gool AR, Fekkes D, Vrolijk JM, Hansen BE, Janssen HL, de Knegt RJ: Psychiatric side effects and fluctuations in serotonergic parameters in the treatment of chronic hepatitis C infection. Neuropsychobiology 2012;65:126–132.
130 Lotrich FE, Ferrell RE, Rabinovitz M, Pollock BG: Labile anger during interferon alfa treatment is associated with a polymorphism in tumor necrosis factor alpha. Clin Neuropharmacol 2010;33:191–197.
131 Musselman DL, Lawson DH, Gumnick JF, Manatunga AK, Penna S, Goodkin RS, Greiner K, Nemeroff CB, Miller AH: Paroxetine for the prevention of depression induced by high-dose interferon alfa. N Engl J Med 2001;344:961–966.
132 Raison CL, Borisov AS, Broadwell SD, Capuron L, Woolwine BJ, Jacobson IM, Nemeroff CB, Miller AH: Depression during pegylated interferon-alpha plus ribavirin therapy: prevalence and prediction. J Clin Psychiatry 2005;66:41–48.
133 Wichers MC, Koek GH, Robaeys G, Verkerk R, Scharpe S, Maes M: IDO and interferon-alpha-induced depressive symptoms: a shift in hypothesis from tryptophan depletion to neurotoxicity. Mol Psychiatry 2005;10:538–544.
134 Raison CL, Borisov AS, Majer M, Drake DF, Pagnoni G, Woolwine BJ, Vogt GJ, Massung B, Miller AH: Activation of central nervous system inflammatory pathways by interferon-alpha: relationship to monoamines and depression. Biol Psychiatry 2009;65:296–303.
135 Dieperink E, Ho SB, Tetrick L, Thuras P, Dua K, Willenbring ML: Suicidal ideation during interferon-alpha2b and ribavirin treatment of patients with chronic hepatitis C. Gen Hosp Psychiatry 2004;26:237–240.
136 Evon DM, Ramcharran D, Belle SH, Terrault NA, Fontana RJ, Fried MW: Prospective analysis of depression during peginterferon and ribavirin therapy of chronic hepatitis C: results of the Virahep-C study. Am J Gastroenterol 2009;104:2949–2958.
137 Janssen HL, Brouwer JT, van der Mast RC, Schalm SW: Suicide associated with alfa-interferon therapy for chronic viral hepatitis. J Hepatol 1994;21:241–243.
138 Fragoso YD, Frota ER, Lopes JS, Noal JS, Giacomo MC, Gomes S, Goncalves MV, da Gama PD, Finkelsztejn A: Severe depression, suicide attempts, and ideation during the use of interferon beta by patients with multiple sclerosis. Clin Neuropharmacol 2010;33:312–316.
139 Holgate ST, Polosa R: Treatment strategies for allergy and asthma. Nat Rev Immunol 2008;8:218–230.
140 Anderson GP: Endotyping asthma: new insights into key pathogenic mechanisms in a complex, heterogeneous disease. Lancet 2008;372:1107–1119.
141 Holt PG, Sly PD: Viral infections and atopy in asthma pathogenesis: new rationales for asthma prevention and treatment. Nat Med 2012;18:726–735.

142 Goodwin RD: Asthma and suicide: current knowledge and future directions. Curr Psychiatry Rep 2012;14:30–35.
143 Kuo CJ, Chen VC, Lee WC, Chen WJ, Ferri CP, Stewart R, Lai TJ, Chen CC, Wang TN, Ko YC: Asthma and suicide mortality in young people: a 12-year follow-up study. Am J Psychiatry 2010;167:1092–1099.
144 Goodwin RD, Demmer RT, Galea S, Lemeshow AR, Ortega AN, Beautrais A: Asthma and suicide behaviors: results from the Third National Health and Nutrition Examination Survey (NHANES III). J Psychiatr Res 2012;46:1002–1007.
145 Philip G, Hustad C, Noonan G, Malice MP, Ezekowitz A, Reiss TF, Knorr B: Reports of suicidality in clinical trials of montelukast. J Allergy Clin Immunol 2009;124:691–696.
146 Favreau H, Bacon SL, Joseph M, Labrecque M, Lavoie KL: Association between asthma medications and suicidal ideation in adult asthmatics. Respir Med 2012;106:933–941.
147 Postolache TT, Stiller JW, Herrell R, Goldstein MA, Shreeram SS, Zebrak R, Thrower CM, Volkov J, No MJ, Volkov I, Rohan KJ, Redditt J, Parmar M, Mohyuddin F, Olsen C, Moca M, Tonelli LH, Merikangas K, Komarow HD: Tree pollen peaks are associated with increased nonviolent suicide in women. Mol Psychiatry 2005;10:232–235.
148 Woo JM, Gibbons RD, Qin P, Komarow H, Kim JB, Rogers CA, Mann JJ, Postolache TT: Suicide and prescription rates of intranasal corticosteroids and nonsedating antihistamines for allergic rhinitis: an ecological study. J Clin Psychiatry 2011;72:1423–1428.
149 Agmon-Levin N, Mosca M, Petri M, Shoenfeld Y: Systemic lupus erythematosus one disease or many? Autoimmun Rev 2012;11:593–595.
150 Meszaros ZS, Perl A, Faraone SV: Psychiatric symptoms in systemic lupus erythematosus: a systematic review. J Clin Psychiatry 2012;73:993–1001.
151 Lauvsnes MB, Omdal R: Systemic lupus erythematosus, the brain, and anti-NR2 antibodies. J Neurol 2012;259:622–629.
152 Karassa FB, Magliano M, Isenberg DA: Suicide attempts in patients with systemic lupus erythematosus. Ann Rheum Dis 2003;62:58–60.
153 Xie LF, Chen PL, Pan HF, Tao JH, Li XP, Zhang YJ, Zhai Y, Ye DQ: Prevalence and correlates of suicidal ideation in SLE inpatients: Chinese experience. Rheumatol Int 2012;32:2707–2714.
154 Nylander A, Hafler DA: Multiple sclerosis. J Clin Invest 2012;122:1180–1188.
155 Chwastiak L, Ehde DM, Gibbons LE, Sullivan M, Bowen JD, Kraft GH: Depressive symptoms and severity of illness in multiple sclerosis: epidemiologic study of a large community sample. Am J Psychiatry 2002;159:1862–1868.
156 Fredrikson S, Cheng Q, Jiang GX, Wasserman D: Elevated suicide risk among patients with multiple sclerosis in Sweden. Neuroepidemiology 2003;22:146–152.
157 Bronnum-Hansen H, Stenager E, Nylev Stenager E, Koch-Henriksen N: Suicide among Danes with multiple sclerosis. J Neurol Neurosurg Psychiatry 2005;76:1457–1459.
158 Vezzani A, Aronica E, Mazarati A, Pittman QJ: Epilepsy and brain inflammation. Exp Neurol e-pub ahead of print 1 Oct 2011; doi: 10.1016/j.expneurol.2011.09.033.
159 Hesdorffer DC, Ishihara L, Mynepalli L, Webb DJ, Weil J, Hauser WA: Epilepsy, suicidality, and psychiatric disorders: a bidirectional association. Ann Neurol 2012;72:184–191.
160 Beck H, Yaari Y: Antiepileptogenesis, plasticity of AED targets, drug resistance, and targeting the immature brain; in Noebels JL, Avoli M, Rogawski MA, Olsen RW, Delgado-Escueta AV (eds): Jasper's Basic Mechanisms of the Epilepsies, ed 4. Bethesda, National Center for Biotechnology Information, 2012.
161 Heyes MP, Saito K, Crowley JS, Davis LE, Demitrack MA, Der M, Dilling LA, Elia J, Kruesi MJ, Lackner A, et al: Quinolinic acid and kynurenine pathway metabolism in inflammatory and non-inflammatory neurological disease. Brain 1992;115:1249–1273.
162 Catalan J, Harding R, Sibley E, Clucas C, Croome N, Sherr L: HIV infection and mental health: suicidal behaviour – systematic review. Psychol Health Med 2011;16:588–611.
163 Ling VJ, Lester D, Mortensen PB, Langenberg PW, Postolache TT: *Toxoplasma gondii* seropositivity and suicide rates in women. J Nerv Ment Dis 2011;199:440–444.
164 Arling TA, Yolken RH, Lapidus M, Langenberg P, Dickerson FB, Zimmerman SA, Balis T, Cabassa JA, Scrandis DA, Tonelli LH, Postolache TT: *Toxoplasma gondii* antibody titers and history of suicide attempts in patients with recurrent mood disorders. J Nerv Ment Dis 2009;197:905–908.
165 Okusaga O, Langenberg P, Sleemi A, Vaswani D, Giegling I, Hartmann AM, Konte B, Friedl M, Groer MW, Yolken RH, Rujescu D, Postolache TT: *Toxoplasma gondii* antibody titers and history of suicide attempts in patients with schizophrenia. Schizophr Res 2011;133:150–155.
166 Zhang Y, Traskman-Bendz L, Janelidze S, Langenberg P, Saleh A, Constantine N, Okusaga O, Bay-Richter C, Brundin L, Postolache TT: *Toxoplasma gondii* immunoglobulin G antibodies and nonfatal suicidal self-directed violence. J Clin Psychiatry 2012;73:1069–1076.
167 Pedersen MG, Mortensen PB, Norgaard-Pedersen B, Postolache TT: *Toxoplasma gondii* infection and self-directed violence in mothers. Arch Gen Psychiatry 2012:1–8.
168 Okusaga O, Postolache TT: Toxoplasma gondii, the Immune System, and Suicidal Behavior; in Dwivedi Y (ed): The Neurobiological Basis of Suicide. Boca Raton (FL), CRC Press, 2012.
169 Vyas A, Kim SK, Giacomini N, Boothroyd JC, Sapolsky RM: Behavioral changes induced by *Toxoplasma* infection of rodents are highly specific to aversion of cat odors. Proc Natl Acad Sci USA 2007;104:6442–6447.
170 Gaskell EA, Smith JE, Pinney JW, Westhead DR, McConkey GA: A unique dual activity amino acid hydroxylase in *Toxoplasma gondii*. PLoS One 2009;4:e4801.

Lena Brundin, MD, PhD
Department of Translational Science and Molecular Medicine
Michigan State University, Van Andel Research Institute
333 Bostwick Avenue NE, Grand Rapids, MI 49503 (USA)
E-Mail Lena.Brundin@hc.msu.edu

Author Index

Subject Index